An Agrarian History of Portugal, 1000–2000

Library of Economic History

General Editor

Peer Vries (*University of Vienna*)

VOLUME 7

The titles published in this series are listed at *brill.com/lehi*

An Agrarian History of Portugal, 1000–2000

Economic Development on the European Frontier

Edited by

Dulce Freire
Pedro Lains

BRILL

LEIDEN | BOSTON

Cover illustration: Ceramic tiles with a market scene at the wall of the Mercado dos Lavradores, Funchal, Madeira (early 20th century).

Library of Congress Cataloging-in-Publication Data

Names: Freire, Dulce, 1969- editor of compilation. | Lains, Pedro, editor of compilation.
Title: An agrarian history of Portugal, 1000-2000 : economic development on the European frontier / edited by Dulce Freire, Pedro Lains.
Description: Leiden ; Boston : Brill, [2017] | Series: Library of economic history, ISSN 1877-3206 ; VOLUME 7 | Includes bibliographical references and index.
Identifiers: LCCN 2016041110 (print) | LCCN 2016042110 (ebook) | ISBN 9789004311534 (hardback : acid-free paper) | ISBN 9789004311527 (e-book) | ISBN 9789004311527 (e-book)
Subjects: LCSH: Agriculture--Economic aspects--Portugal--History. | Economic development--Portugal--History. | Portugal--Economic conditions. | Frontier and pioneer life--Europe. | Europe--Economic conditions. | Portugal--Relations--Europe. | Europe--Relations--Portugal.
Classification: LCC HD2027 .A57 2017 (print) | LCC HD2027 (ebook) | DDC 330.9469--dc23
LC record available at https://lccn.loc.gov/2016041110

Typeface for the Latin, Greek, and Cyrillic scripts: "Brill". See and download: brill.com/brill-typeface.

ISSN 1877-3206
ISBN 978-90-04-31153-4 (hardback)
ISBN 978-90-04-31152-7 (e-book)

This book is printed on acid-free paper and produced in a sustainable manner.

Contents

PART 2
New Frontiers, Crisis and Growth

Acknowledgements

We would like to acknowledge helpful comments from the participants in the two preparatory conferences held in Lisbon and in particular from Giovanni Federico, Cormac O'Grada, Vicente Pinilla and Rui Santos, as well as useful comments from two anonymous referees. We would also like to express special thanks to Conceição Andrade Martins for her contribution to the conception of the book. For the preparation of the book we had the invaluable collaboration of Cláudia Viana, who prepared the maps in the Appendix, and Marta Castelo Branco, Bárbara Direito and Ana Margarida Rodrigues. Finally, we aknowledge funding from the Portuguese Science Foundation under the project Portuguese Agriculture: Food, Development and Sustainability, 1870–2010 (FCT/PTDC/HIS-HIS/122589/2010).

List of Illustrations

Figures

Maps

Tables

List of Contributors

Luciano Amaral
Professor of Economic History at Nova School of Business and Economics, Lisbon.

Amélia Branco
Assistant Professor of Economic and Business History at Lisbon School of Economics and Management, University of Lisbon.

Dulce Freire
Research Fellow at the Institute of Social Sciences, University of Lisbon.

António Henriques
Assistant Professor at the Faculty of Economics, University of Porto.

Pedro Lains
Research Professor at the Institute of Social Sciences, University of Lisbon.

Susana Münch Miranda
Senior Researcher at the Institute of History, Leiden University.

Margarida Sobral Neto
Professor of Early Modern History at University of Coimbra and member of the Portuguese Academy of History.

Jaime Reis
Retired Research Professor of the University of Lisbon.

Ana Maria S.A. Rodrigues
Associate Professor at the University of Lisbon.

José Vicente Serrão
Associate Professor of History at ISCTE/University Insitute of Lisbon.

Ester Gomes da Silva
Assistant Professor, scientific coordinator of Institute of Sociology and researcher at the Centre of Economics and Finance at the University of Porto.

Introduction

Dulce Freire and Pedro Lains

This book follows the path of the renovation of European economic history which has led to a more unified investigation on endogenous sources of growth and stagnation, in the core and the periphery of the Continent, and moved away from the concentration on the study of the causes of the British Industrial Revolution.[1] In order to better understand why European economies expanded in many different patterns and speeds, we need to respond to a wider variety of questions and explore the centuries before industrialization. Possibly the single most relevant enquiry driven from the new research is whether the forerunners, such as England, Holland and France, were already ahead in terms of development when industrialization gained momentum, or whether differences in terms of development were by then relatively small.[2] Was there a "little divergence" of income levels within the European economies, and a "great divergence" between Europe and the rest of the world, dating back as far as the Medieval and the Early Modern periods?[3] This interrogation is directly related to the investigation about the "ultimate sources of economic growth".[4]

Questions about how economic agents responded to the constraints and opportunities in domestic and international markets are complementary to the investigation on cases of success or failure in the adoption of the technological developments of the more advanced economies in northern Europe. In order to test the validity of hypotheses about the basis of economic success, we need to ascertain not only why some countries succeeded, but also why others failed and did not follow the leaders closely. Economic history is also about peripheral and apparently unsuccessful stories of those that remained behind.

The European periphery is often taken as a single historical unit and yet there are relevant differences between or even within nations such as Portugal and Catalonia or regions such as Andalusia and may be even more pronounced ones between northern and southern Italy, or between present-day Hungary and the Balkan Peninsula. Thus, to understand the emergence of industrialization,

1 This renovation is patent in the works of Crouzet (2000), Pomeranz (2000), Maddison (2001), Clark (2007), Allen (2009), van Zanden (2009) and Broadberry *et al.* (2015).

2 See the opposing views in Clark (2007) and Allen (2009).

3 See Pomeranz (2000), van Zanden (2009) and Broadberry *et al.* (2015).

4 See Maddison (2001).

 | DOI 10.1163/9789004311527_002

we gain by looking with more detail at economic development in the European periphery so that we can explore the endogenous reasons behind economic performance there.[5] Due to the lasting dominance of agriculture in these economies, the investigation of the fundamentals of economic growth in the more backward regions of Europe needs to pay particular attention to the primary sector. That is the main purpose of the present book.

Studies on the European economy in the long run depend to a large extent on indirect measures for the evolution of income per capita and factor productivity. That information comes from the analysis of trends in land rents, wages and prices, as well as indicators for population growth, urbanization and other structural transformations. There are now estimates for long-term trends of national income and productivity levels for a wide range of European countries, such as England or Great Britain, the Netherlands, Italy, Spain, Turkey and Portugal.[6] The availability and the accuracy of quantitative data are, however, limited and the resulting indicators need to be compared with qualitative data in order to ascertain error margins. By the same token, the choice and definition of the indicators depend on an accurate evaluation of the historical narrative on the main characteristics of economic activity.

As we advance towards a better evaluation of the long-term evolution of income and productivity levels, we need to go back to the interpretation of institutional developments and other aspects of the economies involved, in order to have more complete answers to the questions about convergence and divergence. The dialogue between different types of sources is crucial, as it helps to point out possible contradictions and suggest new lines of both qualitative and quantitative evidence. That dialogue is, after all, at the root of the field of economic history itself. Long-running studies on growth thus greatly benefit from the in-depth knowledge of the economic and institutional developments of each geographical area under analysis. In turn, the interpretation of long-running developments is crucially dependent on the study of social, political and economic institutions, the structure of land ownership and tenure, the establishment of urbanization and settlement patterns, the introduction of new crops, land tax regimes, income and domestic and foreign trade, the formation and consolidation of political boundaries, among other factors.

---

5 See Federico and Malanima (2004), Federico (2005), Lains and Pinilla (Eds.) (2009), Pinilla (Ed.) (2009), Vivier (Ed.) (2010), Olsson *et al.* (Eds.) (2011) and Béaur *et al.* (Eds.) (2013).

6 See Malanima (2011), Schön and Krantz (2012), Álvarez and Prados (2013), Broadberry (2013), van Zanden and van Leeuwen (2012) and Palma and Reis (2016).

Our aim here is precisely to put together in a coherent way the historical evidence regarding the development of the agricultural sector in Portugal from just before its political independence to the present time. The new quantitative evidence raises questions that have been in the background of historical research on Portuguese agriculture for a long time. First of all, as we searched for a global picture of the evolution of the sector, we found that research on the different centuries or periods was to a large extent dependent on the evaluation of the performance of agriculture in the neighbouring periods. For example, the interpretation of events after the Black Death depends on the perception of what happened in the previous period, whether or not the tragic epidemic was set in a context of growing population and pressure on land resources. By the same token, the interpretation of the more positive view of 19th century agriculture, given by the new quantitative evidence, depends on our view of what happened in the previous century. Thus, we quickly found out that the quest for elements of continuity and discontinuity, which lay at the heart of the long-term analysis, was already embedded in the historiographical debate.

Portugal is a relevant case to study because of its secular borders, long experience with openness to the World – it was a pioneer explorer of overseas territories from the early 15th century – and its pattern of divergence and backwardness, which ultimately meant that it was the last country in Western Europe to industrialize and reduce the significance of the agricultural sector. Fortunately, the study of Portuguese agriculture in the very long run is made possible by the availability of a number of historiographical works on the topic, although with different levels of intensity over time.[7] It is now time to gather all this information and provide a long-term view of Portuguese agriculture in the European and global contexts.

European economic history has shown that, to a certain extent, the continent has been integrated since earlier periods in time, and the agricultural sector is no exception. The links between developments in agriculture and industrialization also need to be studied in the period before the British Industrial Revolution gained momentum. Economic growth is not only about technical and institutional innovation, but also about the development of conditions that allow for the introduction of the new techniques in agriculture and about incremental improvements in the way production factors are used. In earlier phases of development, natural resources, such as fertile lands and water, are more relevant. For agriculture, in particular, it is unlikely that output will

7 For a review of the literature see Costa, Lains and Miranda (2016).

increase in a context of institutional retrenchment or disaggregation of population settlements. Moreover, in earlier periods the structure and development of demand patterns may be of paramount relevance, allowing for changes in the structure of output, notwithstanding the importance of favorable conditions for the accumulation of physical and human capital.[8]

The study of trends and fluctuations in agricultural output and productivity in Portugal over 1000 years implies different sorts of approaches due to the type and availability of data and existing scholarship. As we embarked on this project, it became evident that the quantitative analysis was relatively advanced for the period after 1850, as there are output indices for the use of labor and land and levels of output, as well as capital. There was, however, very little quantitative data available for earlier periods, particularly at the macro level. This has changed recently, and we shall use the new evidence in the present analysis. Moreover, this line of research was globally absent, as economic historians were more concerned with understanding how the economy functioned at the individual and institutional level, rather than with trends in output and productivity, particularly during the Medieval and Early Modern periods. We nevertheless realized that there was enough material to probe the agrarian economy while dealing with the questions of growth and fluctuations. To a certain extent, these were questions that remained implicit in much of the economic historical research. Moreover, the fact that the quest for growth and productivity was kept for the whole period provided a valuable framework for the analysis. Thus, for instance, the understanding of the *Reconquista*, particularly during its heyday, may benefit from being investigated together with the question whether it took place in a period of agricultural growth or stagnation. The set of institutional transformations that for some time now have been identified in the agricultural sector during the 18th century are also better understood if we take into consideration the framework regarding growth and change that we are using in the current investigation. Aside from looking at continuities and discontinuities and growth of output and productivity, this book also looks at the transformation of institutions related to land, labor and domestic and foreign trade, and in these studies we have also found that there are many converging aspects across the whole period analysed here.

8 For these topics related particularly with the agricultural sector in the long run see Hoffman (2000), Bieleman (2010), Bavel and Hoyle (2010), Fonseca and Reis (2011), Myrdal and Morell (Eds.) (2011), Vanhaute *et al.* (Eds.) (2012) and Molle and Segers (2013). See also van Bath (1963), Bairoch (1989), O'Brien and Prados (1992) and Federico (2005).

The fact that we managed to define a set of relevant questions, anchored in previous studies on Portugal's agricultural sector, and closely related to a broader international concern, was a strong motivation for conducting the project. The analysis in each chapter in the book will vary, as each period has to be approached differently because of the different source materials and different types of problems. The chapters about earlier periods need to be more descriptive, whereas quantitative data for the 19th and 20th centuries allow for a closer look at yearly output and factor productivity growth. Foreign trade, technological change and policy are relevant only for the later years too. Despite these differences, the chapters will deal with a set of questions that are common throughout book and are crucial for its main purpose, which is to provide an interpretation of long-term agricultural growth in the southern periphery of Europe. Each chapter will therefore deal with topics such as demographic change and urbanization; land ownership and tenure; settlement patterns, and agricultural output; development of domestic and foreign markets; innovation and technological change; agricultural policy.

Following a period of territorial expansion to the south and the recovery from the Black Death epidemic, the Portuguese polity began a period of overseas expansion that led to the creation of an empire. The empire had different territorial configurations and levels of economic activity, and to a great extent marked the structure of the metropolitan economy. Such a unique feature has led many to question the extent to which the Portuguese economy was interfered with by the existence of empire. The consequences of the empire will be reflected in the conditions of agricultural development, either indirectly, through eventual changes in the supply of labor or directly, through competition from abroad in agricultural produce or the creation of an export market for agriculture. The empire was also a source of opportunities, and it is still unclear whether the domestic economic activity, particularly in the primary sector, responded positively to overseas expansion. Agriculture is also a good field in which to assess the impact of changes in the institutional setting regarding the expansion and consolidation of the state, changes in land ownership legislation, or monetary arrangements. Institutional change was intensified in the modern period, and we need to consider how those changes affected the agricultural sector.

The chapters in the book attempt to address a common set of questions, although with variations across time. In this regard, we will take into account aspects related to institutional developments affecting the structure of land tenancy, labor legislation, as well as domestic and international trade. The chapters will also take into account, albeit in a different fashion, according to the availability of data, the evolution of land, labor and capital supply, the growth of output and its structure, trends in factor productivity, foreign trade, and technical innovation and research. We will also use the study of the broader institutional framework, namely regarding issues concerning the occurrence of wars and major conflicts, major diseases, the relation between the state, the aristocracy and the church, the evolution of different forms of central government, the imposition of the rule of law, and taxation, property rights and economic policy.

The rest of the book is divided into four parts. Part 1 deals with the formation of the Portuguese state and Malthusian growth during the medieval period, a period for which there still is little quantitative information concerning output and productivity trends. It includes Chapter 1, by António Henriques, that studies the economics behind the expansion or the Christian *Reconquista*, from *circa* 1000 to the arrival of the Black Death in 1348; and Chapter 2, by Ana Maria S.A. Rodrigues, that addresses the consequences of the great plague and the slow recovery that followed during most of the 14th century, and shows that recovery was accompanied by relevant structural changes in the agricultural sector. Part 2 deals with the period of opening borders and extensive growth during the modern period, for which new quantitative evidence on output growth is presented. It comprises Chapter 3, by Susana Münch Miranda, that focuses on another period of slow but consistent expansion, from 1500 to 1620, during which Portuguese agriculture was for the first time affected, albeit only marginally, by international markets; Chapter 4, by Margarida Sobral Neto, reveals the years of slowdown and contraction of the agrarian sector, starting in what was identified as a peak of activity, around 1620, and that lasted for the remainder of the war with Spain until the 17th century; Chapter 5, by José Vicente Serrão, deals with the signs of institutional and overall expansion from the new international setting that stemmed from the Anglo-Portuguese commercial and political treaties of 1703, pointing to the increasing role of market relations in agricultural activity, and ends with the study of the reversal provoked by the wars of the turn of the century until 1820; finally, Chapter 6, by Jaime Reis, provides news indices for output and productivity growth for the period between 1500 and 1850, depicting phases of growth and stagnation, and serves as the basis for the more detailed discussions about growth trends and fluctuations in the previous chapters.

Part 3 deals with the late modern period in which sustained structural change occurred for the first time and economic policy gained momentum and achieved a crucial role. It includes Chapter 7, by Amélia Branco and Ester Gomes da Silva, which deals with the origins and developments of what was the first period of continued, though relatively mild, growth of agricultural output and productivity, lasting from 1820 to 1914, in which Portuguese agriculture was substantially transformed while still remaining a backward sector in the European context; and Chapter 8, by Luciano Amaral and Dulce Freire, which deals with the last period of growth of agriculture and its subsequent demise, as industrialization gained momentum, and as the integration in the European markets led to an overhaul of its traditional relevance in Portugal. Finally, Part 4 includes Chapter 9, by Pedro Lains, which concludes by providing a global interpretation of 1000 years of agricultural development in the European periphery. The book ends with an Appendix with maps and graphs that helps to put in context some of the features regarding Portugal's geography and institutional divisions.

References

Allen, R.C. (2009). *The British Industrial Revolution in Global Perspective*. Cambridge: Cambridge University Press.

Álvarez-Nogal, C. and Prados de la Escosura, L. (2013). "The rise and fall of Spain, 1270–1850". *Economic History Review*, 66 (1), pp. 1–37.

Bairoch, Paul (1989). "Les trois revolutions agricoles du monde développé: rendements et productivité de 1800 a 1985". *Annales* économies, sociétés, civilisations, 44 (2), pp. 317–353.

Bath, D.H. Slicher van (1963). *The Agrarian History of Western Europe A.D. 500–1850*. London: Edward Arnold.

Bavel, Bas van and Hoyle, Richard (Eds.) (2010). *Social Relations: Property and Power*. Turnhout: Brepols.

Béaur, G., Schofield, P.R., Chevet, J.-M. and Perez-Picazo, M.-T. (Eds.) (2013). *Property Rights, Land Markets and Economic Growth in the European Countryside (13th–20th centuries)*. Turnhout: Brepols.

Bieleman, J. (2010). *Five Centuries of Farming: A Short History of Dutch Agriculture, 1500–2000*. Wageningen: Academic Publishers.

Broadberry, S. (2013). *Accounting for the Great Divergence. Economic History Working Papers*. London: London School of Economics and Political Science, No. 184.

Broadberry, S., Campbell, B.M.S., Klein, A., Overton, M. and Leeuwen, B. van (2015). *British Economic Growth, 1270–1870*. Cambridge: Cambridge University Press.

Clark, Gregory (2007). *A Farewell to Alms: A Brief Economic History of the World*. Princeton: Princeton University Press.

Costa, Leonor Freire, Lains, Pedro and Miranda, Susana Münch (2016). *An Economic History of Portugal, 1143–2010*. Cambridge: Cambridge University Press.

Crouzet, François (2000). *Histoire de l'économie européenne, 1000–2000.* Paris: Albin Michel.

Federico, Giovanni (2005). *Feeding the World: an Economic History of Agriculture, 1800–2000.* Princeton: Princeton University Press.

Federico, Giovanni and Malanima, Paolo (2004). "Progress, decline, growth: product and productivity in Italian agriculture, 1000–2000". *Economic History Review*, 58 (3), pp. 437–464.

Fonseca, Hélder A. and Reis, Jaime (2011). "The limits of agricultural growth in a fragile eco-system. Total factor productivity in Alentejo, 1750–1850". In M. Olsson and P. Svensson (Eds.), *Growth and Stagnation in European Historical Agriculture.* Turnhout: Brepols, pp. 167–194.

Hoffman, Philip T. (2000). *Growth in a Traditional Society. The French Countryside, 1450–1815.* Princeton: Princeton University Press.

Lains, Pedro and Pinilla, Vicente (Eds.) (2009). *Agriculture and Economic Development in Europe since 1870.* London: Routledge.

Maddison, Angus (2001). *The World Economy. A Millennial Perspective.* Paris: OECD.

Malanima, Paolo (2011). "The long decline of a leading economy: GDP in central and northern Italy, 1300–1913". *European Review of Economic History*, 15 (2), pp. 169–219.

Molle, Leen van and Segers, Yves (Eds.) (2013). *The Agro-food Market: Production, Distribution and Consumption.* Turnhout: Brepols.

Myrdal, J. and Morell, Matts (Eds.) (2011). *The Agrarian History of Sweden: from 4000 BC to AD 2000.* Lund: Nordic Academic Press.

O'Brien, Patrick K. and Prados de la Escosura, Leandro (1992). "Agricultural productivity and European industrialization". *Economic History Review*, 45 (3), pp. 514–536.

Olsson, Mats and Svensson, Peter (Eds.) (2011). *Growth and Stagnation in European Historical Agriculture.* Turnhout: Brepols.

Palma, Nuno and Reis, Jaime (2016). "From convergence to divergence: Portuguese Demography and Economic Growth". Groningen Growth and Development Centre Research Memorandum, 161.

Pinilla, Vicente (Ed.) (2009). *Markets and Agricultural Change in Europe from the 13th to the 20th Century.* Turnhout: Brepols.

Pomeranz, Kenneth (2000). *The Great Divergence. China, Europe, and the Making of the Modern World Economy.* Princeton: Princeton University Press.

Schön, Lennart and Krantz, Olle (2012). "The Swedish economy in the Early Modern Period: constructing historical national accounts". *European Review of Economic History*, 16 (4), pp. 529–549.

Vanhaute, Eric, Devos, Isabelle and Lambrecht, Thijs (Eds.) (2012). *Making a Living: Family, Income and Labour.* Turnhout: Brepols.

Vivier, Nadine (Ed.) (2010). *The State and Rural Societies.* Turnhout: Brepols.

Zanden, Jan Luiten van (2009). *The Long Road to the Industrial Revolution. The European Economy in a Global Perspective, 1000–1800.* Leiden: Brill.

Zanden, J.L. van and Leeuwen, B. van (2012). "Persistent but not consistent: the growth of national income in Holland, 1347–1807". *Explorations in Economic History*, 49 (2), pp. 119–130.

PART 1

State Formation and Malthusian Growth

CHAPTER 1

The *Reconquista* and Its Legacy, 1000–1348

António Henriques

Introduction[1]

This chapter covers the period from circa 1000 to the Black Death. Throughout Europe, this long era witnessed major changes that transformed agrarian life: the "commercial revolution", the "urban revolution", the "renaissance of the 12th century" or the "feudal revolution".[2] In Portugal, these transformations were compounded by a great demographic and territorial disruption: the *Reconquista*.[3] Like the German *Drang nach Östen* or the Hungarian *Honfoglalás*, this process led to drastic changes in borders, populations, land use and institutions (Bartlett 1993). Political boundaries, settlement patterns and some internal differences in land use and property distribution become recognizable in this period. The very existence of Portugal as an independent state with its stable borders and distinctive features is inextricably linked to this process of territorial conquest.

The *Reconquista* and the subsequent decades were also a formative period that saw the emergence of institutions that would endure for many centuries and, thus, influence Portuguese agriculture in later periods. However, according to late 18th and 19th century reformers and historians, these institutions placed a heavy burden on agriculture in that period. Concentration of land ownership, one of the most maligned ills of Portuguese agriculture, was blamed upon the *Reconquista* since the 18th century (Fernandes 1991). In fact, "enlightened" and "liberal" authors took a dim view of "feudal" agriculture. In their view, idle, absentee lords concentrated land ownership, and farmers were oppressed by excessive rents, while mortmain institutions and long-term tenure blocked the development of land markets (Lobão 1814; Silveira 1989). In short, this long period created a powerful path-dependence that only "enlightened" or "liberal" reforms could overcome.

This largely negative appraisal of the institutions born out of the *Reconquista* was toned down during the central decades of the 19th century. The towering

1 The author wishes to thank André O. Marques for his valuable criticisms.

2 For a review of the classic topics in European history see Power (2006).

3 *Reconquista* is used here as shorthand for the conquest of the Muslim polities by the Christian kingdoms. See map A3 in the Appendix.

 | DOI 10.1163/9789004311527_003

works of Alexandre Herculano (1846: 153) claimed that the conquest of Muslim territory and the emergence of Portugal coincided with the development of benign, self-governing institutions that protected the interests of freeholders from the ravages of "feudal" lords.[4] Agriculture benefitted from this favorable framework. A pioneering essay on the history of agriculture by Rebello da Silva even claimed that the period from 1250 to 1325 was the high-water mark of Portuguese agriculture, whereas in the subsequent periods farming succumbed to "oppression" (Silva 1868: 37). This view proved very popular: in 1918, a well-known agrarian reformer urged his countrymen to "complete the *Reconquista*" (Campos 1918: 21).

Twentieth-century historiography did not fully break with these two traditions and their meta-historical concerns.[5] Legal historians continued to regard the institutions developed during the 12th and 13th centuries as the polar opposite of the horrors of European "feudalism" (Costa 1957; Rau 1982). In contrast, the most innovative authors in the fields of agrarian and rural history, Oliveira Marques, Iria Gonçalves and Cruz Coelho, reinstated the old, negative views on Portuguese farming.[6] In fact, their research in agrarian history was much influenced by human geography and by *structuralisme*. The historiographical perspective on agrarian matters was dominated by *longue durée* studies with great emphasis on structures and geographic conditions and which downplayed dynamics and change. Thus, the then-contemporary backwardness of Portuguese agriculture was projected back to earlier ages. Also, the re-introduction of the concept of "feudalism" in the 1970s and 1980s led historians to interpret institutions in purely distributive terms (*i.e.* lords vs. peasants) and misinterpret or neglect issues like living standards, the degree of economic freedom, cooperation and property rights. The argument put forward by these authors was that the country was trapped by poor ecological conditions, namely poor soils, which entailed low agrarian productivity. Accordingly, low productivity levels combined with the predatory nature of the "feudal" system to keep the Portuguese peasantry in poor living conditions which, allied to poor natural conditions and resources, prevented the development of internal markets and limited specialization.[7]

4 See also Herculano (1981: Vol. 4 Appendix).

5 See Amaral (2011) and Barata and Henriques (2011).

6 Tellingly, the first edition of Oliveira Marques' crucial work, the *História da Agricultura em Portugal*, was never on sale for political reasons and the author ended up being exiled in the United States by the Portuguese dictatorial regime.

7 See Marques (1978 and 1986), Coelho (1989 and 1990) and Gonçalves (1989). For a critique of these authors see also Barata and Henriques (2011).

Although the influence of these grand narratives is still felt, recent research on agrarian history is successfully sidelining such meta-historical issues and bringing forward different methods.[8] The present chapter draws on both old and recent research and discusses whether natural endowment or institutions worked as stumbling blocks for Portuguese farmers. This chapter is divided into four sections. As the *Reconquista* is crucial to the understanding of the first centuries of Portuguese agrarian history, the first section provides a brief background to this event and deals with the underlying forces in this process. The second chronicles the emergence of institutions adapted to conditions wrought by the *Reconquista*, namely the relative abundance of land in relation to labor. Then, in section three, we will analyse the role of market mechanisms in agrarian production. The fourth section deals with issues of production and productivity, and the fifth concludes.

The *Reconquista*

By 1000, the County of Portucale, the polity that eventually became Portugal, was but a small patch on the coast with some 12,000 km^2 on the right bank of the Douro. In the following decades, it started to annex Muslim-controlled territory and initiated a long, stop-and-go advance. Expansionist bursts in 1057–64, 1147–69, 1187–1191 and 1235–49 alternated with periods of stability in 1112–35, 1169–87 and 1191–1217, and at some points the Muslim or "Moorish" polities recovered lost ground, such as in 987, 1087–1112, 1169–70 and 1191. Eventually, in 1249, after two centuries of southward and eastward expansion, Portugal acquired its continental borders and a territory of about 90,000 km^2.

The long process of the Christian *Reconquista* occurred at the expense of Islamic states and populations. Muslim political and administrative structures were dissolved and the population was relocated and/or deprived of its status in earlier years. Settlements were wiped out (Boone 1996; Catarino 2002; Tente 2010) and there was a tangible, if unquantifiable, exodus to Muslim-held territory, namely to Andalusia or North Africa. This was detected in the archaeological record. In different settings, like the Serra da Estrela, *c.* 1150, the *Reconquista* coincided with sites where violence occurred being burnt to ashes or with the forsaking of prior settlements, although not all were necessarily Muslim.[9] Some elements of continuity cannot however be put aside: the northern conquerors duly adopted the, often intact, physical capital and

8 See Amaral (2007), Viana (2007), Gonçalves (2011) and Marques (2014).

9 See Tente (2010: Appendix).

resorted to Muslim workers, products and techniques (Barbosa 1992; Fernandes 2000; Boissellier 2003; Viana 2007; Gonçalves 2011); in a few specific cases, the conquerors found it convenient to keep some taxes and other aspects of communal organization (Barros 2007). While such aspects of continuity cannot be ignored, there was one radical change that not even continuity-oriented historians can deny: the substitution of Muslim customs and property rights by radically different ones brought in by the conquerors. This must be emphasized: unlike the Barbarian Invasions in 476 or the Muslim Conquest in 711, the *Reconquista* did not simply mean a change in the ruling elite or in religious denomination; it was an outright invasion which led to the replacement of one society by another. Even when the Christian conquest did not lead to the expulsion or the destruction of the existing communities, it always involved the redefinition of property rights and land use. In fact, in Christian society, military conquest was even regarded as the ultimate arbiter of entitlement to land ownership (Rau 1982: 29).

The year 1000 provides an apt starting point for understanding the forces that have affected both sides of the political and religious divide in present-day Portugal. By then, a long period of Islamic hegemony under the powerful Caliphate of Córdoba appeared more likely than a steady advance of the puny northern states. The army of the Caliphate had in 987 recovered Coimbra and most of the lands south of the Douro River. Defeated, troubled by internal strife and raided by Norsemen, the county of Portucale appeared vulnerable at the turn of the millennium. However, the 10th century Muslim expansion had stopped at the Douro. Whilst Moorish armies did raid the lands north of that river, the Caliphate did not envisage any further conquest. Muslim conquerors were content to fortify and garrison the strategic centres south of the Douro and to obtain the allegiance of the Christian communities. They did not organize the settlement of the conquered territory and, despite some cases of deportation and enslaving, Christian populations remained in large numbers and kept their property rights (Azevedo 1993). Picard (2005: 127–129) goes on to conclude that the Muslims had a structural demographic disadvantage vis-à-vis their Christian foes.

By 1000, the military organization of Córdoba still allowed the Muslims to punch above their demographic weight. However, when the Caliphate collapsed, in 1031, the western lands of modern-day Portugal ceased to be a heavily-guarded frontier of a solid empire and became a distant periphery for the warring kingdoms centred in Toledo, Badajoz or Seville. After some earlier exploits, local nobles started in the 1030s to encroach on lands south of the Douro: Arouca, Montemor-o-Velho and Paiva (Amaral 2007: 136–137; Marques 2008). The campaigns that led to the conquest of Lamego, in 1057, Viseu, in 1058, and Coimbra, in 1064, by King Ferdinand I of Leon placed the lands between the Douro and the Mondego Rivers (some 16,000 km^2) in Christian hands. The contrast between the Muslim (987–1064) and the Christian (since 1064) occupation of Coimbra is telling. Whilst Muslim rule had meant a thin, essentially military occupation, Christian rule led to the large-scale settlement of northern migrants (Amaral 2007: 138–142). King Ferdinand I placed these newly-conquered lands under the rule of a local magnate who was entrusted with the task of, in the words of a contemporary charter, "settling and founding villages" (Rau 1982: 32, note 3).

Settlement did not always wait for military advance. When the Almoravid Empire (1086 to 1140) halted Portuguese progression, settlers continued to claim forest and heathland south of Coimbra (Gomes 2009). After this area was secured by the double conquest of Santarém and Lisbon in 1147, these lands saw the emergence of a string of towns built around existing castles: Leiria, Tomar, Torres Vedras, Torres Novas, Alenquer, Óbidos, Ourém, Atouguia, Lourinhã. All these settlements grew rapidly, as reflected in the building of long walls and the multiplication of urban parishes (Silva 1996: 33, 56–57). By 1200, these towns were surrounded by new villages, established near the best soils. This intense process contrasts with the Muslim neglect of this region in the previous centuries.

The region of Leiria, the southernmost frontier of Portugal c. 1135, provides a case in point.[10] Part of it was claimed by a powerful magnate (Barbosa 1992), whilst the rest was cleared by townsmen (Gomes 2009). Early 12th century documents presented this area as a *terra deserta* (literally, "no man's land"), although archaeology detected some marginal settlements from the Roman period that remained inhabited (Bernardes 2005). The term *terra deserta*, which cannot always be taken at face value, indicates how the Christian northerners perceived the South: as a land up for grabs. The imagery of deserts legitimized the land-grabbing efforts of lords, individuals and institutions and sanctioned the appropriation of land without any regard for previous rights.

The Portuguese *Reconquista* culminated in the capture in 1249 of the last Muslim strongholds in the Algarve (the castles of Albufeira and Alzejur and the cities of Loulé and Faro). In the hundred years, between 1147 and the conquest of Faro the area under Portuguese rule grew by about 35 percent. This time, however, settlement lagged behind military advance. The wide frontier opened in the South was too large to be occupied as intensively as the lands between Coimbra and Lisbon. Whereas north of Santarém there were practically no Muslim communities, the lands to the south of the Tejo River retained some pockets of Moorish population. Still, the link between demography and military advance is very impressive, as shown by the cases of Coruche and Loulé.

10 See map A3 in the Appendix.

Coruche, a precarious military outpost on the south bank of the Tejo was occupied shortly after the fall of Santarém in 1147. About a generation later, in 1181, when it operated as a bridgehead south of the Tejo it was already inhabited by about 400 families, as we learn from a Muslim-chronicler (Boissellier 1999: 113). Loulé was captured in 1249 and two-thirds of its lands distributed among Christian settlers. Fifteen years later, knights from Loulé were claiming lands in the recently-conquered Jerez de la Frontera. The fact that families could benefit from two land distributions in less than one generation was a powerful incentive for moving southwards.

Kings Afonso I (1140–85) and Sancho I (1185–1211) as well the Knights Templar attracted settlers from Northern Europe to the southern frontier. However, we do not find in Portugal the contemporary large-scale operation by which thousands of Frisians, Saxons and Germans from the Rhine were relocated to the Hungarian plains or to the shores of the Baltic. The settlement of the conquered territory was largely spontaneous (Conde 2000: 69–71), apart from later operations, mainly on the northern and eastern borders with Castile. Here, in the second half of the 13th century, there was still room for new sizeable settlements. Indeed, in the north east Afonso III created Vila Real in 1273 for 1000 families. In the same year, the fortified towns of Montalegre and Monforte were built ex novo for 100 families both in the bleakest spots of Trás-os-Montes. In the south, these institutional, large-scale settlements were still underway in 1300, namely Arraiolos, in 1290, Quarteira, in 1297, and Lavre, in 1304.

The reasons for this expansion are the decisive issue for historians studying this period. Clearly, the outcome of the *Reconquista* cannot be regarded solely as the result of the superior military power or better political organization of the Christians. This was a long-term conflict fought by two agrarian societies in which factors like demographic pressures, land-hunger and the fight for rents and revenues were more powerful influences than ideological drive, ambitious strategies or military prowess. In order to make sense of the *Reconquista*, it is imperative to consider agrarian "push" (notably, population outgrowing resources) and "pull" factors (wealthier Muslims with higher land/labor ratios). These forces, however, are extremely hard to disentangle in the few written documents available. Thus, research on the *Reconquista* tends to resort to other disciplines such as archaeology, history of art, palynology, climate history and linguistics. Regional archaeological studies, in particular, provide a useful complement to the written sources as, unlike those, they focus on continuity rather than change (Vieira 2000; Cosme 2002; Tente 2010).

As seen, it is not possible to dissociate population pressure and land-hunger up north from the *Reconquista*. Nevertheless, demography cannot be used as a *deus ex machina* to explain the expansionism of the Christian kingdoms. Instead, the population growth on the Christian side should be regarded as a proximate cause. Ultimately, the reasons for the differences in the population of the two sides should be sought in two different but complementary factors: climate change and institutions. These differences help to explain how a rural society pressed by decreasing returns eventually defeated and ultimately replaced a more urbanized and sophisticated society. The rest of this section explains how these factors had differing impacts on these two rural societies.

The first major factor to consider is climate. Climate specialists agree about the existence of a "Medieval Warm Period" from c. 900 to c. 1300 in the Northern Hemisphere (Esper, Cook and Schweingruber 2002; Brázdil 2005). A positive variation in temperature has different implications for northern and southern territories. In the high, mountainous lands that formed the core of the Christian territories, warming neutralizes the cooling effect of altitude, making plateaus and highlands more hospitable. Also, the high gradients of these rugged territories imply that increases in temperature will allow for extending the grown area and diversifying cultures. In concrete terms, warming opens up the possibility of growing more valuable and nutritional species like vines and olive trees. Although sources are unclear, these effects were confirmed by empirical research on 10th century Portugal: better ecological conditions for men, animals and plants allowed large villages to be settled at altitudes that were not met again in any historical period (Henriques 2003; Fontes and Roriz 2007: 10); likewise, vines reached altitudes that would not be seen again (Durand 1982; Henriques 2003: map 1.1); wine replaced cider as the dominant beverage in the Christian north by 1100, with apple orchards probably being replaced by vineyards (Gouveia 2012); in mountainous areas there was a change from extensive to intensive cattle-rearing somewhere between the 9th and the 13th centuries (this was possible due to the *lameiros*, small irrigated plots with a combination of sown fodder and natural pasture (Henriques 2007)); 11th century charters provide new names for agrarian plots, hinting at the allocation of plots for specific cultures and a more diversified use of the soil (Amaral 2007: 163).

These changes allowed for both the intensification and the extensification of agrarian production. That is to say, farmers could both produce more (in either calorific or monetary terms) using the same amount of land or expand the quantity of land put to use, as warming made the cultivation of hitherto unusable land possible. This probably induced a demographic expansion of this rural society. In contrast, warming could not have had a similar effect on the dry, low-lying southern plains. If anything, it made them more arid. This agrees with some recent research suggesting that warming even reversed endogenous demographic growth in the Muslim area (Drake, Hanson and Boone 2012).

Human societies, however, are not determined by demographic dynamics in the same way as animal populations. They develop rules, some of which are designed to avoid or accommodate growth in numbers. Also, some rules may unwittingly prevent or foster the growth of population. This brings us to the second major factor of our argument: institutions. It can be said that, unlike what happened in the south, institutions in the Christian north were not designed to thwart this demographic growth. In contrast, it can be argued that the strength of fiscal and communal demands on Muslim families deterred sustained demographic growth. The existing 10th and 11th century charters transferring land, by donation, by exchange or by sale, are clear about the rules that affected the distribution of property. Land ownership was divisible, marketable and distributed equally among the existing sons. On the whole, these sources testify to a clear prevalence of individual over communal rights. Typical communal rights, like access to water or forest, appear to have already been privatized in the 11th century and were also divisible, marketable and inherited along with property (Marques 2008: 211–216). By then, there is no indication that public authorities, i.e. the counts of Portucale or the kings of León, were tapping a significant proportion of the existing resources with something resembling a fiscal system, as was the case in the Muslim south.

Researchers detected a strong expansion of the farming area in the north during the 10th and 11th centuries (Amaral 2007 and 2009). The place names in the existing charters indicate that settlement was already pushing towards sub-optimal areas by 1000. In broad terms, it is possible to speak of uphill settlement. An impressive list of parochial churches compiled in Braga in 1085–91 shows that the dense settlement did not stop in the fertile western valleys like Braga or Guimarães, and had already reached the plateaus of Lanhoso and Vieira (Amaral 2007: 345, map 15). In the well-documented district of Penafiel de Sousa, there is evidence that 11th century settlement neared its ecological limits (Santos 2004: vol. 1, 143–144). In the highlands of Barroso, distinctively older place names occur in the favorable valleys, whereas place names implying deforestation are located uphill in worse soils (Henriques 2007).

The old Portucale region already featured its trademark dense network of small settlements (*villae*) (Costa 1997: vol. I, 295). These *villae* brought together one or two dozen families. While many had at some point been owned by a single lord, existing charters show that this was seldom the case by the 11th century. Inheritance laws favored the partition of property and, as land sales show, small properties predominated (Garcia Alvarez 1971–2). By 1000, the *villa* was organized into family-size units (*casais*). The *casal* was an elaborate combination of specialized plots and rights over communal goods, including wasteland, woods and waters. Some of the earliest charters relative to this region suggest that these units were already being parcelled and divided amongst heirs (Marques 2008: 61–65, 91, 234).

The morcellation of land asset shows that institutions were not able to contain the effects of demographic growth on property. This process can be read in a rare quantitative indicator: the price of agrarian land. The high grain price of land by surface before 1060 (see Figure 1.1) reflects this pressure on increasingly scarce land. The conquest of Coimbra and the ensuing southward migration greatly released the pressure on natural resources.

An 11th century list of parochial churches already indicates the dense network of parishes that remains typical of the region (Costa [1959] 1997–2000). Costa (1981b) showed that all the 82 settlements that became parishes in the area of Guimarães were already in existence in the late 10th century. Whilst most types of sources are inconclusive, the emergence of parishes demonstrates that settlement was dense and, ultimately, that these small communities could produce enough surplus to sustain one non-producer (Amaral 2007: 176) and meet the material demands of the cult (candles, lights, works, adornments, ecclesiastical taxes and fees). As churches were founded by local wealthy laymen or by a group of locally important families, they are indicators that the surplus was not captured as taxation and sent to cities or to central authorities. This starkly contrasts with what happened south of the Tejo

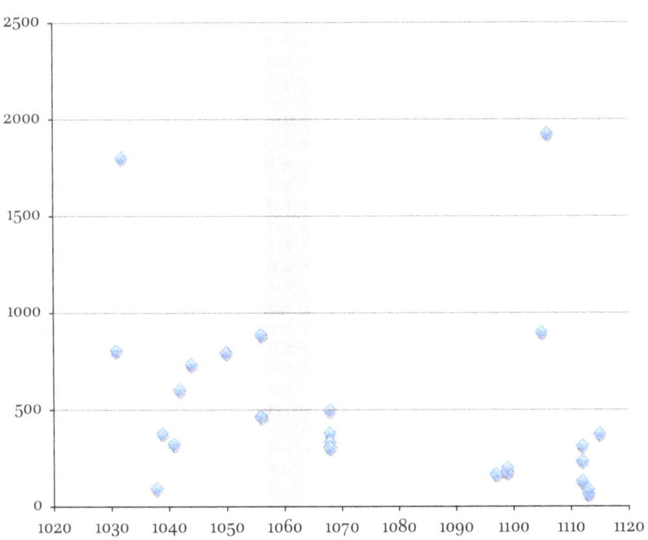

FIGURE 1.1 *Grain price of land in northern Portugal* (*100 = 990* AD)
The shaded area indicates the years between 1057 and 1064 that saw the conquest of Lamego, Viseu and Coimbra

where religious buildings appear confined to the cities. The concentration of religious works in the cities tallies well with the channelling of Muslim agricultural surplus to urban elites or even directly to Córdoba (Picard 2005). As far as existing sources indicate, the north was in sharp contrast with the Moors. Though in the absence of charters nothing of substance can be argued about inheritance or property rights, archaeologists and historians tend to consider that collective interests prevailed. Communal ties appear decisive in the small, nondescript settlements excavated by archaeology (Boone 1996; Catarino 2002; Picard 2005). These communities had to keep their flocks, manage their irrigation as well as man their defences in case of danger. This was compounded by state interest, as Córdoba, according to Picard (2005), drained resources from the countryside to the cities in the form of taxation.

The lack of charters makes the knowledge of agrarian life in the southernmost lands under Islamic domination by 1000 somewhat sketchy. Apart from archaeology, anecdotes recorded in chronicles or in terse geographical tracts remain the main source of information (Boissellier 1999: 452). Still, it appears likely that large property was common in the Muslim-dominated part of later-day Portugal. Eighth-century Muslim conquerors wrestled large properties from the Visigoth nobility, at least when the latter did not submit or convert in order to keep their lands (Chalmeta 1994). Some properties surrounding the main Andalusian cities counted on servile or enslaved workforce. The analysis of the Christian sources relating to Santarém led Viana (2007) to argue that large, villa-type properties dominated the fields around the Tejo. Likewise, archaeology detected the continual occupation of many Roman *latifundia* in the Algarve during the Islamic period. The continuity of the Roman large-scale property tallies well with two other features of Muslim society: pronounced inequality and the dominance of cities over the countryside. Important social differences are confirmed by archaeological research (Gomes 2012). The presence of large properties was not the single source of inequality. Chronicles refer to lively social conflicts in cities like Beja (Alentejo) and Silves (Algarve) setting the urban and literate aristocracy against small landowners or farm workers (Fernandes 2000). Muslim aristocracy combined landownership with income from public offices. This income, in turn, was generated by a fiscal system that diverted resources from the countryside to the cities, namely through the *as-sukhra* or *sofra*, a tax aid in the form of days of forced labor in city works (Guichard 2003).

With their grand public buildings, Muslim cities cut an imposing figure. However, in the spaces between these cities, archaeologists have unearthed a network of small, nondescript villages. At least, some of these settlements were inhabited by clan-type societies of North African stock, as suggested by linguistic and archaeological evidence (Boone 1996: 80). These small villages were most likely organized along communitarian lines.[11] As archaeologist Christophe Picard noted, the archaeological record shows little sign of economic integration. Notably, in stark contrast with its northern equivalents, coinage is noticeable for its absence, and mosques are either modest or non-existent. In the urbanized areas, where land ownership was more concentrated and social differences were steeper, the local elites and the state were able to finance the building of "public works" like castles, palaces and mosques (Picard 2005: 135). The strength of collective interests, in the form of either taxation or communal obligations, likely hampered the dynamism of the Muslim agriculture.

In sum, it can be argued that a long-term ecological improvement combined with institutions that led to the fragmentation of property created a strong push factor in the north. Coupled with a high level of fertility, induced by the expansion of agrarian production, the intense division of property threatened to stretch natural resources. By contrast, warming in a less diversified ecology with a seemingly smaller dynamic distribution of property did not lead to similar effects. At any rate, with the Tejo basin firmly in Christian hands, Portuguese farmers had vast room for expansion within the borders of the kingdom. Thus, the main legacy of the *Reconquista* was that, unlike other European countries, early 14th century Portugal was hardly a *mond plein* struggling with land scarcity and decreasing returns. The expansion of the agrarian surface meant an increase in population. Signs of growing pressure on land are visible in the steady increase of the rate of returns from land (see Figure 1.2).

Nevertheless, this pressure could still be absorbed by Portugal's large stock of land. In the outskirts of Leiria or Santarém, the 1340s, *i.e.*, two centuries after the *Reconquista*, was still a period of forest clearances and drainage of swamps (Gonçalves 1989; Viana 2007; Gomes 2009). From 1306 to 1348 the heathlands around Coina (south of Lisbon) was gradually converted into vineyards and grain fields, until the Plague imposed a sudden halt on the work (Pontes 2006). The vast, loosely settled lands south of the Tejo still allowed for the foundation of new settlements in the limits between villages (Viana 2007) in the 14th century. Thus, in 1329, roughly one century after the conquest, a knight-friar could still describe this territory as a "new land, with much scope for improvement".[12]

---

11 For an alternative interpretation see Boone (1996).

12 See at the Torre do Tombo (Portuguese National Archives): TT, Ordem de Santiago/Convento de Palmela, DR, 1, 25.

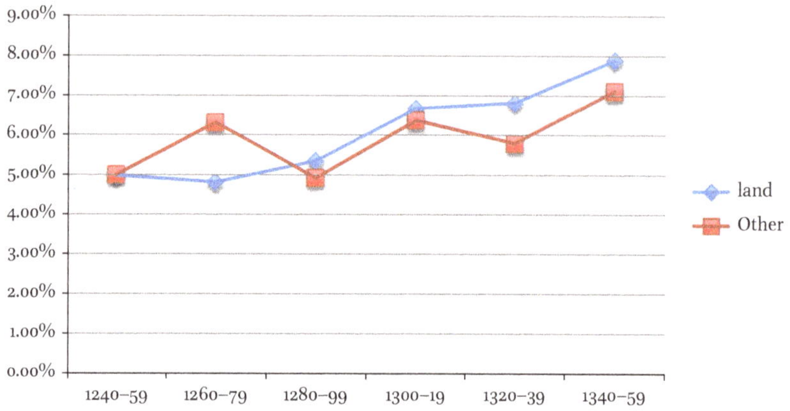

FIGURE 1.2 *Nominal rates of return*

Institutional Developments

Among the institutions that emerged along with the *Reconquista*, tenure was the most maligned from the Enlightenment onwards. Tenure was regarded as an intrinsically unbalanced agreement that encouraged both the exploitation of farmers and the indifference of owners regarding productivity (Lobão 1814). Unlike freehold, tenure discouraged investment and contributed to keepin productivity stagnant or declining. Later authors compounded this negative perspective by interpreting tenure in reference to classical "feudalism" or to "feudal relations", in which lords concentrated a large share of production factors and controlled the fiscal and judicial systems and thus seized an important share of farming output (Coelho 1989). In doing so, they deprived farmers and landowners of the right incentives to bring out the best use of the lands they tilled. Is this correct? This section will assess the intensity of the "feudal" restrictions in three different planes: capital, labor and land.

In "feudal" settings, lords ban farmers from building mills or other capital goods in order to preserve their rents or, alternatively, overinvest in these in order to exploit their monopoly rights, as occurred in England (Campbell 2005: 44). These typical "feudal" settings do not apply to the present case. From the outset, Portuguese common law was against such monopolies (Marques 1978), although some lords secured limited privileges. The custom followed in Estremadura, the *foro de Santarém*, clearly stated the freedom to build water mills and olive or wine presses (Gonçalves 1989: 189). By an undated edict from Afonso III (1248–79), this liberty to erect mills became the norm throughout the kingdom. Even the most powerful abbey of the country, the Cistercian monastery of Alcobaça (close to the city of Leiria, Estremadura), had to comply with this law and allow individual farmers to build new water mills. On the whole, the feudal monopolies did not stand in the way of enterprising farmers, with a few specific exceptions.

In the present state of research, it is hard to find empirical arguments for a structural absence of investment in the countryside. Portuguese farmers were, for instance, early adopters of wind mills, which have been documented in the country since 1182 (Marques 1978: 195). Tidal water mills, a very innovative technology, were also adopted relatively early and became widespread, without major institutional obstacles (Silveira 2009). Another important capital good was the wine press, which allowed the production of a market-oriented commodity. Privately owned, often in shares, wine presses were common throughout the country. Their large numbers and availability are reflected in the small fees (*lagaradigas*) paid for using them (Viana 1998; Henriques 2013). As countless sales contracts testify, the property of a given vineyard, even if relatively small, granted the owner access to one share of a local wine press.

This evidence is coherent with a more systematic indicator: the rate of return. The rate of return on investment was low in comparison with those of other European countries, implying that capital was not in severe want. This can be observed in interest rates implicit in the contract known as *censo* (see Table 1.1). A *censo* was a perpetual interest paid yearly in a given agrarian property in exchange for a lump sum advanced to the owner of that property. It could work as a form of raising capital at the cost of a fraction of future

TABLE 1.1 *Nominal interest rates in rent charges in Europe (percent)*

	1251–1300		1301–1350	
	rate	n	rate	n
England	10.2	43	11.2	3
France	11.1	88		
Germany	10.8	5	10.1	56
Holland	10	5	11.2	20
Italy			7.3	3
Portugal	4.8	29	6.2	12

returns from a given piece of land. As similar contracts existed throughout Europe (Clark 1988; Zuijderduijn 2009: 20, 283), it is possible to compare Portugal's relative availability of capital. Considering the values at Table 1.1, landowners could not complain about the cost of raising capital. Thus, enterprising farmers not only were free to build physical capital as they could do it with moderate costs.

In theory, by enlarging the amount of available land, the *Reconquista* allowed Portuguese farmers to abandon territories where land-labor ratios were lower for the newly-conquered areas. In other words, the *Reconquista* created the conditions for a large North–south migration. However, it is possible to imagine feudal restrictions to the mobility of labor leading to excessive labor in the north and scarce labor in the south. Did this happen? It did not. Feudal rules could not avoid a strong North–south migration. Tenure relations did not prove binding enough to avoid the mobility of labor. As Viana (2007) argued, this can be demonstrated in the outright relinquishing of family-size agrarian units (*casais*) owned by monasteries and other lords in the northernmost areas of the country by their tenants in 1258. The same can also be detected in the centre of the country, where lords acknowledged that they could not halt this flow and were content with merely collecting modest exit fees for tenants. It is thus not surprising to read in the text of a 14th century plea: "that it is the law of the king of Portugal that each man can earn his living wherever he wants".[13]

The ideology, the customs and the institutions in the time of the *Reconquista* favoured the acquisition of land through individual effort. As Barros (1945–1954) and Rau (1965; 1982) pointed out, the old Visigoth law that assigned untilled land to those who grew it was deeply ingrained within the Christian half of the Peninsula. According to Costa (1982), the legal principle of the "conquest of ownership through work" was already firmly established in 12th century common law. The *presúria*, the conqueror's right to dispose of the lands wrested from the infidel, was even older. Immediately after the conquest of Coimbra in 1064, the local count granted free lands to all the incoming settlers (Coelho 1989: vol. 2, 7). The *presúria* benefitted the monarchy and a few enterprising religious institutions and noblemen, which amassed vast lands. However, lesser players could also earn their share through the redistribution of lands by the king and other lords. Furthermore, the *presúria* evolved into the appropriation of any untilled land simply by ploughing for one day. The same principle was at stake in the land considered to be within the limits of any giving municipality in the central and southern lands. Here municipalities distributed unoccupied land to incomers (the practice known as *sesmaria*),

13 See at the Torre do Tombo (Portuguese National Archives): TT, São Jorge de Coimbra, M. 9, n. 32 (1369).

seemingly on a first-come first-served basis (Rau 1965). This practice continued throughout the 14th century (Fernandes 1991).

The Muslim system of property and settlement was destroyed by the conquest. The extent of this change was such that the Moors of Loulé, who had capitulated in the hope of obtaining better terms, had to apply for land along with the incoming Christian settlers. Ignoring all previous property divisions, the Moors of Loulé received one fourth of the arable land. The effects of such a radical redefinition of property rights can also be observed in the fertile alluvial plains between Santarém and Lisbon, as studied by Viana (2007). Under Muslim rule, these plains had belonged to major landowners living on their estates ("*aldeias*") or in the city itself. The Portuguese occupation in 1147 led to the extinction of this class with its rural and urban properties being shared by the victors. Property was distributed from scratch: *terra de novo dividebatur inter populatores* (Costa 1981a). The new division led to the foundation of new quarters and new villages in the periphery of Santarém and, as a result, a few decades after the conquest, settlement had assumed the dispersed pattern typical of the northern and central lands (Viana 2007: 162). The same can be found in the region of Beja, a decayed city that was abandoned by the Moors in the 1170s and occupied by Christian settlers at the turn of the 13th century. Here, the old Muslim villages were eventually occupied, but settlement did not stop there, as newly settled villages emerged alongside them (Fernandes 1991: 95). Even archaeologists and authors defending some measure of continuity acknowledge that pre-conquest settlement patterns changed dramatically (Boissellier 2003: 45–50).

Portuguese settlers could, and did, obtain property rights in the newly conquered South. The impact of this land distribution in the concentration of property is still to be measured. Large-scale quantification has not yet been attempted, though a survey by Durand (1982), using the owners of the contiguous properties, indicates that lay, private owners were dominant even in a region with extremely powerful ecclesiastical lords such as Coimbra. Up to the mid 14th century, land was still being distributed to settlers. As such Beja, like other southern municipalities, kept an officer, the *sesmeiro*, charged with assigning unoccupied land (*sesmo*) to incoming settlers.

Abundance of land in private hands, either through ownership or long-term tenure, was correlative with the scarcity of labor for the lords and great landowners. The more radical forms of forced labor i.e. slavery and serfdom, were rare. Early 10th century charters document the presence of serf families, some of them tied to specific properties. Nevertheless, there are reasons to consider that in the following century serfdom was on the wane, even among the most powerful families (Gouveia 2012). As elsewhere in Europe (namely France, Sweden and the Low Countries), servile and slave labor became rare in the

13th century (Coelho 1989: 276–279; Rodrigues 1996: 582–585). The raids and conquests in Muslim territory provided the monarch and some lords with captives. Some were tied to the seigniorial reserve (Coelho 1989: 279) but eventually became tenants, as suggested by some cases; by 1250, captive moors brought by force to the Northern lands had already merged in the general population (Barros 2005).

The Muslim population that remained in the conquered lands could have provided a vast low-cost workforce for the conquerors. Nonetheless, the Moors remained free. Their status remained inferior, as they bore higher taxes than the Christians, including workdays in crown lands. But they could nevertheless own and let lands and, despite their inferior status, they were eventually integrated into the economy as smallholders and craftsmen. The political interests of the monarchy were perhaps important in keeping some Muslims free, but there were others factors at play. This is illustrated by the fate of the Moors ruled by the Military Order of Santiago. From 1200 to 1245, the friar-knights of Santiago forcefully conquered nearly half of the Alentejo and came to dominate a considerable Moorish population.[14] In the first decades under Christian occupation, the Muslims were considered captives and therefore worked under coercion. But a generation later, they were accepted as tenants for some of the Order's lands. This reflects their growing bargaining power in an underpopulated frontier society.

The scarcity of labor is well reflected in the laws and ordinances issued by the crown. Such rulings show to what extent the lack of laborers grieved the lords. The Laws of 1211 ordered that each man had to live "under a lord", an order that reads as an attempt to halt the deterioration of seigniorial ties and, accordingly, to avoid paying wages. In 1253, a new ordinance tried to fix the yearly wages of the workers in the animal husbandry sector. In 1340, in the height of Pre-Plague demographic growth, a fresh ordinance tried to reduce laborers' wages. Indeed, the landowning opportunities offered by the *Reconquista* certainly reduced the availability of wage workers.

In the absence of better data, such laws indicate high labor costs. For great landowners, it was expensive to internalize the cost of labor and directly farm their lands. Many of them, namely monasteries, parish churches and cathedrals, were not allowed to sell lands. Thus, along with the *Reconquista* and the ensuing migrations, the 12th century witnessed the emergence of a new land tenure institution: the *enfiteuse*. This form of tenure transferred some of the property rights to the tenant either perpetually (*aforamento*) or not (*emprazamento*) (see Table 1.2).

14 See map A3 in the Appendix.

TABLE 1.2 *The main forms of land tenure*

Forms	Sharing property rights?	Subtypes	Periodicity	Rent	Tenant
Enfiteuse	Yes	*Aforamento*	Perpetual	Share	Collective or family
		Emprazamento	lives (typically 3; 1, 2, 4 and 6 also used)	Share or fixed	Family
Arrendamento	No		years (between 2 and 10)	Fixed	Family or individual

In both forms, the *emphyteusis* can be read as an answer to the freedom and scarcity of labor. According to the legal historian Costa (1957: 72–73), the prevailing conditions in the 12th century "led the farmer to strengthen his claims over the tenured land to the extent that it acquired rights that had to be respected by the owner". The suitability of long-term tenures to the conditions in Portuguese agriculture can be observed *a contrario* by the failure of the rival institution: the *complantio*. Under the *complantio*, property was equally divided between the farmer and the owner, after the labor of the former made the necessary changes to make it more productive (as for instance transforming heathland into a vineyard) (Costa 1957). The former would receive a payment for his effort that could be transmitted to his heirs, whilst the owner saw the productive capacity of his property improved. It was a common institution throughout Europe, most of all where labor-intensive cultures like vines were important. Sharing land was appropriate for the interests of families and individuals but not for perpetual entities, like monasteries and cathedrals, which did not intend to give up their patrimony. Also, it was not that attractive to farmers who had cheaper ways of accessing land, either as owners in the centre and south or as long-term tenants anywhere in the country.

Thus, the 12th century saw a shift from direct exploitation to tenure which has been observed in nearly all of the institutions studied.[15] Tellingly, this

15 The exception being the monastery of Aguiar, in the Beira district (Vicente 1996).

shift can be detected even in the Cistercian houses, whose labor needs were potentially lower thanks to the non-waged manpower provided by the *conversi* (devout laymen who retired to the monastery to work and pray). 12th and 13th century Alcobaça, the largest abbey of the realm, provides a case in point. Its ever-growing and widely dispersed estates could not be adequately exploited by the *conversi* and gradually the abbots granted their lands to tenants for long periods (Barbosa 1992: 139).

Many of the lands owned by the monasteries had been at some point private land that had been bestowed on, or bought by, those institutions up to the mid 13th century. Thus, monasteries were great landowners in the sense that they owned many properties; they, however, had few large properties. As lands were seldom brought together through a deliberate, coherent process of acquisition, the landed patrimony of the monasteries appears as a collection of scattered properties. Moreover, these properties had been accumulated by private individuals, who had built the structures they thought necessary for their production choices. Thus, large-scale management was seldom possible for monasteries. Family-sized property required family-scale management, and thus long-term tenure, in the form of either *aforamento* or *emprazamento* remained the typical form of management. Several monographs found that *arrendamentos*, the most commercial and profit-oriented type of tenure, are rare in the archives of monasteries and cathedrals.

The predominance of the *emphyteusis* affected large landowners but it did not stand in the way of a proper market for land. In fact, the holder of land under the title of an *aforamento and emprazamento* was entitled to act as an intermediary and let the property or part of it. Seemingly, the relative scarcity of labor meant that tenants had the upper hand when it came to negotiating terms with the major landowners. Thus, the long-term stability of rents in the region of Coimbra (see Figure 1.3) should not come as a surprise. Landlords appear to have been unable to change the customary rents collected and were not able to profit from the demographic growth that occurred throughout the period studied.

The weakness of feudal bonds could be adequately measured by comparing the average income of a peasant family with the total feudal rent, as was done elsewhere (Kitsikopoulos 2000). Nevertheless, there is still a long way to go before research on Portugal can quantify the overall burden of the various "feudal" rules and customs, along with taxes. Naturally, this burden varied considerably within the country as in some regions, notably Coimbra (Coelho 1989) and the North West (Mattoso 2001), lords appear more powerful than in the rest of the country. What is clear is that two of the more onerous forms of feudal extraction, namely monopolies and seigniorage, were the exception

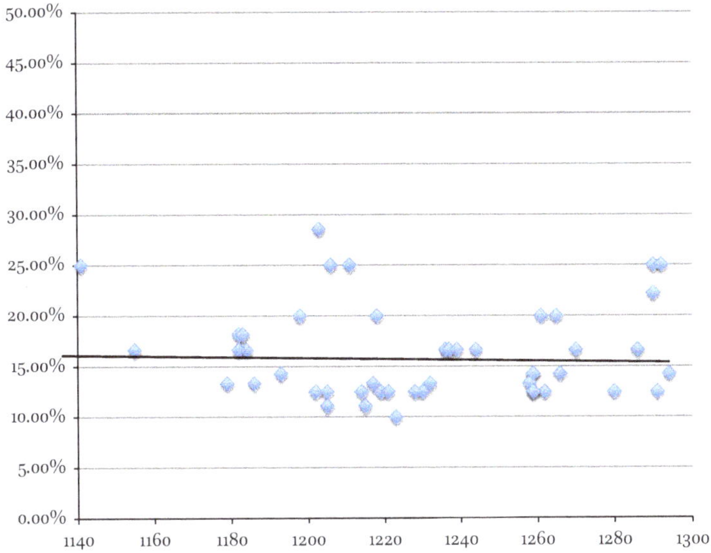

FIGURE 1.3 *Rent agreed in collective contracts in the region of Coimbra (percent of total output) (1135–1300)*

rather than the rule. Only the few border districts where Leonese coinage circulated were liable to the tax corresponding to the seigniorage (*moneda*). As mentioned, monopolies were the exception, with the *nullius diocesis* territory of Tomar under the Knights Templar being the most noteworthy case. State taxation, in the form of extraordinary aids (*serviços*), was hardly important, with typical rates varying between two and four day wages per household, as indicated in contracts. As these aids were rather infrequent, this did not represent a heavy burden. Also a few days, often one per year, of unpaid work (*geirus*) were sometimes demanded in tenure contracts in the northern half of the country. By contrast, in Lisbon or further south, such clauses very seldom appear in contracts.

As elsewhere in Europe, tithes placed a heavy burden upon the peasantry. Existing norms were favorable to the church, which, in the case of grain, olive oil and wine, kept between 11 and 13.3 percent of the gross product. Nonetheless, the livestock sector largely evaded tithes as, according to recent calculations, this tax attached to only between 3 and 5 percent of the gross pastoral product (Henriques 2013: 84). Rent was equally focused on arable agriculture: the grain and wine output was heavily taxed (with rates as high as 1/3), while payments in livestock or animal products were negligible. Nevertheless, as Figure 1.3 indicates, average shares were lower than 15 percent, even in a region where ecclesiastical lords held considerable power (Coelho 1989). With regular conditions 1/5 and 1/6 were more common (Durand 1982) and, even so, livestock commodities were almost always exempt from such taxes.

Cities, Markets and Agro-towns

During the 13th and early 14th centuries the growing Portuguese population relied on agriculture primarily to subsist. Given that there is no evidence of regular large-scale food imports (Barbosa 1992), Portuguese agriculture could be relied upon to satisfy the food needs of the population. The import of grain by Lisbon and coastal towns would become frequent from the early 15th century onwards. Before the Plague, however, there is no indication that such trade existed. In fact, a law in 1272 actually forbade the export of cereals, hinting at a more favorable position. However, this remained "calamity-sensitive" (Campbell 2005) agriculture that was vulnerable to natural shocks. Narrative sources and scattered charters provide occasional mentions of "famine", *i.e.* crop failures, although there is no way of assessing the magnitude of such events. Some major landowners, like the king and the bishops, kept important granaries to store wheat. In the south of the realm, including Lisbon, grain was stored underground in pits dug in rocks and sealed (*covas de pão*). There are indications that this type of storage was aimed at maximizing profit, and speculation in grain was not unheard of. The episcopal granary of Braga, for instance, sold shares of its future reference (Costa 1981b: doc. 42).

Despite evidence of trade in agricultural products throughout the country, the market potential of Portuguese agriculture remained to be fulfilled. Though geographical conditions were not especially favorable, the key problem appears to be the connection between town and country. In theory, town and country have complementary interests, with the former consuming the surplus of the latter and selling the production of its specialized craftsmen. This perfect symbiosis is contradicted by existing sources, which indicate a strong tension. Farmers could, and did, make their own tools and townsmen could, and did, grow their own wine and grain. From the 12th century the municipality of Coimbra pressed neighbouring farmers not to make their own tools but to buy them in the city (Durand 1982: 205). Fourteenth-century Porto was a proper city, in the sense that it specialized in maritime freight and in manufactures like ironworking. Yet, its comparatively wealthy citizens faced severe supply problems and had to resort to price fixing on essential foodstuffs. They also had to force the farmers under their jurisdiction to bring their surpluses in grain, fish and salt to be sold in the city (Melo 2009: 210). The risk of shortages and the use of coercion were a price that Porto had to pay for its specialization.

Proper cities in the economic sense were thus rare (see Table 1.3). By contrast, towns that started their lives as castles were common.[16] The foundation of towns by kings and great lords (mainly ecclesiastical) reflected their strategic interests more than spontaneous concentrations of population. Most of these can be regarded as "agro-towns" (*agrovilas*), that is relatively large settlements where agriculture is the main economic activity (Viana 2007:100). Oliveira Marques described Arruda, in the hinterland of Lisbon, as an "eminently rural [town], dedicated to the tending of its vineyards and on the labor of the fields" (Marques 1980). In some Portuguese townships between one-fourth and one-third were non-agricultural households, with craftsmen growing part of their food themselves (Rodrigues 1996: 65; Viana 1998: 96). Thus, the concentration of population in towns did not necessarily reflect or stimulate productivity increases in agriculture.

Migration and political integration eventually led conquering and conquered regions to form a cohesive unit in terms of language and laws. Still, conquered lands were different in one decisive aspect: the degree of urbanization. By 1050, Portucale was entirely rural, except for the small boroughs of Braga, Guimarães and Porto. The Christian conquest of Coimbra (1064) endowed Portucale with a major city. Between 1147 and 1249, the country acquired at least three cities (Santarém, Lisbon and Évora) along with a few recently conquered urban centres like Beja, Silves, Tavira and Faro. Of these, all but Silves had been Roman cities that were part of Lusitania, an urbanized province. The difference in the degree of urbanization, with all its important agrarian

TABLE 1.3 *Urbanization rate by region by 1300*

Region	**Urbanization rate**		**Cities considered**
	(> 5000 inhabitants)	**(> 10000 inhabitants)**	**(population in thousands)**
North	1.4	0	Porto (8)
Centre	2.0	0	Coimbra (6)
South	10.6	5.6	Lisbon (16); Évora (8); Santarém (7); Elvas (5)
Portugal	5.0	1.6	All above (50)

16 See map A9 in the Appendix.

consequences, was one that neither Barbarians nor Muslims erased, or that Christian conquest would change.

Institutional strictures contributed to thwart trade within the country. One feature of the Portuguese town was jurisdiction over the countryside. Accordingly, the municipal government (the *concelho*) had an important regulatory role in agriculture, as well as in the transport and trading of products. Landowning municipal elites aimed to market the product of their lands at favorable prices (Barata and Henriques 2011: 269–270). The written customs of the towns limited trade. Throughout the country, municipalities prohibited the import of wine from outside until all the local wine had been sold (Viana 1998: 186–187). Towns aspiring to autarky created a "protectionist" background that prevented the emergence of comparative advantages, as first suggested by Marques (1978: 129–131) and confirmed by Viana (2009).

The rugged terrain and natural accidents have often been invoked as reasons for the belated development of internal markets (Marques 1978: 175). Despite some known efforts in keeping old routes and opening new ones in the 12th century, rugged mountains that ran from north east to south continued to make trade difficult in the interior.[17] Nevertheless, in this period, ruggedness was less of a problem than in later eras as, by 1300, some rivers were still navigable, allowing some interior towns ready access to the sea: Ponte de Lima, Leiria, Aljezur, Mértola, Odemira and Silves. With the sea providing more efficient transport, natural obstacles were less influential in this period.

Despite these domestic strictures, Portuguese agriculture received market stimuli. The *Reconquista* had bequeathed yet another gift: a long strip of Atlantic coast with three prime seaports (Porto, Lisbon and Setúbal) and a network of small harbours.[18] This coast gave Portugal a relevant place in intra-European trade. This is important for present purposes, because Portuguese agriculture was efficient enough to provide all of the country's known exports. Data is hard to come by, but the combination of disparate references confirms that Portuguese agricultural products were exported. In 1194, a Portuguese ship loaded with wine, honey and wood sank off Nieuwpoort, in Flanders. A Flemish text from the same period informs us that "the realm of Portugal provides honey, leather, wax, rawhides, grain, grease, oil, figs, raisins and whale products" (Answaarden 1991). In the 13th and 14th centuries, England, Flanders

17 See maps A2 and A14 in the Appendix.

18 Valença, Viana, Vila do Conde, Gaia, Ovar, Estarreja, Vagos, Buarcos, Paredes, Pederneira, S. Martinho, Alfeizeirão, Salir, Atouguia, Lourinhã, Ericeira, Almada, Alcácer, Odemira, Odeceixe, Arrifana, Lagos, Alvor, Silves, Portimão, Albufeira, Faro, e Tavira. (Marques 1996: 490). See map A13 in the Appendix.

and Normandy imported wine, olive oil, honey, salt, figs, raisins, wax and fruit from Portugal (Miranda 2012: 75). Further south, Lisbon, Setúbal and the ports of the Algarve were also exporting wine as well as other agricultural produce, such as dried fruit and olive oil. The Atlantic also gave Portugal the opportunity to import other goods. Wheat imports, which would later become very important, appear to have been neither significant nor permanent. Manufactures, rather than agrarian goods, were most likely the main imports.

Production and Productivity

The previous sections have assessed economic forces, institutions and market stimuli affecting Portuguese agriculture. It is now time to determine how all these elements impacted on agricultural production and productivity. In theory, the space obtained by the *Reconquista* allowed for increasing both agrarian production and population throughout this period. Land-hungry northern farmers could escape decreasing returns by moving south. The availability of land ensured that arable production could increase without the need for rising per capita productivity. As was seen in the previous sections, 13th or 14th century farmers could also raise production levels by increasing the use of arable and pastoral land or even the amount of capital, the cost of which appears to have remained moderate.

Palynology provides less circumstantial evidence of the extensification process brought about by the *Reconquista*. The data obtained from pollen sites in the centre of the country indicates that the expansion of cereal, sheep-rearing and olive trees at the cost of the older forests started in 1041(±40 years) (Knaap and Leeuwen 1995). This date closely matches the Christian conquest of Lamego, Viseu and Coimbra (1057–64). This finding is perfectly compatible with what was argued earlier about the dramatic changes brought about by the *Reconquista*, at least in areas that were little exploited during the Muslim occupation. The conquest by the Christian kingdoms meant a displacement of populations that resulted in the expansion of the arable and pastoral area.

The *Reconquista* was followed by southward and eastward migration from the densely settled north west. The greater availability of land in the south attracted farmers who struggled with high rents. The forsaking of agrarian units belonging to the monasteries by the tenants documented in a royal enquiry held in 1258 (Viana 2007) illustrates this tendency. This movement had two beneficial consequences for productivity: first, it alleviated demographic pressure in the north and, second, it allowed one to tap yet unused natural resources in the south. This is visible in an estimation of the per capita agrarian production for the first quarter of the 14th century, when this internal migration was at an advanced stage. As Map 1.1 shows, the farmers established in the southern dioceses of Lisbon and Évora (Silves/Algarve was left blank due to the lack of data) were among the most productive in the country, despite their drier climate. Nevertheless, the home of most migrants, the north-western dioceses of Braga (Western half) and Porto, was at a similar level. The less productive agriculture was located inland, most likely because, as seen in the previous section, it did not have the same commercial stimuli as the coastal areas. The composition of Portuguese agrarian output has only recently been addressed. The combination of supply-side sources, like tithe revenues, with demand-side information allows for roughly outlining the shares of the different agricultural sectors in the first half of the 14th century. The allowances in Table 1.4 reflect median consumption as they are taken from the records few belonging to charitable houses for urban and rural impoverished people. The estimated output in the areas of Tomar and Avis is relevant because these are vast areas.

Demand- and supply-side indicators appear broadly consistent. They agree that the grain sector was the most important in money terms. It was not, however, the most efficient. Considering animal prices and consumption patterns present in allowances, livestock appears to have been relatively abundant. The poor people who lived off allowances such as those depicted at Table 1.4 could count on a respectable calorific intake through animal products or wine. This contrasts favorably with the diets of later periods, namely those of the 19th century. Combining the consumption of wheat implied in those estimates (4.7 hl of grain per caput) with a 1276 estimation of production per surface for wheat and second-rate cereal in optimal land (20 hectolitres/ha), Portugal required some 241.935 hectares to provide the grain required for its 1300 population (estimated at around 1 million inhabitants).[19] This is less than one-third of the 850,000 hectares of arable land put to use in 1636, according to Silva (1868: 322). In this context, it is not surprising to find areas specializing in tradable products, namely olive oil and wine. Tithe accounts for Avis (Alentejo) and Tomar (Estremadura), for instance, show that there was some specialization in olive oil (Henriques 2013). Nevertheless, these were wealthy areas and do not necessarily reflect the country as a whole.

A tax assessment of all churches in the realm made in 1320 on behalf of the papacy provided the basis for a reconstruction of the country's gross agricultural product (Table 1.5). As the bulk of the parish churches' revenue

---

19 See source at TT, Chanc. Afonso III, Lv. 1, fl. 139. The calculation refers to the high-yielding alluvial fields of the Tejo (the *lezírias*), which were certainly well above the country average.

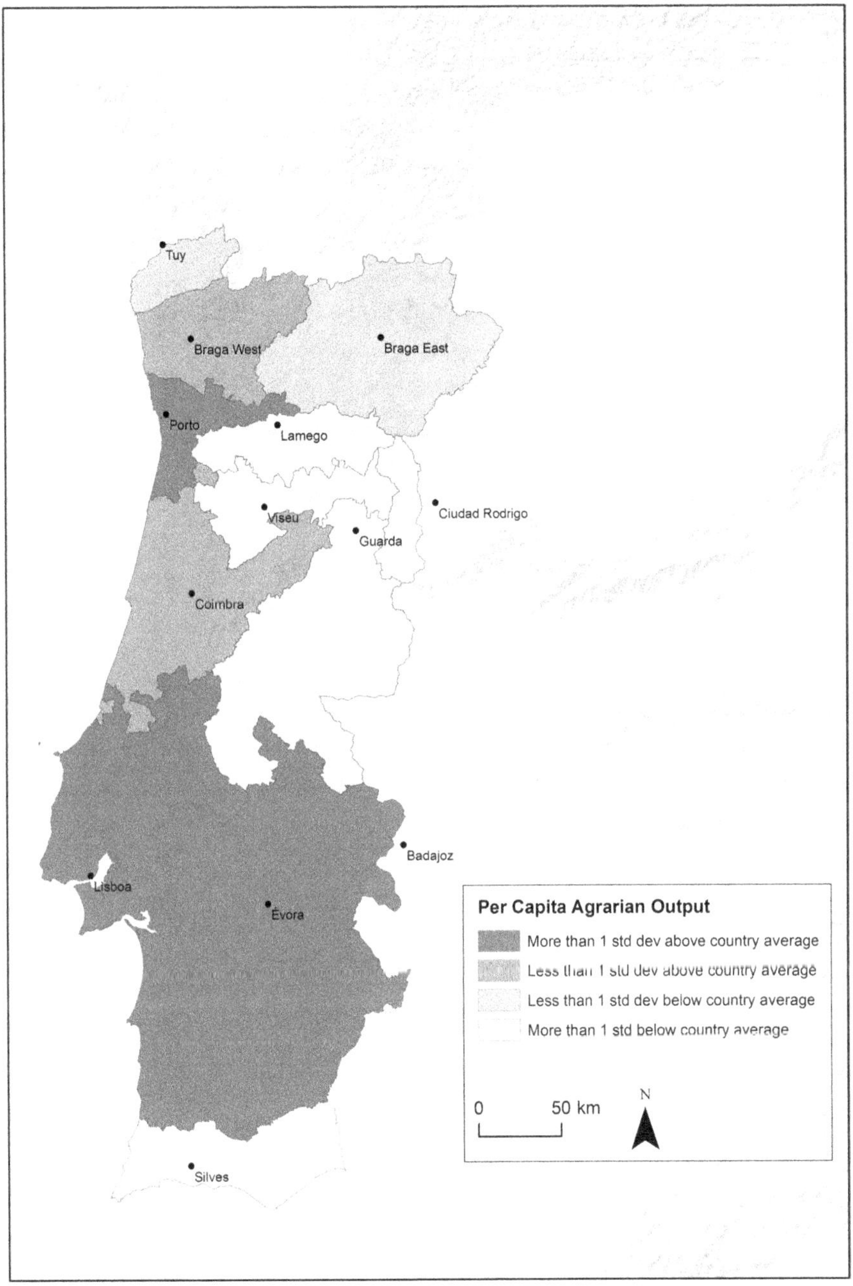

MAP 1.1 *Agricultural production per capita by diocese (c. 1320)*
MAP DRAWN BY THE AUTHOR, ANTÓNIO HENRIQUES

TABLE 1.4 *Data on the composition of the agrarian product (1316–1349)*

	Allowance for urban poors in Coimbra, 1328	Allowance for poor farmers in Gulfar, 1349	Output in Tomar (1316–26)	Output in Avis -1327
Grain	52%	57%	56%	30%
Meat (fish)	27% (4%)	33% (1%)	17%	15%
Wine	17%	9%	11%	
Olive oil	–	–	13%	55%
Other	–	–	2%	1%

TABLE 1.5 *Agrarian output in Portugal and England (c. 1300)*

	Portugal 1311–20	England 1300
Population	1.0 million	4.25 million
g of pure silver per capita	168	218
Real per capita output (in Portuguese *libras*)	14	10.4

was provided by the tithe and as this tax caught nearly all agrarian outputs, this source allowed the estimation of the totality of the agrarian production (Henriques 2015). The reasonable living conditions afforded by Portuguese agriculture can be observed in a recent comparison with England, a country with different features. The estimates confirm that Portugal was not struggling with excess population and could assign enough land to the pastoral sector. Endowed with larger natural resources, Portuguese agriculture had the conditions to be more productive; hence, some exportation of agricultural products (wine and olive oil) and the high animal protein intake referred to earlier.

Portuguese per capita agrarian production was higher than that of England when measured in real terms, but not when considered in metallic terms. This likely reflects the higher degree of commercialization of English agriculture. It can be argued that Portuguese low prices reflected the abundance of land relative to labor. Widespread land ownership and/or long-term tenure meant that land was relatively well distributed, and thus Portuguese agriculture could not count on important demand-side stimuli. Coupled with low urbanization(Table 1.3), the ample provision of land in 14th-century Portugal implied an important degree of self-consumption. In fact, for England and the Netherlands, the existence of "proletarians", i.e. landless laborers who lived off wages, was an important factor in developing tradable sectors within agriculture (Campbell 2005; Bavel 2010). By contrast, Portugal, with its low urbanization and an apparent predominance of family-sized exploitation, seems to have underdeveloped its agrarian and industrial sectors. This trap of excessive resources is the negative legacy of the *Reconquista.*

Conclusion

As the previous pages have suggested, recent research paints a relatively favorable picture of Portuguese agriculture in these three and half centuries. In fact, this period witnessed a steady expansion of the agricultural surface and production, which, in turn, accommodated a roughly commensurate population growth. This agrarian expansion gained momentum from the second half of the 11th century, when the conquest and claiming of land went side by side, and did not cease until the middle of the 14th century. Available data suggests that this growth was not accompanied by decreasing returns and that, barring natural shocks, Portuguese farmers could ensure a high level of food production.

Throughout this period Portuguese agriculture successfully adapted to the existing conditions and sustained a growing population. Feudal institutions did not prove strong enough to stifle powerful forces, namely the north–south migration of land-seeking farmers. Nevertheless, a relatively low level of urbanization and the prevalence of self-consumption meant that demand-side stimuli were weak. The conditions imposed by the *Reconquista* protected Portugal from the *mond plein* of, for instance, contemporary England, but did not favor an even development of product markets throughout the country.

This expansionist period came to a halt because of a single event, the Black Death of 1348–9. While this was a profound shock, it did not wipe out the legacy of the *Reconquista*, as the settlement pattern, the composition of the product, the distribution of property rights and the institutions that emerged in this period remained in place. Whether this legacy outlived its utility is a question to be answered in the following chapters.

References

Amaral, Luís Carlos (2007). *Formação e Desenvolvimento do Domínio da Diocese de Braga no Período da Reconquista (Século IX-1137)*. Porto: Universidade do Porto.

Amaral, Luís Carlos (2009). "O povoamento da terra bracarense durante o século X". *Revista da Faculdade de Letras – História*, 10 (3), pp. 13–127.

Amaral, Luís Carlos (2011). "Half a century of rural history of the Middle Ages in Portugal: a possible overview". In José Mattoso (Ed.), *The Historiography of Medieval Portugal, c.1950–2010*. Lisbon: Instituto de Estudos Medievais, pp. 303–322.

Answaarden, R.V. (1991). *Les Portugais devant le Grand Conseil du Pay Bas*. Paris: Fondation Calouste Gulbenkian.

Azevedo, Ruy de (1993). *O Mosteiro do Lorvão na Reconquista Cristã*. Lisbon: Bertrand.

Barata, Filipe Themudo and Henriques, António (2011). "Economic and fiscal history". In José Mattoso (Ed.), *The Historiography of Medieval Portugal*. Lisbon: Instituto de Estudos Medievais, pp. 261–282.

Barbosa, Pedro G. (1992). *Povoamento e Estrutura Agrícola na Extremadura Central. Século XII a 1325*. Lisbon: Instituto Nacional de Investigação Científica.

Barros, Henrique Gama (1945–1954). *História da Administração Pública em Portugal*. Lisbon: Sá da Costa.

Barros, Maria Filomena L. (2005). *Mouros da Terra e Terra de Mouros*. Palmela: Câmara Municipal de Palmela.

Barros, Maria Filomena L. (2007). *Tempos e Espaços de Mouros: a Minoria Muçulmana no Reino Português (Séculos XII a XV)*. Lisbon: FCT/Fundação Calouste Gulbenkian.

Bartlett, R. (1993). *The Making of Europe. Conquest, Colonization and Cultural Change. 950–1350*. London: Penguin.

Bavel, Bas van (2010). "The medieval origins of capitalism in the Netherlands". *BMGN-Low Countries Historical Review*, 125 (2–3), pp. 45–79.

Bernardes, J.P. (2005). "Entre romanos e medievos. O problema do povoamento da região de Leiria durante a Alta Idade Média". *Arquipélago – História*, 9, pp. 563–576.

Boissellier, Stéphane (1999). *Naissance d'une identité portugaise. La vie rurale entre Tage et Guadiana de l'Islam à la reconquête (Xe–XIVe siècles)*. Lisbon: Imprensa Nacional-Casa da Moeda.

Boissellier, Stéphane (2003). *Le Peuplemente médiéval dans le sud du Portugal*. Paris: Fundação Calouste Gulbenkian.

Boone, James (1996). "Uma sociedade tribal no baixo Alentejo medieval?" *Arqueologia Medieval*, 4, pp. 25–35.

Brázdil, Rudolf, Pfister, Christian *et al.* (2005). "Historical climatology in Europe – the state of the art". *Climatic Change*, 70, pp. 363–430.

Campbell, B.M.S. (2005). "The agrarian problem in the early fourteenth century". *Past and Present*, 188 (1), pp. 3–70.

Campos, Ezequiel de (1918). *A Evolução e a Revolução Agrária*. Porto: Renascença Portuguesa.

Catarino, Helena (2002). "Chateaux et peuplement rural islamique et de la reconquête en Algarve Oriental. Les données archeologiques". In M. Bourin and S. Boissellier, (Eds). *L'Espace rural au Moyen Âge. Portugal, Espagne, France (XIIe–XIVe siècle)*. Rennes: Presses Universitaires de Rennes, pp. 41–56.

Chalmeta Gendron, Pedro (1994). *Invasión e islamización: la sumisión de hispania y la formación de al-andalus*. Madrid: MAPFRE.

Clark, Gregory (1988). "The cost of capital and medieval agricultural technique". *Explorations in Economic History*, 25 (3), pp. 265–294.

Coelho, Maria Helena da Cruz (1989). *O Baixo Mondego nos Finais da Idade Média*. Lisbon: Imprensa Nacional-Casa da Moeda.

Coelho, Maria Helena da Cruz (1990). "Apontamentos sobre a comida e a bebida do campesinato coimbrão em tempos medievos". In M. Helena da Cruz Coelho (Ed.). *Homens, Espaços e Poderes: Séculos XI–XVI*. Lisbon: Livros Horizonte, vol. 1, pp. 9–22.

Conde, Manuel Sílvio Alves (2000). *Uma Paisagem Humanizada. O Médio Tejo nos Finais da Idade Média. A Terra e as Gentes*. Cascais: Patrimonia Historica.

Cosme, Susana (2002). *Entre o Côa e o Águeda. Povoamento nas Épocas Romana e Alti-medieval*. Porto: Universidade do Porto.

Costa, Mário Júlio de Almeida (1957). *As Origens da Enfiteuse no Direito Português*. Coimbra: Coimbra Editora.

Costa, Avelino de Jesus da (1981a). "Inventário dos bens e obituário de Santa Maria de Alcáçova de Santarém". *Boletim da Biblioteca Universitária de Coimbra*, 36, pp. 1–29.

Costa, Avelino de Jesus da (1981b). *Povoamento e Colonização do Território Vimaranense nos Séculos IX a XI*. Guimarães: s.n.

Costa, Mário Júlio de Almeida (1982). *Os Contratos Agrários e a Vida Económica na Idade Média*. Guimarães: s.n.

Costa, Avelino de Jesus da (1997–2000 [1959]). *O Bispo D. Pedro e a Organização da Diocese de Braga*. Braga: Irmandade de S. Bento da Porta Aberta.

Drake, B.L., Hanson D.T., and Boone, J.L. (2012). "The use of radiocarbon-derivedD13C as a paleoclimate indicator: applications in the lower Alentejo of Portugal". *Journal of Archaeological Science*, 39 (9), pp. 2888–2896.

Durand, Robert (1982). *Les campagnes Portugaises entre Douro et Tage aux XIIe et XIIIe siècles*. Paris: Fundação Calouste Gulbenkian.

Esper, J., Cook, E.R. and Schweingruber, F.H. (2002). "Low-frequency signals in long tree-ring chronologies for reconstructing past temperature variability". *Science*, 295, pp. 2250–2253.

Fernandes, Hermenegildo (1991). *Organização Social do Espaço e Sistema Social no Alentejo Medievo. O Caso de Beja*. Lisbon: Universidade Nova de Lisboa (Master's dissertation).

Fernandes, Hermenegildo (2000). *Entre Mouros e Cristãos. A Sociedade de Fronteira no Sudoeste Peninsular Interior (sécs. XII–XIII)*. Lisbon: Universidade de Lisboa (PhD dissertation).

Fontes, Luís, and Roriz, A. (2007). *Património Arquitectónico e Arqueológico de Vieira do Minho*. Vieira do Minho: Câmara Municipal.

García Álvarez, M. Rubén (1971–1972). "Algunos aspectos de la economía estructural básica bracarense en los siglos X y XI". *Bracara Augusta*. XXV–XXVI, pp. 38–124.

Gomes, Saúl António (2009). *Vinhos e História na Alta-Estremadura. Entre os Séculos XII e XVI*. Leiria: Folheto.

Gomes, Rosa Varela (2012). "El mundo rural, en el sur del actual territorio portugués (siglos XII–XIII)". *Arqueologia Medieval – Els Espais de Secà*, 4, pp. 99–116.

Gonçalves, Iria (1989). *O Património do Mosteiro de Alcobaça nos séculos XIV e XV*. Lisbon: FCSH-UNL.

Gonçalves, Luís (2011). *Sistemas de Povoamento e Organização Territorial: Dois Vales na Periferia de Lisboa (Séculos IX–XIV)*. Lisbon: Universidade de Lisboa.

Gouveia, Mário de (2012). *Ilduara Mendes e a vila de Nogueiró. Gestão Feminina de uma Exploração Agrária (Século XI)*. Lisbon: s.n.

Guichard, P. (2003). "Quelques réflexions sur la société urbaine en Al-Andalus". *Mainake*, 25, pp. 7–20.

Henriques, António (2003). *O Rei e a Terra de Barroso. Poder Régio, Montanha e Periferia*. Lisbon: Universidade Nova de Lisboa (Master's dissertation).

Henriques, António (2007). *As Origens do Lameiro Barrosão. Ecologia e Transformação Social*. Lisbon: Centro de Estudos Históricos.

Henriques, António (2013). "O 'fruto' e o 'produto'. Do dízimo eclesiástico às contas nacionais". In M. Viana and J.A. Solorzano Telechea (Eds). *Portugal Século XIV, Economia e Instituições na Idade Média. Novas Abordagens*. Ponta Delgada: Centro de Estudos Gaspar Frutuoso, pp. 66–94.

Henriques, António (2015). "Plenty of land, land of plenty: The agrarian output of Portugal (1311–20)". *European Review of Economic History*, 19 (2), pp. 149–170.

Herculano, Alexandre (1846–1853). *História de Portugal*. Lisbon: Viúva Bertrand e Filhos.

Herculano, Alexandre (1981). *História de Portugal: Desde o Começo da Monarquia até ao Fim do Reinado de Afonso III*. Lisbon: Bertrand.

Kitsikopoulos, Harry (2000). "Standards of living and capital Formation in pre-plague England: a peasant budget model". *Economic History Review*, 53 (2), pp. 237–261.

Knaap, W.O. van den, and Leeuwen, J.F.N. van (1995). "Holocene vegetation succession and degradation as responses to climatic change and human activity in the Serra da Estrela, Portugal". *Review of Palaeobotany and Palynology*, 89, pp. 153–211.

Lobão, Manuel Almeida e Sousa (1814). *Tractado Pratico e Crítico de Todo o Direito Emphyteutico, Conforme a Legislação e Costumes deste Reino e Uso Actual das Naçoens*. Lisbon: Impressão Régia.

Marques, A.H. de Oliveira (1978). *Introdução à História da Agricultura em Portugal. A Questão Cerealífera em Portugal na Idade Média*. Lisbon: Cosmos.

Marques, A.H. de Oliveira (1980). *Estratificação Económico-Social de uma Vila Portuguesa na Idade Média*. Lisbon: Vega.

Marques, A.H. de Oliveira (1986). *Portugal na Crise dos Séculos XIV e XV. Nova História de Portugal*. Lisbon: Presença.

Marques, Mário Gomes (1996). *História da Moeda Medieval Portuguesa*. Sintra: Instituto de Sintra.

Marques, André E. (2008). *O Casal. Uma Unidade de Organização Social do Espaço no Entre Douro e Lima (906–1200)*. Noia: Toxosoutos.

Marques, André E. (2014). *Da Representação Documental à Materialidade do Espaço. Território da Diocese de Braga (séculos IX–XI)*. Porto: Afrontamento.

Mattoso, José (2001). *A Identificação de um País. Ensaio sobre as Origens de Portugal (1096–1325)*. Lisbon: Círculo de Leitores.

Melo, Arnaldo de Sousa (2009). *Trabalho e Produção em Portugal na Idade Média: o Porto, c. 1320–c. 1415*. Braga: Universidade do Minho.

Miranda, Flávio (2012). *Portugal and the Medieval Atlantic. Commercial Diplomacy, Merchants, and Trade, 1143–1488*. Porto: Universidade do Porto (PhD dissertation).

Picard, Christophe (2005). "Le changement du paysage urbain dans le Gharb al-Andalus (X–XIIe siècle): les signes d'une dynamique". In J. Barroca and I.C. Fernandes (Eds.) *Mouros e Cristãos entre o Tejo e o Douro (sécs. VIII–XIII)*. Palmela: Câmara Municipal de Palmela, pp. 129–143.

Pontes, Rui Manuel (2006). "Povoamento e desenvolvimento económico do senhorio de Coina estuário do Tejo) nos séculos XIII e XVI: a construção de uma paisagem rural". In Iria Gonçalves (Ed.), *Paisagens Rurais e Urbanas. Fontes, Metodologias, Problemáticas. Actas das Segundas Jornadas*. Lisboa: Centro de Estudos Históricos, pp. 213–239.

Power, Daniel (Ed.) (2006). *The Central Middle Ages: 950–1320*. In *The Short Oxford History of Europe*. Oxford: Oxford University Press, vol. III.

Rau, Virgínia (1965). *Presúrias e Sesmos no Povoamento de Portugal*. Barcelona: s.n.

Rau, Virgínia (1982). *Sesmarias Medievais Portuguesas*. Lisbon: Presença.

Rodrigues, Ana Maria S.A. (1996). "A população de Torres Vedras em 1381". In Ana Maria S.A. Rodrigues (Ed.). *Espaços, Gente e Sociedade no Oeste. Estudos sobre Torres Vedras Medieval*. Cascais: Patrimonia, pp. 45–84.

Santos, M.J. Ferreira dos (2004). *A Terra de Penafiel na Idade Média. Estratégias de Ocupação do Território*. Porto: Universidade do Porto.

Silva, Luiz Augusto Rebello da (1868). *Memoria sobre a População e a Agricultura de Portugal desde a Fundação da Monarchia até 1865: Parte 1*. Lisbon: Imprensa Nacional-Casa da Moeda.

Silva, Manuela Santos (1996). *Óbidos e a sua Região na Baixa Idade Média*. Lisbon: Universidade de Lisboa.

Silveira, X. Mouzinho da (1989). *Obras*. Lisbon: Fundação Calouste Gulbenkian.

Silveira, Ana Cláudia (2009). "Novos contributos para o estudo dos moinhos de maré no estuário do Tejo: empreendimentos e protagonistas (séculos XIII–XVI)". In Amélia A. Andrade, Hermenegildo Fernandes and João Luís Fontes (Eds.), *Olhares sobre a História. Estudos Oferecidos a Iria Gonçalves*. Lisbon: Caleidoscópio, pp. 581–610.

Tente, Catarina (2010). *Arqueologia Medieval Cristã no Alto Mondego. Ocupação e Exploração do Espaço nos Séculos V a XI*. Lisbon: Universidade Nova de Lisboa.

Viana, Mário (1998). *Os Vinhedos Medievais de Santarém*. Cascais: Patrimonia Histórica.

Viana, Mário (2007). *Espaço e Povoamento numa Vila Portuguesa: Santarém*. Lisbon: Caleidoscópio.

Viana, Mário (2009). "A viticultura nas cidades portuguesas". In J. Solórzano Telechea and A. Arízaga Bolumburu (Eds.), *Alimentar La Ciudad En La Edad Media. Nájera Encuentros Internacionales Del Medioevo 2008. Actas*. Logrono, pp. 87–110.

Vicente, António. B. (1996). *Santa Maria de Aguiar, um Mosteiro de Fronteira. Património Rural e Paisagem Agrícola. Séculos XII–XIV*. Lisbon: Universidade de Lisboa.

Vieira, Marina (2000). *Alto Paiva. O Povoamento nas Épocas Romana e Alti-medieval*. Coimbra: Universidade de Coimbra.

Zuijderduijn, C.J. (2009). *Medieval Capital Markets: Markets for Renten, State Formation and Private Investment in Holland (1300–1550)*. Leiden: Brill.

CHAPTER 2

The Black Death and Recovery, 1348–1500

Ana Maria S.A. Rodrigues

Introduction

Until recent times, the period we will be addressing in this chapter, from the outbreak of the Black Death to the beginning of a new era of overseas expansion, has been widely considered as an era of prolonged economic crisis in Portugal and the Iberian Peninsula. In fact, this interpretation of events is still present in works published recently.[1] That view is somehow related to similar interpretations regarding the rest of Europe. New research, particularly for England, has however challenged those results and explored new paths, paying more attention to positive terms in the evolution of demand, as well as the links between the conditions of the agrarian sector and the remaining sectors, namely commercial activity and monetary circulation, which have developed in a more favorable way.[2] Historians have also been debating the meanings of the concept of *crisis* and asking questions about the chronological delimitation, the geographical extension and the very existence of a *crisis* across the Mediterranean region during the 14th century.[3]

Until recently, Portuguese historians have remained absent from these debates, but that has changed substantially, firstly with the work of Barata and Henriques (2011). Indeed, the authors question the theoretical relevance of the Malthusian model that dominated medieval economic research, suggesting alternative interpretations based on a revision of data previously known, to which they have added new evidence. This new perspective is also present in Henriques (2015) and Chapter 1 (in this book). Duarte (2012) also questions the Black Death as a main element in explaining the evolution of the economy throughout the period analysed in this chapter, and points to the relevance of the effects of wars, particularly those fought during the reigns of King Fernando I (1367–1383) and João I (1385–1433) of Portugal. Muhaj (2013) also argues that the swerve in commercial routes caused by the Hundred Years War strongly

1 See Moreno (2009). For an older view see also Marques (1978 and 1987).

2 See Britnell (1993), Britnell and Campbell (1995), Masschaele (1997) and Epstein (2000).

3 See Oliva Herrer and Benito (2007), Bourin and Menant (2011), Bourin, Carocci, Menant and To Figueras (2011) and Benito i Monclús (2013).

 | DOI 10.1163/9789004311527_004

benefited Portugal, opening an era of prosperity and expansion rather than contraction and decline, particularly in the later part of the 15th century. Yet Portugal's economic conditions worsened considerably during the 14th century.

The chapter is structured as follows. In the first section we discuss the combined effects of the shocks of the plague and wars on agriculture, as well as the institutional responses that ensued. The crisis was however restricted to a relatively short phase and did not cause adverse structural changes in rural society. The second section therefore analyses the changes through which lords, landowners and farmers adapted to new demographic and economic conditions, setting the foundations for the following growth period. The third section shows how internal and external trade provided stimuli for those changes. A fourth and last section links the early Portuguese overseas expansion to North Africa and the Atlantic islands, with the search for new supply areas for much demanded commodities, both inside and outside the kingdom's borders.

Plague, War and Demographic Decline

The Black Death reached Portugal at the end of September 1348 and took its toll until the end of January of the following year. The few contemporary documents that report on the plague's effects speak of an extremely high mortality rate. In fact, as described in a number of ecclesiastic registers, "the prior and the singer and all the canons" of the church of S. Pedro of Almedina, near the city of Coimbra, died, as did "most of the nuns" of the nearby monastery of Lorvão. The archdiocese of Braga, a vast and densely populated rural area in the north of the country, was "deprived of the consolation of its ministers" (Marques 1987: 20–21).[4] Other local studies lead to similar conclusions in two towns near Lisbon. In Torres Vedras, the plague was responsible for 38 percent of deaths among the collegiate churches' members, and for 62.5 percent of deaths among the town's public notaries (Rodrigues 1995: 80); and in Sintra, the public notaries died in the same proportion (Carvalho 1985: 130–131). There is no quantitative evidence for exclusively rural areas. However, it is known that in urban centres and crowded monasteries contagion spread more quickly and mortality was higher. It is thus estimated that roughly one third of the population died as a result of the pestilence (Rodrigues 2009: 113). Accepting the estimates according to which before the Black Death the kingdom's population

4 See maps in the appendix.

was roughly one million people, the plague would have caused the disappearance of more than 300,000 souls.

This demographic collapse had an immediate impact on land distribution. A great number of manors and a multitude of small freeholdings were bequeathed to parish churches, convents and hospices by their dying owners in the hope of gaining "eternal salvation".[5] Yet the Church could not find either peasants to rent them or even rural workers to till them, because they also died during the plague or had become wealthy and did not want to return to their old occupations, as many survivors received unexpected inheritances. This led previously landless people, who usually worked for wages, to leave their employers and start a new life as tillers of their own lands. Others workers were still available, but claimed higher wages and shorter agreements, for weeks or months, instead of years. In other cases, there were also able men and women that preferred living on charity to working. These were, at least, the arguments used by King Afonso IV (1325–1357) to justify an ordinance from 1349, quite similar to the English Ordinance of laborers of the same year. The Portuguese king wished to compel those who had been journeymen before the plague and all the landless men and women to work for a master for fixed terms and wages. The ordinance exempted from this obligation those that had recently amassed some wealth, allowing them to request the privileged status of *peões* or *cavaleiros-vilãos*, who farmed their own lands and performed their military duties on foot (the former) or on horseback (the latter) (Rau 1982: 80–82).

For many lords and landowners, scarcity of labor was not new. As shown in the previous chapter of this book, during the *Reconquista* northern villains of customary lands could migrate to southern regions and become freeholders, as there was a vast amount of newly conquered land to distribute. At the beginning of the 14th century, when several regions in England, France and the Low Countries were a *monde plein*, there were still many barely populated areas in Portugal. Thus the Black Death did not create, but rather dramatically worsened, the already existing imbalance between labor and land. The situation deteriorated as a result of new plague outbreaks in 1356, 1361–63, 1365 and 1374 (Rodrigues 2009: 114).

In 1369, King Fernando I decided to reclaim the throne of Castile, after the murder of Pedro I, *the Cruel*, by his half-brother Enrique of Trastámara. The Portuguese king invaded the province of Galicia and, in return, Castilian military expeditions entered Portugal in 1369–71 and 1372–73 (Marques 1987: 511–516). Wars did not cause as many deaths as pestilence, but set men away from their work and provoked extensive destruction. Contenders set fire to wheat fields,

5 See Ramos (1963: 221–222), Sousa (1990: 42) and Rodrigues (1995: 376–378).

vineyards, olive groves and orchards, demolished granaries, stables, mills, and pillaged draught animals and other cattle with long-lasting effects. The quarters outside the walls of besieged towns were razed, which implied the destruction of the plots that supplied the urban market with fresh vegetables and fruit (Martins 2006).

A number of historical studies focusing on Portuguese rural areas and Church estates in the later Middle Ages confirm the desolation in which they lived by the end of the 14th century.[6] In wetlands, the lack of manpower hampered the continuation of drainage activities, and as a result swamps took over cultivated land. Vineyards, olive and fruit trees were no longer tended to as necessary, and productivity was substantially reduced. Moreover, an increasing share of arable lands was kept fallow for long periods, ultimately becoming covered by brushwood (*mato*), and turning into a refuge for wild animals, which in turn destroyed the harvests and threatened the cattle of the neighboring farmers. As a consequence of this general abandonment of farming activities, there was a corresponding increase in the uncultivated fields that were turned into commons (*baldios*), used by rural communities for pasturage and firewood. This transformation allowed for an increase in livestock husbandry, but agricultural output suffered a severe breakdown (Marques 1987: 104). In an attempt to stop the negative consequences of this situation for the agrarian sector, a royal ordinance from 1375 dealt mainly with the large extension of land that was left uncultivated and the corresponding lack of grain to supply the kingdom's inhabitants. According to the ordinance, all those who possessed or held lands, including lords, villains and freeholders alike, were required to keep the land productive. Otherwise, after a certain period of time, the lands would be expropriated and given to whoever would guarantee its cultivation. This was the so-called *Sesmarias* Law, a designation that was borrowed from the system that had been used during the *Reconquista* period to share conquered lands among new settlers (Rau 1982: 50–57).

The ordinance however went even further. Since landowners would not be able to avoid expropriation if they could not find workers and working animals at a reasonable cost, King Fernando I also determined that the municipalities would fix realistic cattle prices and rural wages, and laid down penalties for those who dared to pay more. The *Sesmarias* Law also, once again, compelled several social categories to work for wages: former journeymen who no longer wanted to be rural workers, because they could find better jobs elsewhere; able indigents, who lived on alms; fake friars and all kinds of vagrants

6 See Coelho (1983), Gonçalves (1989), Sousa (1990), Amaral (1994) and Rodrigues (1995).

that went from one town to another, living by their wits and causing trouble (Rau 1982: 90–91).

Yet still worse was to come. In 1381–1382 Castilian incursions into Portuguese territory took place, and English troops, called to help the Portuguese military, increased the distress, acting as if they were in occupied land. When Fernando I died, in 1383, leaving as his successor a ten-year-old daughter, married to the Castilian King Juan I, there was a risk that the Portuguese crown would be occupied by that king. A popular upheaval around the choice of another king, João I, a half-brother of the deceased king, displeased the Castilians and Portugal was invaded again. The wars ceased in 1402, after several victories by the Portuguese army eventually forced Castile to sign a truce. Meanwhile, plague struck in the years 1383 to 1385, 1389 and 1400. At the turn of the century, the kingdom was largely exhausted and its economy partially destroyed.[7]

In the following decades, although wars ceased within Portuguese borders, pestilence made its appearance in several occasions and the number of deaths continued to increase in impressive numbers (Rodrigues 2009: 114–115). Since the first population estimate for the whole kingdom dates only from 1527–1532, when it is believed that the pre-plague population level had already been surpassed (Miranda, Chapter 3 in this book), it is difficult to provide an estimate for the overall decline. We can however use as a proxy the number of cross-bowmen that each municipality (*concelhos*) had to deliver to the royal army, which was by law directly related to the size of the population, and thus we can assume that the total are proportionally related to each municipality's totals.

As shown in Table 2.1, the number of cross-bowmen fell everywhere in the country, though with important regional differences. For most municipalities, there are only figures for 1385 and 1422. Furthermore, the two southern administrative divisions (*comarcas*), the Alentejo and the Algarve, were less hit by the problem. During the period under analysis, the decline in the number of cross-bowmen varied from the minimum of 9 percent in Faro, 11 percent in Veiros, and 12 percent in Tavira and Setúbal, to a maximum of 52 percent in Sines. But if we take into account the information regarding some *comarcas* for which there is data up to 1439–45 or 1458–68, like Marvão, the decline reaches 60 percent. Likewise, in Estremadura, Beira and Entre-Douro-e-Minho, the decline in the number of officers was about 50 percent from 1422 to 1458–68.[8]

The demographic depression that the above figures depict impacted significantly on the patterns of population settlement. In the northern *comarca*

7 See Marques (1987: 519–538) and Rodrigues (2009: 114).
8 For the administrative divisions (*comarcas*) see Map A5 in the Appendix.

TABLE 2.1 *Numbers of cross-bowmen required from some municipalities to serve the royal army (1385–1468)*

Region / Municipality	1385	1422	1439–45	1458–68	Downfall
Minho					
Braga	–	50	35	25	50%
Porto	–	40	25	–	37.5%
Beira					
Aveiro	–	13	7	–	44%
Coimbra	–	86	–	60	31%
Leiria	–	40	–	20	50%
Linhares	–	30	25	–	17%
Marialva	–	14	–	10	29%
Estremadura					
Abrantes	–	30	12	–	60%
Lourinhã	–	10	–	6	40%
Sintra	–	20	–	12	40%
Torres Vedras	–	50	40?	30	40%
Além-Tejo					
Setúbal	34	30	–	–	12%
Sines	21	10	–	–	52%
Crato	–	20	–	10	40%
Marvão	–	25	15?	12	60%
Nisa	–	25	–	14	44%
Veiros	18	16	–	–	11%
Évora	150	100	–	–	33%
Montemor–o–Novo	32	30	20	–	37.5%
Mourão	16	10	–	–	37.5%
Alcáçovas	16	10	–	–	37.5%
Aljustrel	13	10	–	–	23%
Messejana	16	12	–	–	25%
Almodôvar	16	11	–	–	31%
Alvito	14	12	–	–	14%
Castro Verde	14	12	–	–	14%
Odemira	21	12	–	–	43%
Garvão	14	8	–	–	43%
Algarve					
Aljezur	15	10	–	–	33%
Albufeira	12	10	–	–	17%

Region / Municipality	1385	1422	1439–45	1458–68	Downfall
Faro	33	30	–	–	9%
Tavira	34	30	–	–	12%
Castro Marim	20	16	–	–	20%

SOURCE: MARQUES 1987: VOL 4, 24–25 (ADAPTED).

of Entre-Douro-e-Minho, where settlement was very dense and scattered, a number of small villages (*aldeias*) and family-size agrarian units (*casais*) were abandoned, as their inhabitants died or migrated into urban centres (Marques 1988: 276–285). In Estremadura, where manors (*quintãs*), hamlets and isolated *casais* had started to spread in the hinterland of small towns and large villages, there was a halt in this dispersive movement and many of the smallest units disappeared (Rodrigues 1995: 65–70; Viana 2007: 170–173). In the South, however, the decline in population led to the decline in importance of some larger villages, whereas smaller settlements increased in size, which implied an increase in the dispersion of the population across the land (Boissellier 2003: 138–139). These transformations in the size and distribution of population necessarily had an impact on productivity and levels of commercialization of the agrarian sector.

Yet, after the mid-15th century onwards, the level of population increased again. According to one source, the inversion of the trend happened in the 1460s, and the rate of increase reached a new level at the turn of the century (Sousa 1993: 335). These trends are confirmed in the case of the municipalities for which there are population counts for the years before 1527. For example, in Alenquer, an agro-town located in the hinterland of Lisbon, records show that the population increased by about 55 percent between 1497 and 1527 (Ferro 1996: 96). Such increases were more pronounced in the municipalities of Beira, in the centre of Portugal, and Braga, in the north, where the population increased almost two-fold, in the periods from 1496–1527 and 1477–1527, respectively (Dias 1988: 50–51; Marques 1988: 290).

Did the *Sesmarias* Law succeed in reversing the disastrous effects of the population loss on agrarian activities? There is evidence that at least part of the law was actually enforced, as abandoned lands kept being expropriated and given to other cultivators until the end of the 15th century. These procedures led to the privatization of some uncultivated lands that had been in common use for centuries (*baldios*), a movement that would accelerate in the following decades (Rau 1982: 104–141). Yet, the fact that the same fields were expropriated over and over again, together with the persistence of references to idle lands and ruins during the whole period under study, shows that attempts made to recover the pre-Plague cultivated area suffered several setbacks (Coelho 1983: vol. 1, pp. 69–70; Rodrigues 1995: 7). The frequency of complaints presented in the Parliament *Cortes* during the 15th century about the excessively high level of wages, as well as the information that the sons of rural workers did not want to continue in their ancestors' profession, suggest the inefficiency of the law in supplying journeymen and forcing down wages (Marques 1987: 275). But were those complaints justified?

Due to the scarcity of historical sources, it is not possible to build continuous series of prices and wages for such an earlier period. Yet the fragmentary information at our disposal indicates a rise in the real wages of rural laborers until the end of the 14th century, then followed by a period of stagnation and, later, decline, during the 15th century. We also have evidence pointing to the fact that the earnings of craftsmen remained higher than those of rural laborers, which created an incentive for the population to leave the rural areas and activities, although we cannot ascertain the significance of that effect. Moreover, stagnation and decline in those wages have also been noted (Ferreira 2007: 168–215). Such trends in wages were met by a large degree of instability in the prices of grain, particularly wheat, but also in millet, rye, oats and barley. The prices of wine and olive oil did not fluctuate as much, but experienced a decreasing trend during most of the 15th century.[9]

The Reversal of Fortunes

With such depressed market conditions, the low level of wages was not necessarily sufficient to favor landowners and the overall development of the agrarian sector. The substantial decrease in the population and the workforce had greater consequences for the estates of the lords and large landowners, as they were more dependent on hired work for cultivating their demesnes. In Portugal, labor services (*geiras*) performed by peasants with a form of customary land that gave them access to the demesnes were usually quite limited, as they meant an average of three to four days' work each year (Durand 1982: 375–377; Gonçalves 1989: 144). Additionally, in copyhold tenures, handed over to free peasants, laboring duties could not be imposed. Therefore, in the middle of the 14th century, most of the demesnes were cultivated using waged labor. When the availability of manpower was reduced, competition increased, hence the

9 See Marques (1978: 218–224), Coelho (1983: vol. 1, 428–436), Viana (1998, 2001 and 2008) and Ferreira (2007: 28–49, 59–66, 94–96). See also Reis, Chapter 6 in this book.

privileges some lords tried to obtain from the monarchy to recruit journeymen anywhere in the kingdom, and the efforts of certain municipalities to reserve the workforce of their inhabitants for only local landowners. These measures were however insufficient to prevent all of them from experiencing difficulties in paying the rising wages seen in the second half of the 14th century. Especially because the revenue coming from leased lands was declining at the same time: not only did land rents decrease after the plague and stagnate in the first decades of the 15th century, but there was also a clear build-up of arrears and refusals of payment during that period, and landlords repeatedly sued their tenants in the urban courts of justice to force them to meet their obligations.[10]

We do not know much about what happened to lay landowners' estates, because most of their cartularies, deeds and account books have disappeared. When by chance any evidence has survived, it reveals that the area under direct exploitation by the lord had usually become extremely small by the end of the 14th century. Most of the manors were divided into *casais* leased to free peasants for terms of lives (Johnson 2002). The same happened to smaller ecclesiastical lords' estates. As for the estates of great monasteries, such as São Salvador of Grijó, in the Porto area, Santa Maria of Alcobaça and Santa Cruz of Coimbra, located in the geographical centre of the kingdom, or São Vicente de Fora, in Lisbon, they still included vast demesnes, but few were still farmed under the monks' supervision. Several other demesnes were being leased to individual tenants for terms of lives or years, sometimes together with the corresponding legal jurisdictions, a way of externalizing labor costs. The lessees were mainly intermediaries (noblemen, the king's officials, members of the urban aristocracy, clerics), who had the means to pay the rents in cash once a year, to manage the demesnes and to wait for the sale of the produce to receive the return (Rodrigues 1998: 13–16). This might have been considered by the monks as a temporary situation to deal with hard times, until they were able to return to direct farming.

Yet, historical evidence shows that, if they took back any demesnes, it was because of the failure of the lessees to keep them under cultivation and pay the rents (Gonçalves 1982). Indeed, during these uncertain decades, investing in agriculture was hazardous and many a lessee lost his belongings for having underestimated the difficulties involved. The larger monasteries also continued to alienate parcels of their large demesnes or to divide them into *casais* and hand them over to peasants through leasehold contracts.[11] Therefore, despite the great flow of donations to the Church and the concentration of real estate

10 See Coelho (1983: vol. 1, 326–332) and Rodrigues (1998: 38–41).

11 See Coelho (1983: vol. 1, 286–289) and Gonçalves (1989: 157–159).

in the hands of fewer owners after the Black Death, there were no signs of the formation of larger estates. As we saw before, in fact, some of the previously existing ones actually got smaller. The estates of lords and great landowners, including those amassed during this period, remained divided into a multitude of *casais* and plots dispersed over vast areas.[12] Only small farm areas (*quinta* or *granja*), or a few of them in the case of the greater lords, were reserved for direct farming. Thus, tenancy, as a typical management strategy, became even more generalized than before.

The usual types of tenancy were the same as in the previous period: perpetual *aforamentos*, *emprazamentos* for terms of lives and *arrendamentos* for terms of years.[13] The former allowed the transmission of the land to the holder's descendants and even its sale to someone else, subject to the owner's permission and the payment of a fee (*laudémio*). The latter only assured to the holder of the land usufruct for a limited number of years. Not many of these have been found in historical archives because they were short-lived: they could be oral arrangements and, even if they were written on parchment or paper, they were possibly destroyed after they came to an end.[14] They were particularly numerous, though, in the monastery of Grijó's lands, in the hinterland of the city of Porto. The inventory dating back to 1366 informs us that 44.9 percent of the land was under *arrendamento* contracts, many of which were made "until a settler shows up" (*ata que lhe saia pobrador*) (Amaral 1994: 79–81). Landowners most likely used *arrendamentos* in the worst periods of labor shortage to prevent more lands from being abandoned. However, they preferred longer arrangements, which provided more guarantees that the soil would be carefully treated and not be exhausted by overexploitation, and so did tenants who wanted to enjoy the returns of investments in work and capital made in the holdings.

During the *Reconquista* period, perpetual leases had been the typical landlords' concessions to attract peasants to their estates and ensure that these lands remained populated and cultivated. In the late 14th and 15th centuries, *aforamentos* were once again one of the preferred tenure contracts to stimulate new holders to perform the more demanding tasks (clearing, canal digging, vine planting, etc.) necessary to make lands that had been abandoned productive again. Additionally, in customary lands, vacant holdings had to be handed over to peasants under the same conditions as before. Thus, in some

12 See Cunha (1990: 79–123), Rosa (1995), Oliveira (1999: 71–94 and 2005: 171–183).

13 For a definition of these types of tenancy regimes see Table 1.2 and Chapter 1 above.

14 See Coelho (1983: vol. 1, 298–299), Gonçalves (1989: 198–200) and Rodrigues (1995: 434–435).

regions, about 60 percent of the countryside leases made throughout this period were still perpetual, while in others they did not exceed 30 percent. Even if in the long term the trend was for a decrease in perpetual contracts, it appears that there were some short-term increases, which correspond to phases of intensification of the activity.[15] Nonetheless, as leases for terms of lives also gave holders return guarantees for their investments, this is the type of contract that gained ground during those decades. A trend towards the predominance of leases for "three lives" is perceptible not only in the outskirts of urban centres, where they were the majority, but also in rural areas. Stable tenancy was thus sought by tenants and promoted by landowners.

Rents paid for these leases were divided into two parts: a main rent (*cânon*) and accessory rents (*direituras*, *miunças* or *foragens*) made of small quantities of grain or bread, poultry, eggs or a variety of other agrarian products. The Church also collected tithes, which amounted to 10 percent of output. In the countryside, sharecropping remained the prevailing way of paying the main rent. It is possible to calculate that it was used in 50 to 60 percent of contracts. Although the shares varied from one region to another, the information available indicates that the most usual ones were ¼ of grain, ⅙ of wine and ⅛ of vegetables and fruit.[16] Money rents were also used. They became very popular in the small plots surrounding towns and were starting to advance to more distant countryside areas when, during the reigns of King Fernando I and King João I, war efforts forced the monarchy to devalue the currency several times (Marques 1996: 169–183). This led to a considerable devaluation of money rents that the Ordinances of Equivalence, issued by the crown to define the correspondence between the old nominal units and the money in circulation, compensated for only in part (Henriques 2008: 195–205). As a consequence, a long stagnation in the number of cash rents ensued.[17] Eventually, due to the combined efforts of lords and peasants, houses and granaries were rebuilt, swamps were drained once more and abandoned fields were gradually returned to cultivation. There were no dramatic improvements in agrarian techniques, which remained the same as in the period referred to in Chapter 1 of this book (Marques 1987: 46–49). However, many of the recovered fields were converted into vineyards and olive groves, especially in the outskirts of

15 See Coelho (1983: vol. 1, 297–298), Gonçalves (1989: 189–193) and Rodrigues (1995: 419–425).

16 See Coelho (1983: vol. 1, 309–314 and 319–321), Gonçalves (1989: 205–208) and Rodrigues (1995: 437–443).

17 See Coelho (1983: vol. 1, 327–329), Gonçalves (1989: 208–213) and Rodrigues (1995: 445–446).

urban centres. Vast areas of wasteland were furthermore carefully preserved to allow for the extraction of timber for naval and urban construction, and to feed the flocks of sheep and cattle that had become an important source of revenue. Those changes were a consequence of the increasing commercialization of agricultural produce.[18]

Commercialization

Although, as we have seen before, commodity prices revealed a declining trend during the decades under analysis, there were periods in which they rose strongly. For example, historical evidence has been found for a dearth (*carestia*) of grain for 67 years between 1350 and 1500 (Marques 1987: 257–282). Historians are however still discussing the meaning of the term *carestia*: for some it derives from the Latin *carere* (to be wanting), for others it derives from *carus* (expensive) (Bourin and Menant 2011: 15–16). This allows for two different perspectives: did cereals become too expensive because of a decrease in production or for other reasons? According to what was argued in Chapter 1 regarding the continuous migration of people from the over-populated North to the loosely-settled South, there are no reasons to believe that, before the Great Plague, demographic pressure on land caused soil exhaustion and thus declining returns, and even less so afterwards. The usual Malthusian explanation for food subsistence crises should therefore be dismissed.

As for environmental factors, though a change in climate actually took place in Europe in this period, the cold and rain that it brought does not seem to have had calamitous consequences in the agrarian economy of the Mediterranean (Campbell 2010: 15–20).[19] Only in very few cases do we know that a dearth resulted from poor crops caused by bad weather, drought (in 1355, 1375, 1412) being more harmful than excessive rainfall (in 1372). War was also to blame in some of these periods, as we have seen: the dearths of 1371–72, 1374–76 and 1384–87 coincide with or follow Castilian military incursions and the destruction they caused. This does not however mean that those ravages affected the whole kingdom. In fact, most of the evidence collected on *carestias* concerns urban centres – not only cities like Lisbon, Porto or Évora, but also smaller towns such as Elvas, Chaves, Loulé – that could not find grain at reasonable prices to feed all their

18 See Coelho (1983: vol. 1, 200–201), Rodrigues (1995: 254–255), Viana (1998: 48–51), Conde (2000: 162) and Oliveira (2008: 237–243).

19 For temperatures in Portugal during this period see Figure A1 in the Appendix.

inhabitants (Duarte 2012: 248–251). This may indicate that the problem was not – or at least not only – a shortage in production, but rather an excessive rise in prices. For other European regions, it has been suggested that "the essential factors of the gravity of famine were speculation on commodities, wars, authorities' errors in prediction and administration, irresponsible, sectarian, and corrupt practices of power, and a ruinous tax regime for tax-payers" (Bourin, Carocci, Menant and To Figueras 2011: 673). Was this the case in Portugal?

Despite the fact that they also lost considerable numbers of their inhabitants during the outbreaks of pestilence and military maneuvers, towns kept receiving migrants from the countryside during this period and most of them even grew (Rodrigues 2009: 123).[20] Lisbon became the kingdom's capital, the place where the king and the court sojourned for longer periods, and where most of the monarchy's administrative institutions were based. It was also an important centre for crafts and trade. Its attraction was so intense that it caused the shrinking of a few neighboring towns, such as Torres Vedras, Arruda and even Santarém.[21] At the beginning of the 16th century, Lisbon was by far the kingdom's largest city and could compare favorably with some of the busiest cities in Western Europe, such as Florence, London, Bruges or Antwerp, although it was still far from equal to Venice and Paris (Bairoch, Batou and Chèvre 1988) (see Table 2.2).

While a few urban dwellers still cultivated small gardens on the outskirts of Lisbon (Barros 1998: 90–91; Oliveira 1999: 124–126) and some lords brought produce from their distant manors for their own families' consumption, most of the inhabitants needed to buy food, and the city depended on an ever-growing supply area. The charter that details the amounts paid at the local toll in c. 1377 (*foral da portagem*) shows that Lisbon received grain, fruit, wine, olive oil, meat, fish and other foodstuffs from across the kingdom.[22] These goods arrived at Lisbon markets mainly by way of seaports, but could also be transported through existing river ports on the Tejo and other smaller rivers, or to a lesser extent by road.[23] In fact, a considerable amount of these commodities was consumed locally, though part of them was exported abroad. Historians have usually interpreted the integration in 1385, by King João I, of the neighboring towns of Sintra, Mafra, Alenquer and Torres Vedras into the municipal area of Lisbon, thus annulling their autonomy, as a punishment for their collaboration

20 For the location of the main urban centres see Map A9 in the Appendix.

21 See Marques (1980: 124–125), Rodrigues (1996a: 58–60) and Viana (2007: 175).

22 See Arquivos Nacionais – Torre do Tombo (National Portuguese Archives), Forais Antigos, m. 2, n. 3.

23 See Maps A13 and A14 in the Appendix.

TABLE 2.2 *Population of some European urban centres, 1300–1500 (in thousands)*

Urban centre	1300	1400	1500
Lisbon	35	55	65
Porto	6	18	14
Coimbra	6	–	6
Évora	12	–	13
Santarém	7	–	10
London	35	45	50
Antwerpen	17	5	30
Bruges	40	125	35
Rome	30	33	55
Florence	95	55	55
Genova	100	100	58
Venice	110	100	100
Paris	150	275	225

SOURCE: BAIROCH, BATOU AND CHÈVRE 1988: 11, 28, 33, 43, 47, 49, 57.

with Castilian rule (Marques 1987: 532). However, as these were typical agro-towns, which commanded rich rural areas, this decision might also have been caused by the urgent need to provide the city with foodstuffs when plague and war had interrupted the usual commercial routes.

Something similar happened in Porto. This city, the most important in the North, became a busy industrial and commercial centre, depending as much as Lisbon on provisions that were brought from an expanding supply area, both for local consumption and for export (Melo 2009: vol. 1, 188–196). In other medium-sized towns, such as Santarém, Évora, Ponte de Lima or Guimarães, most inhabitants belonged to the privileged orders or worked in the administration, crafts, and trade, depending on the market for the provisioning of their larders and tables.[24] Even in agro-towns like Torres Vedras and Alenquer, in this period there was a growing diversification and specialization on crafts works (Rodrigues 1995: 305–313; Ferro 1996: 151–157). In this context,

24 See Beirante (1980: 194–222), Andrade (1990: 163–173), Beirante (1995: 507–578) and Ferreira (2010: 460–616).

urban demand for foodstuffs increased and price speculation became more common.

The legal powers conferred by the municipal charters (*forais*) granted by kings to the cities and major towns allowed these to have legal mechanisms to force farmers from the surrounding rural areas to sell their surplus in the urban market. Yet municipal authorities had no influence over those who horded great masses of grain, wine or olive oil. They were often the lessees of ecclesiastical manors and the collectors of the king's rural rents and taxes, many of whom were also merchants, who could wait for prices to rise or send provisions to other places, including abroad, to be sold more profitably. When facing a shortage of grain and rising food prices, municipal authorities therefore usually fixed prices and forbade the grain to be taken beyond municipal borders. When these protectionist measures were taken at the same time by several urban centres, the grain could not flow from the areas where there were surpluses to those where there were shortages. This contributed to amplifying food crises.

The fact that in these depressive contexts the kings also forbade the export of foodstuffs warns us of their importance for Portuguese foreign trade. In 1391, King João I wrote to Porto's municipal authorities to inform them that no grain should be sent out of the kingdom (Ferreira 2007, 33–34). Several years later, in 1414, Porto's municipal authorities forced the grain stationed at the local seaport and ready to be sent abroad to be sold in the city market. King Afonso V also issued a general prohibition on grain export in 1452 (Marques 1978: 265, 273). The same kind of measure was applied to other commodities. In 1394, the Lisbon representatives in the Parliament (*Cortes*) complained that there was no meat in the city because of its export to Castile, repeating this accusation in 1446 and 1448 (Ferreira 2007: 68–69). In 1442, in the face of a wine shortage, the Porto municipality forbade its export to prevent an excessive rise in prices (Ferreira 2007: 62). If on some occasions merchants made higher profits by selling abroad grain, wine and meat that were in shortage inside national borders, from 1370 onwards they imported grain ever more often and in increasing quantities to benefit from the trade tax (*sisa*) payment exemptions that the kings continued to enact in times of food scarcity to ensure the supply of urban centres. Another measure taken by the monarchy or the local authorities to attain this goal was to authorize fish and salt exports only when the ships brought wheat in return (Marques 1978: 157–167, 257–282).

Portuguese external trade was strongly stimulated by the shift in commercial routes caused by the *Hundred Years War*. Because of the war, there was a growth in the volume of trade and contacts with Flanders, England, Brittany, Castile and Aragon were intensified. Portugal also established for the first time commercial relations with the Hanse, the Baltic countries, Ireland, Venice, Florence, Sicily, and the northern and western African coasts (Muhaj 2013: 101–130). Muhaj does not mention which goods were traded with each of these regions, but research carried out by other historians informs us that a considerable part of the traded goods was based on agricultural production. To the North Atlantic regions (France, Ireland, England, Flanders, Zealand and the Baltic Sea) the Portuguese started exporting wine, olive oil, fruit, honey, wax, fish, salt. With the beginning of overseas expansion to the African coast and the Atlantic islands, these products were joined by sugar and ivory (see Figure 2.1). In return, the Portuguese imported cloth, staple goods, metal and shipping equipment (Miranda 2012: 169–174).

To the western Mediterranean areas (Barcelona, Valencia, Majorca and Italy) the Portuguese exported salted fish, leather, later joined by slaves, wheat and sugar. In return they imported cloth, different manufactured goods, raw materials (tar, pitch, metal, etc.), weapons, spices and grain (the last mainly from Sicily). To North Africa Portugal exported fruit, iron and weapons (despite the prohibition of selling them to "enemies of the Christian faith") in exchange for grain, copper, and, later, raw cloth necessary to support trade across the western African coast (Barata 1998: 421–423, 431–434). Thus there was an increasing external demand for foodstuffs, mainly wine and olive oil, whose area of production grew (Rodrigues 1998: 42). Yet the lack of wheat remained a major concern in the kingdom, which found only a partial solution to the problem with the consolidation of the Atlantic islands' settlements.

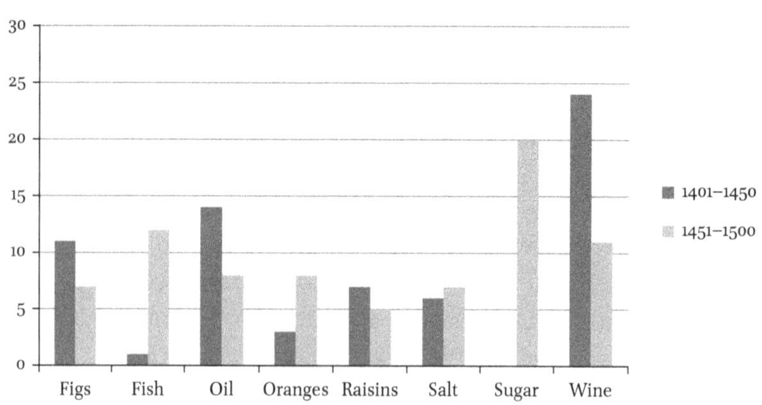

FIGURE 2.1 *Main foodstuffs exported in the 15th century to the North Atlantic (number of mentions in sources)*
SOURCE: MIRANDA 2012: 172 (ADAPTED).

The Impact of the Early Overseas Expansion

The launching of Portuguese overseas expansion in the early 15th century did not bring major changes to national agrarian activities. If one of the main arguments for the military conquest of the town of Ceuta (in North Africa), in 1415, was the local abundance of wheat, which would compensate for the kingdom's inability to produce enough grain to feed all its inhabitants, the fact is that quite the opposite happened. The kingdom had to provide Ceuta with commodities as long as it remained the only Portuguese military fortress in North Africa, because the Portuguese could control neither the hinterland areas, where grain was produced, nor the commercial routes that bypassed the town after the conquest. Even when the Portuguese occupied the towns of Alcácer Ceguer (1458), Arzila and Tangiers (1471) the overall supply situation did not change, as again these remained isolated urban centres among the enemy's large territories.[25] Thus, for Portugal the need to import wheat continued to be an important political matter.

The colonization of the Atlantic islands had more impact on the realm's agrarian activities. Madeira island had been represented in portulans since the 14th century but was rediscovered by the Portuguese c. 1419, and its colonization started in the following decade. Settlers were prisoners to whom freedom was granted by the monarchy in exchange for their settlement overseas (*homiziados*) and common people who were looking for a better life, but also elements of the lower nobility and members of the household of Prince Henry *the Navigator*, who was made lord of the island in 1433. The former received in *sesmaria* only the quantity of land they could look after and had to make it productive for a ten-year term while the latter were favoured with vast estates and not submitted to a time limit for improving them (Rodrigues 1996b: 41–45).

During the first years of colonization, agrarian output was enough to feed the local population, and there is evidence of exports of different products, starting with wood, as the island was densely forested and clearing was intense. Within a few years, wheat became more important due to the high levels of productivity of these virgin soils. An Italian visitor, probably exaggerating, speaks of a seeding ratio of 60:1 in the first years, and between 30:1 and 40:1 in 1455 (Cadamosto 1944: 172). Wine was soon exported as well. After 1460, the expansion of sugar cane was made at the expense of grain cultivation, and later wheat to supply Madeira's inhabitants had to be imported from the Continent and the Azores islands (Rodrigues 1996b: 62–63). In the last quarter of the 15th century, Madeira became the main European producer of sugar,

25 See Braga (1998: 316–317), Schwartz (2002: 22) and Pedreira (2007: 53).

driving out its Mediterranean competitors, like Sicily and Granada. A considerable part of these trade activities was in the hands of foreign merchants (Vieira 2002: 55–68).

The Azores archipelago was discovered between c. 1430 and 1453 (Carita 2008). Colonization started around 1440 with the islands of Santa Maria and São Miguel and, in the following decades, extended to the other seven islands. Early colonization was also led by Prince Henry, and then entrusted to another lord, the duke of Beja and Viseu. Thus, the members of these princes' households became the settler elite. Common inhabitants came not only from Portugal (mainly *homiziados* and Jews), but also from Flanders, France, Castile, Italy, and Morocco (Meneses 2008a). The land was given out again under the *sesmaria* system and settlers accepted the obligation to cultivate it within five years. Some historical evidence suggests that in the beginning land may have been taken freely (*tomada*) by the newcomers (Gregório 2008: 113–114).

Once again, the great fertility provided by these virgin soils permitted high grain yields. Wheat soon became the main agricultural product from the Azores, ensuring local consumption and exports to Madeira Island, Portuguese military fortresses in Morocco, and to the Portuguese European territory. Flemish settlers introduced a dyeing plant, pastel, in the 15th century, but prosperity through its export to Northern Europe would come only at the end of the following century. Vines and sugar cane were also introduced in the islands during those early years, yet they never became important local commodities, due to unfavorable ecological conditions and competition from Madeira Island (Meneses 2008b).

Conclusion

As in other Iberian kingdoms, in Portugal the Black Death and its reappearances aggravated an already existing imbalance between the population and the land. The demographic setback was more strongly felt in the countryside, because of a persistent exodus to towns boosted by unexpected inheritances for some and higher urban wages for others. Hamlets and villages were emptied and many lands abandoned, generating a breach in production that, nevertheless, did not match the population reduction. As a result, commodity prices fell. They rose sharply only when there were poor crops, due to unfavorable weather conditions, or when wars caused monetary devaluation and a disruption of trade. These two factors were almost always felt simultaneously from 1371 to 1402, rendering the provisioning of foodstuffs to cities and towns very difficult.

The demographic decline eventually came to a halt and the population started to grow, slowly at first, but very steadily at the end of the 15th century. An increasing demand from the kingdom's urban centres and also from abroad stimulated the reconstruction of the countryside, as well as the expansion of cattle and agrarian products supported by exports, namely vines and olive groves. Similarly to what happened with wheat, those outputs remained irregular and the search for new food supply areas was one of the most important arguments for the conquest of Ceuta and the settlement of Madeira and the Azores. If Portuguese expansion to Morocco worsened the supply problem instead of solving it, the colonization of these archipelagos at first brought a certain relief. However, the discovery that sugar could be cultivated in Madeira with excellent results soon filled the island with cane to the detriment of grain. Sugar started to be exported along with traditional agricultural products, fish and salt in exchange for the wheat that the kingdom could not produce in sufficient quantity to feed all its inhabitants. The continuation of overseas conquests and discoveries would soon add new commodities to this growing transcontinental trade. The increase in commercial activity, within the country, with the empire and Europe was joined by a sustained recovery of the agricultural sector as a whole.

References

Amaral, Luís (1994). *São Salvador de Grijó na Segunda Metade do Século XIV*. Lisbon: Cosmos.

Andrade, Amélia Aguiar (1990). *Um Espaço Urbano Medieval: Ponte de Lima*. Lisbon: Livros Horizonte.

Bairoch, P., Batou, J. and Chèvre, P. (1988). *La population des villes européennes de 800 à 1850*. Genève: Droz.

Barata, Filipe Themudo (1998). *Navegação, Comércio e Relações Políticas: os Portugueses no Mediterrâneo Ocidental 1385–1466*. Lisbon: Fundação Calouste Gulbenkian – Junta Nacional de Investigação Científica e Tecnológica.

Barata, Filipe Themudo and Henriques, António (2011). "Economic and fiscal history." In José Mattoso (Ed.), *The Historiography of Medieval Portugal*. Lisbon: Instituto de Estudos Medievais, pp. 261–282.

Barros, Maria Filomena Lopes de (1998). *A Comuna Muçulmana de Lisboa (sécs. XIV–XV)*. Lisbon: Hugin.

Beirante, Maria Ângela (1980). *Santarém Medieval*. Lisbon: Faculdade de Ciências Sociais e Humanas – Universidade Nova de Lisboa.

Beirante, Maria Ângela (1995). *Évora na Idade Média*. Lisbon: Fundação Calouste Gulbenkian – Junta Nacional de Investigação Científica e Tecnológica.

Benito i Monclús, Pere (Ed.) (2013). *Crisis alimentarias en la Edad Media: modelos, explicaciones y representaciones*. Lleida: Milenio.

Boissellier, Stéphane (2003). *Le Peuplement médiéval dans le sud du Portugal*. Paris: Centre Culturel Calouste Gulbenkian.

Bourin, Monique and Menant, François (2011). "Les disettes dans la conjoncture de 1300 dans la Méditerranée occidentale". In Monique Bourin, François Menant and John Drendel (Eds.), *Les disettes dans la conjoncture de 1300 en Méditerranée occidentale*. Rome: École Française de Rome, pp. 9–33.

Bourin, Monique, Carocci, Sandro, Menant, François and To Figueras, Lluis (2011). "Les campagnes de la Méditerranée occidentale autour de 1300: tensions destructrices, tensions novatrices". *Annales. Histoire, sciences sociales*, 3 (juillet–septembre), pp. 663–704.

Braga, Paulo Drumond (1998). "A expansão no Norte de África". In A.H. de Oliveira Marques (Ed.). *A Expansão Quatrocentista*. In Joel Serrão and A.H. de Oliveira Marques (Eds.). *Nova História da Expansão Portuguesa*. Lisbon: Estampa, vol. 2, pp. 235–337.

Britnell, Richard H. (1993). *The Commercialization of the English Society, 1000–1500*. Cambridge: Cambridge University Press.

Britnell, Richard H., and Campbell, Bruce M.S. (Eds.) (1995). *A Commercializing Economy. England 1086 to circa 1300*. Manchester: Manchester University Press.

Cadamosto (1944). "Primeira navegação". In João Martins da Silva Marques (Ed.). *Descobrimentos Portugueses – Documentos para a sua História*. Lisbon, Suplemento to vol. 1, Doc. 133, pp. 164–231.

Campbell, Bruce M.S. (2010). "Physical shocks, biological hazards, and human impacts: the crisis of the thirteenth century revisited". In Simonetta Cavaciocchi (Ed.), *Le interazioni fra economia e ambiente biologico nell'Europa preindustriale secc. XIII–XVIII*. Firenze: Firenze University Press, pp. 13–32.

Carita, Rui (2008). "O descobrimento dos Açores". In Artur Teodoro de Matos, Avelino de Freitas de Meneses and José Guilherme Reis Leite (Ed.). *História dos Açores. Do Descobrimento ao Século XX*. Angra do Heroísmo: Instituto Açoreano de Cultura, pp. 49–61.

Carvalho, Sérgio Luís de (1985). "A peste de 1348 em Sintra". In *1383–1385 e a Crise Geral dos Séculos XIV e XV. Actas das Jornadas de História Medieval*. Lisbon: História e Crítica, pp. 129–135.

Coelho, Maria Helena da Cruz (1983). *O Baixo Mondego nos Finais da Idade Média*. Coimbra: Faculdade de Letras da Universidade de Coimbra.

Conde, Manuel Sílvio Alves (2000). *Uma Paisagem Humanizada. O Médio Tejo nos Finais da Idade Média*. Cascais: Patrimonia.

Cunha, Mafalda Soares da (1990). *Linhagem, Parentesco e Poder. A Casa de Bragança (1384–1483)*. Lisbon: Fundação da Casa de Bragança.

Dias, João José Alves (1988). "A Beira interior em 1496 (sociedade, administração e demografia)". In João José Alves Dias (Ed.). *Ensaios de História Moderna*. Lisbon: Presença, pp. 11–102.

Duarte, Luís Miguel (2012). "'Tomar o pão dos coitados'. Para repensar a crise do século XIV em Portugal". In Álvaro Garrido, Leonor Freire Costa and Luís Miguel Duarte (Eds.). *Estudos de Homenagem a Joaquim Romero de Magalhães. Economia, Instituições e Império*. Coimbra: Almedina, pp. 241–261.

Durand, Robert (1982). *Les campagnes portugaises entre Douro et Tage aux XIIe et XIIIe siècles*. Paris: Fundação Calouste Gulbenkian – Centro Cultural Português.

Epstein, Stephen R. (2000). *Freedom and Growth. The Rise of States and Markets in Europe, 1300–1750*. London and New York: Routledge.

Ferreira, Maria da Conceição Falcão (2010). *Guimarães: "Duas Vilas, um Só Povo". Estudo de História Urbana (1250–1389)*. Braga: CITCEM-Universidade do Minho.

Ferreira, Sérgio Carlos (2007). *Preços e Salários em Portugal na Baixa Idade Média*. Porto: Universidade do Porto (PhD dissertation).

Ferro, João Pedro (1996). *Alenquer Medieval (Séculos XII–XV). Subsídios para o seu Estudo*. Cascais: Patrimonia.

Gonçalves, Iria (1982). "Custos de montagem de uma exploração agrícola medieval". In *Estudos de História de Portugal. Homenagem a A.H. de Oliveira Marques*. Lisbon: Estampa, vol. 1, pp. 255–270.

Gonçalves, Iria (1989). *O Património do Mosteiro de Alcobaça nos Séculos XIV e XV*. Lisbon: Faculdade de Ciências Sociais e Humanas – Universidade Nova de Lisboa.

Gregório, Rute Dias (2008). "Formas de organização do espaço". In Artur Teodoro de Matos, Avelino de Freitas de Meneses and José Guilherme Reis Leite (Eds.). *História dos Açores. Do Descobrimento ao Século XX*. Angra do Heroísmo: Instituto Açoreano de Cultura, pp. 111–140.

Henriques, António (2008). *State Finance, War, and Redistribution in Portugal 1249–1527*. York: University of York (PhD dissertation).

Henriques, António (2015). "Plenty of land, land of plenty: The agrarian output of Portugal (1311–20)". *European Review of Economic History*, 19 (2), pp. 149–170.

Johnson, Harold B. (2002). "Uma propriedade portuguesa de finais do século XIV". In Harold B. Johnson (Ed.). *Camponeses e Colonizadores: Estudos de História Luso-Brasileira*. Lisbon: Estampa, pp. 75–88.

Marques, A.H. de Oliveira (1978). *Introdução à História da Agricultura em Portugal*. Lisbon: Cosmos.

Marques, A.H. de Oliveira (1980). "Estratificação económico-social de uma vila portuguesa na Idade Média". In A.H. de Oliveira Marques (Ed.). *Ensaios de História Medieval Portuguesa*. Lisbon: Vega, pp. 121–133.

Marques, A.H. de Oliveira (1987). *Portugal na Crise dos Séculos* XIV *e* XV. In Joel Serrão and A.H. de Oliveira Marques (Eds.). *Nova História de Portugal*. Lisbon: Presença, vol. 4.

Marques, José (1988). *A Arquidiocese de Braga no Séc.* XV. Lisbon: Imprensa Nacional-Casa da Moeda.

Marques, Mário Gomes (1996). *História da Moeda Medieval Portuguesa*. Sintra: Instituto de Sintra.

Martins, Miguel Gomes (2006). "Ficou a terra estragada que maravylhosa cousa era de ver. Guerra e paisagem no Portugal medieval (1336–1400)". In Iria Gonçalves (Ed.). *Paisagens Rurais e Urbanas. Fontes, Metodologias, Problemáticas*. Lisbon: Centro de Estudos Históricos – Universidade Nova de Lisboa, vol. 2, pp. 125–146.

Masschaele, James (1997). *Peasants, Merchants and Market: Inland Trade in Medieval England (1150–1350)*. New York: Saint Martin's Press.

Melo, Arnaldo Rui Azevedo de Sousa (2009). *Trabalho e Produção em Portugal na Idade Média: O Porto (c. 1320–c. 1415)*. Braga: Universidade do Minho (PhD dissertation).

Meneses, Avelino de Freitas de (2008a). "O povoamento". In Artur Teodoro de Matos, Avelino de Freitas de Meneses and José Guilherme Reis Leite (Eds.). *História dos Açores. Do Descobrimento ao Século* XX. Angra do Heroísmo: Instituto Açoreano de Cultura, pp. 63–109.

Meneses, Avelino de Freitas de (2008b). "Recursos, excedentes e carências". In Artur Teodoro de Matos, Avelino de Freitas de Meneses and José Guilherme Reis Leite (Eds.). *História dos Açores. Do Descobrimento ao Século* XX. Angra do Heroísmo: Instituto Açoreano de Cultura, pp. 141–172.

Miranda, Flávio (2012). *Portugal and the Medieval Atlantic. Commercial Diplomacy, Merchants, and Trade, 1143–1488*. Porto: Universidade do Porto (PhD dissertation).

Moreno, Humberto Baquero (2009). "La crisis del siglo XIV en Portugal". In Maria Isabel del Val Valdivieso and Pascual Martínez Sopena (Eds.), *Castilla y el Mundo Feudal. Homenaje al Profesor Julio Valdeón*. Valladolid: Junta de Castilla y León – Universidad de Valladolid, vol. 3, pp. 53–61.

Muhaj, Ardian (2013). *Quando todos os Caminhos Levavam a Portugal. Impacto da Guerra dos Cem anos na Vida Económica e Política de Portugal (Séculos XIV–XV)*. Lisbon: Universidade de Lisboa (PhD dissertation).

Oliva Herrer, H.R. and P. Benito i Monclús (Ed.) (2007). *Crisis de subsistencia y crisis agrarias en la Edad Media*. Sevilla: Universidad de Sevilla-Secretariado de Publicaciones.

Oliveira, José Augusto da Cunha Freitas (1999). *Organização do Espaço e Gestão de Riquezas: Loures nos Séculos XIV e XV*. Lisbon: Centro de Estudos Históricos.

Oliveira, José Augusto da Cunha Freitas (2008). *Na Península de Setúbal, em finais da Idade Média: Organização do Espaço, Aproveitamento dos Recursos e Exercício do Poder*. Lisbon: Universidade Nova de Lisboa (PhD dissertation).

Pedreira, Jorge M. (2007). "Costs and finantial trends 1415–1822". In Francisco Bethencourt and Diogo Ramada Curto (Eds.). *Portuguese Oceanic Expansion (1400–1800)*. Cambridge: Cambridge University Press, pp. 49–87.

Ramos, Luís António de Oliveira (1963). "Consequências económicas da Peste Negra". In Virgínia Rau. "Para o estudo da Peste Negra em Portugal". *Bracara Augusta*, 14–15, pp. 220–229.

Rau, Virgínia (1982). *Sesmarias Medievais Portuguesas*. Lisbon: Presença.

Rodrigues, Ana Maria S.A. (1995). *Torres Vedras. A Vila e o Termo nos Finais da Idade Média*. Lisbon: Fundação Calouste Gulbenkian – Junta Nacional de Investigação Científica e Tecnológica.

Rodrigues, Ana Maria S.A. (1996a). "A população de Torres Vedras em 1381". In Ana Maria S.A. Rodrigues (Ed.). *Espaços, Gente e Sociedade no Oeste. Estudos sobre Torres Vedras Medieval*. Cascais: Patrimonia, pp. 45–84.

Rodrigues, Ana Maria S.A. (1998). "Change in tenurial relations in late medieval Portugal". *Cadernos do Noroeste*, 11 (1), pp. 5–47.

Rodrigues, Miguel Jasmins (1996b). *Organização dos Poderes e Estrutura Social. A Madeira: 1460–1521*. Cascais: Patrimonia.

Rodrigues, Teresa Ferreira (2009). "A crise dos séculos XIV e XV (1325–1480)". In Teresa Ferreira Rodrigues (Ed.). *História da População Portuguesa*. Lisbon: Afrontamento, pp. 101–137.

Rosa, Maria de Lurdes (1995). *Pero Afonso Mealha. Os Bens e a Gestão da Riqueza de um Proprietário Leigo do Século XIV*. Redondo: Patrimonia.

Schwartz, Stuart B. (2002). "The economy of the Portuguese empire". In Francisco Bethencourt and Diogo Ramada Curto (Eds.). *Portuguese Oceanic Expansion (1400–1800)*. Cambridge: Cambridge University Press, pp. 19–48.

Sousa, Armindo de (1993). "A população". In José Mattoso (Ed.). *A Monarquia Feudal*. In José Mattoso (Ed.). *História de Portugal*. Lisbon: Círculo de Leitores, vol. 2, pp. 327–360.

Sousa, Bernardo Vasconcelos e (1990). *A Propriedade das Albergarias de Évora nos Finais da Idade Média*. Lisbon: Instituto Nacional da Investigação Científica.

Viana, Mário (1998). *Os Vinhedos Medievais de Santarém*. Cascais: Patrimonia.

Viana, Mário (2001). "Alguns preços do vinho em Portugal". *Arquipélago – História*, 5, pp. 605–625.

Viana, Mário (2007). *Espaço e Povoamento numa Vila Portuguesa (Santarém 1147–1350)*. Lisbon: Caleidoscópio – Centro de História da Universidade de Lisboa.

Viana, Mário (2008). "Alguns preços de cereais em Portugal (séculos XIII–XVI)". *Arquipélago – História*, 11–12, pp. 207–279.

Vieira, Alberto (2002). "A Madeira e o mercado do açúcar. Séculos XV–XVI". In *A História do Açúcar: Rotas e Mercados*. Funchal: Centro de Estudos de História do Atlântico, pp. 55–89.

PART 2

New Frontiers, Crisis and Growth

CHAPTER 3

Coping with Europe and the Empire, 1500–1620

Susana Münch Miranda

Introduction

The period from the mid-1490s to the 1550s is often viewed as a *Golden Age* in Portuguese history as this was when Portugal built an empire comprising settlements in Asia, Africa and the New World and gained a position of leadership in intercontinental trade, particularly with Asia (Disney 2009: 143–145).[1] This new era came about against the background of a long-term cycle of demographic and economic expansion in Europe.[2] Still recovering from the losses caused by the Black Death, the Portuguese population increased steadily throughout the 16th century, as the literature has long stressed. In response, meanwhile, to increasing pressure on food resources, and thanks to the additional labor force available, agricultural output also rose vigorously, at least until the mid-1500s, when investments in land clearings seem to have lost momentum. The prevailing picture depicted by the historiography then points to a depressed phase towards the late 16th or early 17th century, and which ended the long-term cycle of demographic and agricultural expansion.[3] Outbreaks of plague, poor harvests and food shortages subsequently caused death rates to soar and ushered in a recession, although estimations of the timing and depth of this have varied. In the case of population dynamics, the scale of the trend inversion has been debated, especially the question of whether this translated into a sharp fall back to the levels of the 1530s.

Recently assembled data suggest, however, a different narrative. For one, the demographics in the late 16th century are shown to represent only a slow-down in an upward trend and to have lasted approximately twenty years, with growth resuming in the 1610s (Palma and Reis 2016). As for gross agricultural output, recent estimates by Reis revealed three distinct phases in the time span analysed in this chapter: contraction (1500 to mid-1520s), recovery (until the late 1560s) and finally a flattening-out that lasted until the late 1660s. Furthermore, and similar to what has been noted for most countries in Europe,

1 See Map A4 in the Appendix.

2 See Braudel (1992, vol. 2) and Livi-Bacci (1999: 17–23).

3 See Magalhães (1988), Oliveira (1991: 52–56), Santos (2003) Rodrigues (2008) and Costa *et al.* (2016).

 | DOI 10.1163/9789004311527_005

output *per capita* decreased "by more than half from 1500 to around 1625".[4] These estimates clearly tone down the more optimistic view of the agricultural expansion in the 1500s conveyed by the literature.

This chapter examines the agricultural sector between 1500 and 1620 and seeks specifically to highlight the trends in population, trade and industry that boosted demand. The first section presents an overview of the available demographic data on population dynamics and the rate of urbanization. Following the changes taking place on the demand side, the second section examines the signs of output growth and discusses the role of socio-institutional arrangements. The third section evaluates the composition of the sector and indicates the probable effects of the expansion of cropland at the expense of grazing, while the final section discusses the effects of market conditions in agriculture, specifically interrelations with trade and industry.

Population Pressure and Urban Expansion

After the demographic recession attributable to the Black Death in the second half of the 14th century, recovery was well under way in the final decades of 15th century. Indeed, estimates suggest that, by 1500, the population had recovered to its pre-plague levels. By then, Portugal had approximately 906,000 inhabitants, representing less than 2 percent of Western Europe's population and just over 16 percent of the population of the Iberian Peninsula.[5] With a fairly small pool of human resources, Portugal had scope for demographic growth and, thirty years later, the population count of 1527–1532 confirmed that the country was experiencing steady growth in line with the long-term upward trend in Europe. Despite the difficulties arising from the fact that "hearths" rather than souls were counted, this census was the first to cover nearly the entire kingdom. By then, the population had reached an estimated 1,215,000, based on a conversion factor of 4.3 inhabitants per hearth (Table 3.1).[6] This figure implies an average density of 13.6 inhabitants per square kilometre, which was not as low as in Spain (9.9), but clearly below the densities attested for other countries in Western Europe, such as Italy (29.9), England (23.2) or the Netherlands (28.8) (Malanima 2009: 16). However, this comparatively low average density conceals substantial regional differences, which were a legacy of

4 See Reis, Chapter 6, Figure 6.3 in this book.

5 See Álvarez and Prados (2007: 330) and Palma and Reis (2016).

6 A recent estimate by Palma and Reis points to a lower figure of 1,138,000 (Palma and Reis 2016).

settlement patterns dating back to the *Reconquista*.[7] Of the six territorial units (*comarcas*) into which the kingdom was divided for judicial purposes, Entre-Douro-e-Minho, in the north-west, was by far the most densely populated.[8] Although this region accounted for just 8 percent of the kingdom's total area, it housed about one fifth of the country's population and, as such, had a density of 32 inhabitants per square kilometre. The two southern *comarcas* of Alentejo and Algarve had the lowest densities (6.9 and 8.5, respectively). Between these two extremes, the north-west and the south, Beira appears to have been the most balanced *comarca*, with 19.03 inhabitants per square kilometre, despite the irregular intra-regional distribution resulting from the existence of mountainous areas such as the Serra da Estrela. Almost 80 percent of the Portuguese lived in the four *comarcas* located north of the Tejo River and, while accounting for approximately 40 percent of the territory, the two southernmost regions together harboured only one fifth of the total population, which was the same as the number of people living in Entre-Douro-e-Minho. A densely populated north, therefore, contrasted with a high land-labor ratio in the south, where conditions were similar to those of a frontier economy, as also noted for Spain (Álvarez and Prados 2013).

TABLE 3.1 *Regional distribution of Portuguese population (1527–1532)*

Comarca	Number of hearths	Population[1]	Percentage of total population	Density (inhab./ km²)	Area (km²)	Percentage of total area
Entre-Douro-e-Minho	55,016	236,569	19.46	32.62	7,252	8.12
Trás-os-Montes	35,629	153,205	12.60	13.33	11,493	12.87
Beira	67,696	291,093	23.95	19.03	15,298	17.13
Estremadura	65,515	281,715	23.17	14.14	19,930	22.32
Alentejo	48,934	210,416	17.31	6.94	30,319	33.96
Algarve	9,918	42,647	3.51	8.55	4,989	5.59
	282,708	1,215,644	100.00	13.62	89,281	100.00

[1] 4.3 inhabitants/household

SOURCES: SERRÃO 1996: 68; RODRIGUES 2008: 177–178.

7 See Henriques, Chapter 1 in this book.

8 See Map A5 in the Appendix.

The Portuguese population continued to increase throughout the 16th century, although the rate of this growth has been the subject of debate. It is widely accepted that a phase of rapid demographic growth persisted until 1580, followed by a slow-down that lasted until the 1620s.[9] A recent estimate narrowed this slow-down to the time span between the 1590s and the 1610s (Palma and Reis 2016), with data collected from a few regions illustrating these trends and suggesting an average annual growth rate of between 0.4 and 0.8 percent. Among the best-studied cases, the region of Évora (Alentejo) reveals growth rates falling from 0.63 percent for 1527–1589 to 0.42 percent for 1589–1593 and only 0.12 percent thereafter (Santos 2005: 353). However there were also substantial differences in population growth across regions. One of the most dynamic *comarcas* was Beira, where the annual variation of 1 percent between 1496 and 1527 was due mainly to flows of Jewish immigrants expelled from Castile in 1492. Contrasting with the scale of this upsurge, the demographic variation in Estremadura was a modest 0.1 percent in the first quarter of the 16th century.[10] An overall average growth rate of 0.61 percent between 1500 and 1600, which is slightly higher than the 0.58 percent estimated for Spain, puts Portugal as having 1,667,449 inhabitants in 1600 (Palma and Reis 2016).

After 1590, the vigorous expansion was hampered by the accumulated effects of mortality crises. In 1593–1594, and again in 1599, abnormal weather conditions hit not only Iberia, but also France, Germany, Italy and England, and caused severe harvest failures.[11] Famines, coupled with outbreaks of disease, notably plague in 1603–1604, were responsible for soaring death rates across Western Europe. In Portugal, *comarcas* as widely separated as Entre-Douro-e-Minho, Trás-os-Montes, Beira, Alentejo and Algarve experienced serious mortality crises.[12] The overall population declined slightly towards the end of the 1590s and flattened out in the subsequent years. Growth resumed in the late 1610s, when the balance of births and deaths favoured the former, albeit only slightly (Palma and Reis 2016). Owing to changes in the pattern of epidemic-related deaths, outbreaks of bubonic plague became less frequent and had fewer victims.[13] As for emigration, the other variable relevant to demographic evolution, this seems to have played only a secondary role both in the flattening trend and in the overall rate of the population growth in the 16th

9 See Oliveira (1991: 52–56) and Rodrigues (2008: 519).

10 See Pereira (2003: vol. 1, 48–53).

11 See Dupâquier (1997: 20) and Campbell (2010). See also Figures A1 and A2 in the Appendix.

12 See Magalhães (1970), Oliveira (1971–1972), Oliveira (1990), Borges (1996), Amorim (1997) and Santos (2003).

13 See Magalhães (1970: 23–28), Serrão (1993a, 51) and Rodrigues (2008: 182).

century, even though by 1500 the Portuguese were building an empire spread across Africa, South America and Asia.

As the data available are incomplete, estimates for overseas emigration should be taken only as rough orders of magnitude. Throughout the 15th century, an estimated 500 emigrants left the mainland each year for North Africa and the Atlantic islands (Madeira and Azores). By 1500, their number had swelled, as new settlements were formed in Asia and Brazil. According to one estimate, about 3,500 people left the country every year between 1500 and 1580, amounting to a total of 280,000 (Godinho 1978a: 9). However, this estimate does not tally with another estimate, according to which, by 1600, some 100,000 Portuguese were living overseas, and of whom 30,000 were in Brazil.[14] It seems reasonable to conclude, therefore, that the smaller magnitude of emigration flows meant that they did not hinder population growth in the mother country. Indeed, these flows are an indicator of the vigorous long-term demographic growth experienced by Portugal between 1500 and the 1590s. Responding to this general increase and to favourable forces related to overseas expansion, the *long sixteenth century* in Portugal also witnessed urban expansion, just as was happening elsewhere in Western Europe.[15]

If cities are considered to be settlements with at least 10,000 inhabitants (so as to avoid counting people mainly employed in agriculture as belonging to an urban population), the urbanization rate in 1530 stood at 6.7 percent (see Table 3.2). Not surprisingly, this level of urbanization reflects a pre-modern economy, where agriculture was the main source of income. Although this was higher than the average urbanization rate in England (5 percent in 1550), it was lower than in the most advanced urban areas in Europe, such as the southern Low Countries (roughly 20 percent), Italy (12.5–16.6 percent) and even the average in Southern Europe (11.4 percent).[16]

Ideally a benchmark for the urbanization rate in 1600 would allow gauging of the urban expansion of the 1500s. However, it is not until 1695 that the progression of the urban population can be reassessed, by which time urban dwellers comprised 10.3 percent of the total population (see Table 3.2). Knowing that, in the long run, urban expansion in Europe was far from even, scholars usually stress that the 16th-century urban advance in Portugal related mainly to the capital, Lisbon (in Estremadura). Indeed Lisbon grew steadily throughout the 16th century. By 1527, with 56,000 to 60,000 inhabitants, the Portuguese capital was already the seventh largest city in Europe and even

14 See Subrahmanyam (1995: 307–308) and Johnson (1998: 273).

15 See Livi-Bacci (1999: 60) and Clark (2009: 112–119).

16 See Livi-Bacci (1999: 57) and Clark (2009: 120).

TABLE 3.2 *Urbanization in Portugal (1530–1695) (cities with 10,000 or more inhabitants)*

	1530	Percentage of total population	1695	Percentage of total population
Entre-Douro-e-Minho	1	5.5	3	8.6
Trás-os-Montes	–	–	–	–
Beira	–	–	2	3.8
Estremadura	1	19.9	1	23.6
Alentejo	1	5.8	4	15.6
Algarve	–		–	–
Portugal	3	6.7	10	10.3

SOURCES: SERRÃO 1996: 75; RODRIGUES 2008: 193.

larger than Florence, Genoa, Rome, London and Antwerp (Clark 2009: 37). Served by a deep-water port, Lisbon was also the seat of the royal court and the capital of the kingdom and empire, which explains its ability to act as a powerful magnet of immigration. Under the stimuli of the *entrepôt* trade, especially with Asia, its population increased incessantly, reaching about 120,000 inhabitants in the third quarter of the 16th century (Rodrigues 2008: 191). Like other cities in Southern Europe, such as Seville or Naples, Lisbon provides an example of a hypertrophic capital, crowded with casual laborers and ultimately unable to effect major transformations in its hinterland.

In the urban hierarchy, Lisbon was followed, at a substantial distance, by Porto and Évora, each with about 12,500 inhabitants. Despite being served by a port and the Douro River, Porto failed to act as a regional trading centre of the country's northern region, while its growth was also hampered by Lisbon's ability to dominate overseas trade. By 1600, the number of Porto's inhabitants hardly surpassed the level achieved in 1530. In Alentejo, Évora's place in the urban network rested on several factors: firstly on an ancient urban tradition going back to the Muslim occupation, which resulted in a more concentrated pattern of settlement, while secondly on the polarization of the regional agrarian production and its role as a supplier of goods such as wheat and olive oil. Throughout the urban expansion of the 16th century, other cities such as Braga and Guimarães in northern Portugal, as well as Coimbra (Beira) and Elvas (Alentejo), came close to the threshold of 10,000 inhabitants. Nevertheless, none of these cities, including Porto, was able to grow to 20,000 inhabitants, thus stressing Lisbon's dominance of the urban system (Rodrigues 2008: 191). Although the urban network also comprised another handful of small-sized towns, each with roughly 5,000 inhabitants, a good number of these did not function as cities in many respects.[17] A few of them admittedly saw their populations increase in the 16th century. These included the seaports of Aveiro (Beira) and Viana do Castelo (Entre-Douro-e-Minho) in north-west Portugal, which expanded from the 1550s onwards due to their interactions with Northern Europe and the South Atlantic. They also included the coastal areas of Algarve, where dynamic commercial relations with Andalusia, particularly Seville, and the Portuguese fortresses in North Africa, also prompted a brief surge in the numbers of urban dwellers in Lagos and Tavira. Before the end of the 16th century, however, some of these smaller townships had stopped growing as severe mortality crises, problems in agrarian productivity and shifts in trade networks all took their toll. By 1600, for example, the coastal towns of Algarve were losing inhabitants as commercial relations with the Spanish empire, through Seville, came to a halt.[18]

Institutional Framework and Cycles of Agrarian Output

As can be expected in an agrarian-based society, the long-term population growth in the 16th century boosted demand for land, while urban expansion also fostered increased demand for foodstuffs and raw materials. Arable land was consequently extended to meet these needs. Apart from Entre-Douro-e-Minho, where only hilly areas remained uncultivated, and the hinterland of larger cities, Portugal still had significant amounts of land that could be put to agricultural use. At first, the extension of arable land involved taking over previously occupied areas left vacant after 1350. Across the country, lay and ecclesiastical landlords encouraged the recovery of abandoned farming plots by granting tenants lower fees for land use, as in the case of the monastery of Santa Cruz of Coimbra in the central region, which did so in the early 1500s, and the military order of Santiago in its landholdings in Alentejo. The latter recognized in 1510 that it held land that had lain idle for decades and made efforts to bring it back into cultivation.[19] This expansion of farmland also involved clearing land that had never been under cultivation, although the empirical data[20] make it difficult to distinguish between recovery on the one hand and taking over new land on the other. In the case of the Cistercian monastery

17 See Silva (1997: 783–786) and Rodrigues (2008: 350). See Map A10 in the Appendix.
18 See Magalhães (1970: 185–189, 237–242).
19 See Pereira (2003: vol. 2, 259–261).
20 See Devy-Vareta (1986: 12) and Coelho (1989).

of Alcobaça (Estremadura), an estimated 77 percent of farm leases and rents related by the end of the 1400s to newly acquired land (Gonçalves 1989).

Although examples of expanding amounts of arable land can be detected almost everywhere, the main centres of agrarian pressure were in Beira and Estremadura, and included the regions and subregions along the lower Mondego River, the lower Vouga River and the Tejo River valley (from Almada to Abrantes and Tomar) and, closer to Lisbon, around Sintra and Alenquer. Land clearings also took place, albeit with less intensity, further north around Lamego (Beira), in the southernmost regions of Alentejo (Montemor-o-Novo, Évora, Elvas, Arronches, Olivença, Monsaraz and Mértola) and close to a few urban centres in Algarve, notably around Lagos.[21]

This expansion took the form of woodland clearing, marshland draining and the reclamation of pasture land, achieved through the enclosure of fields. Evidence of clearings, dating back to the final two decades of the 1400s and lasting until the 1550s, is available for the lower Mondego valley, as well as for the southern *comarcas* of Alentejo and Algarve. In some regions the clearings led to deforestation. By 1530, for example, the edges of the forest near the town of Alcobaça had retreated as a result of new settlements being created (Gonçalves 1989), while impoverished peasants in the hilly areas of the north-west used burn clearing to gain new land for the cultivation of rye, albeit for only short-term use. In the 1530s, after decades of clearings, also related to intense shipbuilding activities, shortages of wood became obvious around the city of Viana do Castelo (Entre-Douro-e-Minho) (Devy-Vareta 1986). Further south, in Arronches (Alentejo), some of the cork oak and holm oak forests were cut down, thus jeopardizing the traditional balance between arable and pasture lands.[22] Meanwhile in Algarve, the recurrent use of slash-and-burn clearings triggered deforestation of the mountains near Tavira. This prompted an order in 1561 prohibiting the use of the mountains for this purpose and with the aim of protecting the woodlands and the shipbuilding sector (Magalhães 1970: 41–42).

These changes occurred within the institutional framework that took shape during the early Middle Ages and that governed access to land and determined social property relations. Before addressing the main features of this framework it should be noted that no clearly defined property rights to land existed. Instead, property rights to the same plot of land (including rights of use, exploitation and disposal) were often fragmented and held by multiple parties,

21 See Gil (1965: 41–49), Magalhães (1970: 52–53), Gonçalves (1989: 247–251), Rodrigues (1995: 236), Pereira (2003: vol. 1, 329–330), Silva (2006) and Rodrigues (2013).

22 See Pereira (2003: vol. 1, 328).

who could bundle them with similar or other rights to other plots. Although such fragmentation was not unusual in Western Europe, two closely related factors account for the pervasiveness of overlapping property rights to land in Portugal. The first relates to the widespread use of a form of long-term tenure (emphyteusis), whereby limited property rights were transferred to tenant farmers individually or, as in the early days of the *Reconquista*, collectively to entire communities.[23] The second factor relates to the role played by the crown in encouraging collective forms of land settlement, as well as tenant farming. This is best demonstrated by the charters of municipal rights (*cartas de foral*) that granted property rights to a community.[24] These included common land such as *baldios* (wasteland for pasture), as well as forests managed by *câmaras* (municipal governments) on behalf of the community's inhabitants (*concelho*). In return for the property rights transferred in this way, the landowner (the king, a religious institution or a nobleman) claimed various payments or *direitos de foral* from the *concelho* (Costa *et al.* 2016: 28). These payments comprised the equivalent of ground rent, as well as seigniorial rights. In an attempt to standardize the *direitos de foral* collected across the kingdom, King Manuel (1495–1521) enacted a complete revision of the charters of municipal rights granted since the 11th and 12th centuries. After this reform, the charters became an updated list of "royal rights", i.e., land rents and seigniorial rights to be paid to the crown or its grantees (*donatários*). This standardization had long-term effects on the social distribution of agrarian produce, with the impact of this spanning beyond the 16th century and lasting until the liberal revolution of the 19th century.[25]

Against the background of this "imperfect" landownership structure, the most important landowners by the end of the 15th century were the crown and the royal family, the church and other religious institutions, and the nobility. On a much smaller scale, municipal institutions also enjoyed property rights under the charters of municipal rights. Lastly, a few commoners, such as urban elites (merchants and craftsmen) and wealthy rural residents, also owned land.[26] While it is impossible to know exactly how landownership

23 On this, see Costa *et al.* (2016: 28–29).

24 The *cartas de foral* were originally collective agrarian contracts, mostly used in the 11th and 12th centuries as a device for settlement and reoccupation of lands taken from Muslims. Although by the early 14th century collective contracts were no longer granted, the payments due by the *concelhos* of Trás-os-Montes and part of Beira to their respective lords continued to reflect the original collective agreement (Monteiro 2005: 70–71).

25 See Ramos, Sousa and Monteiro (2009: 227–228) and Costa *et al.* (2016: 94).

26 See Magalhães (1970: 228–229) and Rodrigues and Duarte (1998: 83–85).

was distributed, especially among traditional landowners, several features of their respective patrimonies can be highlighted because of their implications for agriculture. The crown owned several landed estates, known as *reguengos*, scattered across the realm and exploited mainly by peasant tenure. Some of these estates were regularly granted to members of the nobility as a reward for military and administrative services, contributing to a high degree of fluidity in the crown's landed property. Less fluid, although not static in the 16th century, was the landed property of ecclesiastical institutions, which was also spread across the six *comarcas* of the kingdom. While exploiting a significant share of their holdings by leasing them to peasants, many of these institutions continued to keep a few estates under direct administration and recruited hired labor for that purpose (Rodrigues and Duarte 1998: 88–91).

As for the aristocracy, research on a handful of houses revealed that, by 1550, a major share of their income derived from crown property (*bens da coroa*), which consisted mainly of seigniorial dues enshrined in the *cartas de foral*.[27] The aristocracy's revenues also extended to tithes (*dízimos*), land rents and seigniorial rights derived from the approximately 600 commanderies of the three military orders (Christ, Avis and Santiago) that the crown regularly bestowed on the aristocracy after 1551.[28] Although the aristocracy held some estates, belonging to their entailed family estates, in full ownership, these represented a comparatively small share of their income. Hence, various circumstances thwarted the aristocracy's direct influence over agriculture. For one, the "royal rights" in the reformed *forais*, which crystallized under the terms set out by King Manuel I, inhibited grantees from arbitrarily raising rents and seigniorial dues.[29] As for the commanderies, their managerial rationale, shaped by the military orders throughout the second half of the 16th century, hampered grantees' ability to interfere with existing tenancy contracts.[30] By the 1580s and in response to this limitation, a growing number of commandery holders had started collecting their income through rent collectors, and

27 See Lobo (1984: 486) and Monteiro and Salvado (2014: 250, 255, 261–263).

28 The Pope granted King John III perpetual control over the commanderies of the three military orders in 1551, thus providing the crown with further resources to redistribute among the nobility in return for military or administrative services. In the late 18th century, the aristocracy derived an average of 30 percent of their income from the commanderies. See Monteiro (2007: 272–273).

29 This reform was part of the crown's interventions designed to control and check the fiscal prerogatives of competing powers. See Hespanha (1994: 141).

30 Although the rules pertaining to the exploitation of the Order of Christ commanderies were formally enacted in 1628, they were already in place in the final decades of the 16th century (Salvado 2009: 270).

this continued throughout the next two centuries.[31] Unlike other countries in Europe, the Portuguese aristocracy was thus mainly a "rent collecting" group (Monteiro 1998: 316).

Within this social and institutional framework, peasant households were the main locus of agrarian production. This is best mirrored in the *casal* (plural *casais*), which had been the major unit for land occupation, farm labor and taxation, from northern to southern Portugal, since medieval times. The *casal* was mostly a small-scale production unit, drawing upon the domestic household for labor and often comprising a dwelling and several plots of land, usually scattered across a wide area and growing a variety of crops to provide for the needs of the family. Agrarian contracts regulated access to a *casal* or other farming plots. Although a variety of arrangements existed, two – perpetual or long-term tenures through emphyteusis, and short-term farm leases – were the most frequently used. It is generally agreed that the balance between the two differed in the northern and southern parts of the country. To the north of the Tejo River, especially in the north-west, long-term tenure dominated as a legacy of secular settlement prior to the formation of the kingdom, while short-term leases were more frequent in Alentejo as a result of the formation of large, market-oriented landholdings (Santos 2005: 350). Nevertheless, the choice between these arrangements was ultimately determined by other variables, and this helps to explain why both were used across the country.

Large and medium-size tenant farms, as in the case – albeit not exclusively – of *casais*, were more commonly exploited under short-term leases, whereby tenants acquired the right to farm the land for a relatively short duration (from two to nine years) in return for a fixed rent. From an income perspective, short-term leases seem to have been more advantageous to the property owner, given that the rent could be renegotiated at the end of every contract term. Contemporary documents, however, record ample cases of land being overworked to the point of soil exhaustion within the few years of a contract's duration (Magalhães 1993: 252). Long-term contracts, on the other hand, seem to have increased the likelihood of properties being preserved in a favourable state. Following a tradition going back to the medieval period, emphyteutic contracts were therefore particularly widespread and also proved more suitable for the expansion of arable land taking place in the early 1500s.

Under an emphyteutic contract, property rights over a plot of land were split into two distinct domains. While the landlord kept the direct domain, the useful domain was transferred to the tenant (*foreiro*), either in perpetuity

31 See Salvado (2011: 50) and Monteiro and Salvado (2014: 254).

(*aforamento* or *prazo enfateusim*) or for a number of lives (*emprazamento*).[32] In return for paying an annual ground rent (*foro*), the *foreiro* acquired a limited array of property rights, including the right to dispose of the useful domain, during his lifetime or after his death, by means of a will.[33] Provided he obtained the consent of the landlord and paid a penalty, the tenant could therefore sell, lease out and even sub-emphyticate the landholding (*subenfiteuse*) (Serrão 2000: 563–564). As a typical form of "imperfect" property, emphyteutic contracts satisfied conditions for both landowner and tenant: the former was guaranteed a source of income with very little risk, while the latter received mitigated property rights and long-term opportunities to farm land in return for investments in labor and capital.[34] Eventually these arrangements encouraged the rise of a secondary land market in useful domains, which may have stimulated cultivators' access to land, as studies for later periods have shown (Santos and Serrão 2011).

Regional studies confirm that emphyteutic contracts for lives were particularly widespread, while those given in perpetuity were used more sparingly.[35] Given that they involved alienation of the useful domain, such *aforamentos* were granted almost exclusively for the cultivation of soils of marginal fertility or for highly demanding clearings. In Palmela town, for example, emphyteusis in perpetuity accounted for only 8 percent of the new contracts established in the landholdings of the military order of Santiago in the early 1500s, while the same relative proportion was also identified for lands belonging to the Monastery of Grijó in northern Portugal.[36] Nevertheless, emphyteusis for lives ended up providing long-term rights to exploit land under the "right of renewal" (*direito de renovação*) that allowed the last tenant to renew the emphyteutic contract for a further three lives, provided he paid ground rent of an additional 25 percent. Despite not being legally recognized until 1769, this informal right was already in place and largely accepted by landlords in the 16th century.[37]

Although emphyteutic contracts were the most common way to bring more land under cultivation, they were not the only way. Rights of exploitation over

---

32 The most common form of *emprazamento* involved three lives (the tenant's, his wife's and an heir's), although cases of two or four lives were not unknown.

33 See Serrão (1993b: 85 and 2000: 426–427).

34 Based on Roman Law, emphyteutic contracts were widely used in continental Europe in several different forms (Clavero 1986; Congost 2003).

35 See Marques (1978: 102), Gonçalves (1989: 248–249), Amorim (1997: 97–100), Rodrigues and Duarte (1998: 91–94), Mata (2007: 348) and Rodrigues (2013: 572–573).

36 See Amorim (199: 97) and Costa (2010: 46–47).

37 See Mata (2007: 347–382) and Rodrigues (2013: 576).

communally owned plots of land were regularly transferred to peasant families under short-term leases, and this fostered the replacement of pasture by arable land (Magalhães 1993: 246). Yet another method comprised the granting of *sesmarias*; in other words, vacant or abandoned land appropriated by the crown and allocated to peasants on the condition that they put it into cultivation.[38] Around Óbidos town (Estremadura), woodlands were still granted in *sesmaria* to promote cereal cultivation in the late 1530s, although at a much slower pace than in the late 15th and early 16th centuries.[39]

The landed classes also played an active role in the expansion of arable land by investing in more expensive and technically demanding ventures, such as the drying of marshlands or the clearing of woodlands. Across the country, but especially in the valley of the Mondego River (Coimbra), around the Tejo River (Santarém, Torres Novas, Tomar) and in the vast lands of Alentejo, religious institutions and members of the nobility channelled resources into such land reclamations.[40] The king's landed estates (*reguengos*) went through similar efforts, as did those held by the queen, whose investments in the draining of swamps around Óbidos (Estremadura) are well known.[41] The ambitious engineering work that changed the course of the Tejo River in 1543–44 involved some of the most expensive investments in agriculture at the time. Designed to protect the fertile farmland in the floodplains around Santarém from the sand brought by the river, this enterprise involved building an artificial riverbed with an extension of 10 kilometres and was paid for by Prince Luís, one of the most influential seigniorial lords of the time (Dias 1984: 68–70).

A few contemporary sources provide evidence of increased agricultural output, at least in some parts of the country. By 1514, improvements in building irrigations ditches and reclaiming swamps around Coimbra allowed wheat production to increase from 15,000 to 20,000 *alqueires*.[42] Similar gains in production were achieved in the valley of the Tejo River and in the fertile wheat lands around Beja (Alentejo). Nevertheless, and contrary to the standard narrative, recently amassed data suggest that gross agricultural production fell by about 40 percent between 1500 and the mid-1520s. Explaining this sharp decline requires further research at a regional level in order to determine the variables that caused it. One hypothesis relies on a succession of poor harvests combined with epidemics, with the bad harvests of 1503 and 1506,

38 See Rau (1982: 120) and Gomes (1994: 28).

39 See Rodrigues (2013: 110, 153, 544).

40 See Gil (1965: 41–49) and Pereira (2003: vol. 1, 327–328).

41 See Magalhães (1993: 247) and Pereira (2003: vol. 1, 327).

42 1 *alqueire* =14–18 litres. See Oliveira (1971–1972: vol. 2, 147).

caused by adverse climatic events, and those of 1521 and 1523 being particularly dramatic.[43]

Agricultural output subsequently recovered in the period to the late 1560s. A steady rise in prices of agricultural produce marked this cycle of growth, creating incentives to invest in land and determining land-use choices. Land rents also increased steadily, an indicator that is consistent with high demand for land as a result of population pressure and increasing demand for food, as happened in Alentejo and in Ourém and Caldas da Rainha (Estremadura agro-towns).[44] This cycle eventually gave way to a protracted stagnation that started around the 1570s and 1580s, signalling problems affecting the sector, and lasted until the 1660s.[45] Although a comprehensive explanation of the turn in this cycle is still lacking, there is ample evidence from regional studies across the country that the final quarter of the 16th century witnessed a succession of failed harvests, which may have been influenced by the strongest cooling phase of the "Little Ice Age" in the northern hemisphere.[46] An in-depth study of the region around Évora (Alentejo) that cross-checked the behaviour of key variables, such as demography, land rents and prices of foodstuffs since the 1590s, offers a few insights. It concluded that the combination of poor harvests and famines led to mortality crises and highly volatile wheat prices in the short term, with the latter adversely affecting waged farm workers and farmers alike. Signalling crop failures, the peaks in prices hit peasant agriculture hard and translated into falling land rents, a sign that most farming units experienced economic hardship and impoverishment (Santos 2005).

Although the problems that affected Portuguese agriculture in the last quarter of the 16th century deserve further investigation, particularly at a regional level, it seems clear that seigniorial and fiscal constraints did not play a significant role in these problems, contrary to what has been noted for Castile. Seigniorial dues crystallized in Portugal in the 1520s in the form of the revised *forais* and remained unchanged until the early 19th century, while the state's fiscal burden stabilized in the 1560s by way of an institutional arrangement with the municipalities.[47]

---

43 See Reis, Chapter 6 in this book and Rodrigues (1993: 215–218).

44 See Santos (2005: 370–371) and Rodrigues (2013: 765, 767, 772–773).

45 See Neto, Chapter 4 in this book.

46 See Magalhães (1970: 76); Oliveira (1971–1972: vol. 2, 168); Oliveira (1990: 38–47, 73–75); Oliveira (2005: 33). On the "Little Ice Age", see Kelly and Ó Gráda (2014) and White (2014).

47 For 16th-century Castile, see Thompson and Yun (Eds.) (1994). For the evolution of the crown's fiscal resources in Portugal, see Costa *et al.* (2016: 97–99).

The Landscape and the Production Structure

Regional data for the 16th century provide a picture of the major features of the structure of agricultural output. Since this information is lacking for most of previous centuries, a comprehensive comparison is not possible and we have to rely on circumstantial evidence. Considering the underlying factors encouraging agriculture (population growth, rising prices and urban expansion), it nevertheless seems reasonable to assume that a few changes took place in the structure of agricultural output during this period.

As elsewhere in Europe, most of the arable land in 16th-century Portugal was allocated to cereal production. Taking Estremadura as an example, monographic studies have shown that between 87 and 95 percent of the arable land was used for grain crops.[48] The cereals most commonly cultivated in the early 1500s were wheat, rye, millet and barley. Of these, wheat was the most widespread, even though only a few areas met the necessary soil and climate conditions to ensure good yields of this demanding crop. Such was the case in the lands around Beja, the red lands around Lisbon and the floodplains of the Tejo River. Even in adverse natural conditions (as in Algarve), wheat was still farmed, and this certainly accounts for overall productivity being among the lowest in Europe (Marques 1978: 22–23). Both the country's cultural preference for white bread and the social significance of its consumption justified its cultivation from north to south.

Rye, rather than wheat, was the leading grain crop in the mountainous area of Trás-os-Montes and in several areas of Beira (Viseu and Serra da Estrela) as it was more resistant and better suited to withstanding the cold winters and hot, dry summers and the region's poor soils. It was thus "poor cereal for paupers".[49] Sorghum or millet, another grain crop considered of inferior quality, could be found in fields of north-west Entre-Douro-e-Minho in Beira and also in Estremadura.[50] As this was a spring harvest, it was commonly grown in combination with wheat, rye or barley. Among the various breadmaking grains, barley was planted throughout the kingdom, but especially in Estremadura and Entre-Douro-e-Minho. It was common on the tables of the poorest populations, but was even more important as a feed grain for horses and mules (Marques 1978: 83, 227).

Two new cereal crops, rice and maize (*zea maïs*), were introduced in early 16th-century Portugal, albeit with mixed results. The former remained confined

48 See Rodrigues (1995: 253 and 2013: 552).

49 See Oliveira (1990: 29–31) and Magalhães (1993: 258).

50 See Gonçalves (1989) and Oliveira (1990: 29–31).

to a few wetlands, mostly in Estremadura (such as the valleys of the Asseca, Ota, and Muge rivers), where the unhealthy conditions of marshy areas, which entailed greater risks of contracting malaria, may have contributed to its limited adoption (Magalhães 1993: 260–261). Maize, on the other hand, presents a different story, and its diffusion in Portugal, especially from the 17th century onwards, has been described as "a revolution".[51] It was first grown in the area of Coimbra, probably between 1515 and 1525. By the 1530s, maize was being successfully combined with other grain crops and livestock in the fields of Entre-Douro-e-Minho, where abundant water reserves provided suitable conditions (Almeida 1992). High yields and a rapid growth cycle (five months from sowing to harvest) explain its fast regional diffusion, such that by the 1600s it was the most popular crop in Minho. Despite maize's success in the north-west, its influence remained regional throughout the 16th century, with the country-wide spread of maize occurring only later, in the 1600s and 1700s.[52]

The other two staples of Mediterranean husbandry – olives and grapes – had been widely grown throughout Portugal since medieval times and were hardly a novelty in the early modern period. During the first phase of agrarian expansion and although land reclamations were mainly intended for the cultivation of grain crops, harvests of both these products kept expanding, thanks to the favourable macroeconomic framework. Vine cultivation benefitted from the natural conditions prevailing throughout the kingdom, thus allowing for abundant production (Marques 1987a: 98). From the early days, winemaking had been a commercial undertaking, targeting urban and overseas markets. Continuing the expansionary efforts that began in the late Middle Ages, new vineyards were planted in Minho (before the 1520s) and in the Beira triangle between Aveiro, Viseu, and Coimbra, as well as in Alentejo (Alcácer do Sal) and Algarve.[53]

As for olive trees, the dissemination of this culture to northern Portugal during the 1500s was underpinned by both urban and foreign demand for olive oil. From Coimbra and Covilhã, the upper-north limit where they could be found in the 14th century, olive trees spread to Trás-os-Montes and Minho.[54] A few landlords, such as church institutions, encouraged the planting of new trees in an attempt to meet the high demand for a product with multiple uses (Mota 2006: 324–325). Besides being part of the Mediterranean diet, olive oil was used in the processing of industrial products such as soap and wool, and

51 See Godinho (1982–1984: vol. 4, 33–38) and Ribeiro (1984: 299).

52 See Neto, Chapter 4 and Serrão, Chapter 5 in this book.

53 See Oliveira (1990), Pereira (2003: vol. 1, 385) and Oliveira (2005).

54 See Ribeiro (1979) and Magalhães (1993: 261–262).

as fuel in lighting. Despite the northern expansion, the main production centres remained in the central and southern provinces of Portugal, where climate conditions provided a more suitable environment. Regional studies show that olive oil was produced in higher amounts in Beira, Estremadura, Alentejo and Algarve, with Coimbra, Tomar, Évora, Elvas and Lisbon being among the most important centres for distributing olive oil across the country. Besides meeting domestic demand, olive oil was also regularly exported, particularly to the Southern Netherlands and to Spain, which indicates that foreign demand may have accounted for the expansion of the areas in which olive trees were grown.[55]

Although having only a minor share in agricultural output, fruit trees played an important role in exports, at least at a regional level. In Algarve, new orchards were planted to boost fig production, following the local seaports' integration into international trade, particularly with neighbouring Spain.[56] Figs were also an integral part of the local diet and replaced bread during times of famine (Magalhães 1970: 125–128). Signs of commercial agriculture around Lisbon in response to the growing openness of the economy can also be seen in the production of oranges, which were exported to London in considerable amounts (Costa *et al.* 2016: 66).

Like any typical agrarian economy of the Old Regime, the agricultural landscape included vast areas of uncultivated land, which played a crucial role in the rural economic system. Forests, natural grasslands, moorlands and scrublands provided a wide range of raw materials and foodstuffs and were also used as pasture for livestock.[57] Given that they were common land stemming from the granting of charters of municipal rights, significant parts of these resources were managed collectively by municipal governments.

Going back to *Reconquista* times, livestock farming played an important role in all six *comarcas* of the kingdom, and particularly in Alentejo, Beira and the mountainous areas of the interior, where poor soils, topography and adverse climate conditions made tilling the land challenging. Transhumant sheep grazed in a long belt running through the central mountainous areas from the Serra da Estrela to the plains in the inner south (Campo de Ourique, Idanha). These large herds, which also comprised animals from Spain, created tensions between owners and municipalities, while also damaging the arboreal

55 See Langhans (1949: 82–93) and Costa *et al.* (2016: 65–66).

56 See Godinho (1982–1984: vol. 3, 187) and Pereira (2003: vol. 1, 340).

57 See Serrão (1993b: 74) and Rodrigues (1998: 165–167).

vegetation and causing soil erosion.[58] Aristocratic houses, meanwhile, often included cattle in their capital goods. A 1565 inventory of the House of Bragança, for example, the richest in the kingdom, lists a herd of 3,000 heads (sheep and goats), a number on a par with the enormous herds of the *Mesta*.[59] The same prevalence of sheep was seen in Minho, although here oxen were reared in higher numbers and more so than anywhere else in the kingdom. In the hinterland of Braga (Minho), a contemporary observer assessed the number of oxen in the mid-century to be as high as 5,000, along with 12,800 sheep, goats and pigs (Barros 1919: 127). Not uncommonly in this north-western *comarca*, irrigated meadows and grasslands sustained cattle breeding, which by the early 16th century was already being confined. Owing to its more intensive animal husbandry, Minho was able to export meat to Coimbra and Lisbon.[60] Mid-16th century descriptions of the northern central region of Lamego estimated a total of 10,200 pigs and 5,100 sows in the town's surrounding countryside, while herds of 100 to 150 cows and oxen were common in the mountains of Montemuro (Viseu) (Coelho 1989: 166). As for the southern province of Algarve, the intra-regional complementarity between the hilly areas of the interior and the sea coast allowed successful livestock production, as evidenced in the 1570s by the regular exports of cows, goats and pigs to Castile (Guerreiro and Magalhães 1983).

This portrait of agricultural production supports the notion that animal husbandry was an integral part of farming practice. In an estimate from the early 1500s, livestock stood out as the largest sub-sector, accounting for 43 percent of total agricultural output. This was followed by cereal crops (38 percent), wine (16 percent) and olive oil (3 percent) (Godinho 1978b: 17–18). The overall value is estimated at 1,634 million *réis*, or the equivalent of 148 tons of silver at a time when average nominal daily earnings for unskilled labor were 3.1 grams of silver. Although there are no other estimates for aggregate output at the end of the century so as to provide an idea of the breakdown of land use, a mid-century inventory of the House of Braganza provides insight into the structure of agricultural capital. In this inventory, cattle alone accounted for 34.7 percent of the total amount of 3,046,160 *réis*, while 53 percent of the value share of land was allocated to grain crops, 10 percent to vineyards and only 1 percent to olive trees (Costa *et al.* 2016: 67).

58 Some 50,000 to 60,000 sheep came from Spain to graze along this belt. See Trindade (1965: 113–114) and Devy-Vareta (1986: 16–17).

59 See Phillips and Phillips (1997: 103–105) and Costa *et al.* (2016: 66).

60 See Magalhães (1993: 267–268) and Pereira (2003: vol. 1, 347).

How the two structures (agricultural output and capital) evolved during the second half of the 16th century remains unresolved. It seems reasonable to assume, however, that population pressure prompted the conversion of pasture, waste and woodland into arable land, with the result that cultivation spread to inferior soils and tended to reduce land available for grazing. Eventually this extensive growth may have upset the balance between cereal cropping and livestock, as the regional studies of conflicts relating to the use of common lands seem to indicate. Indeed, local disputes between crop farmers and cattle raisers became more and more common. These were triggered by the encroachment of agricultural use into communally owned land, which was mainly used as collective pasturage.[61] Examples of these disputes can be seen in the attempts by crop farmers to plant broad beans (*favas*) and peas in plots of land lying fallow (*contrafolha*), as attested to for some communities in central and southern Portugal.[62] Since these attempts reduced the resources available for feeding the livestock needed for manure and draught power, they were generally short-lived, as in Algarve. Like elsewhere in Europe, therefore, technological limitations were still hindering an efficient integration of crops and livestock.

As in medieval times, Portuguese peasants continued to use the same crop rotations when exploiting the land; in other words, the two or three-field system. Throughout the early modern period, the former was employed more extensively since the Mediterranean climate conditions that affected a significant part of the territory did not encourage spring crops (Marques 1978: 91–92). Both systems, however, relied on fallow as a natural fertilizer and involved a permanent division between arable and pasture lands. These technological limitations thus hindered productivity gains in both land and labor during the 1500s.[63] Like in other European countries, the Portuguese population grew faster than the agricultural output in the 16th century, and this translated into a sharp decline, by more than half, in the food supply *per capita* between 1500 and 1625, as a recent study revealed.[64] Against a background of stagnant techniques, the primary sector remained highly vulnerable to contingent factors such as extreme weather conditions, while harvests fluctuated from year to year. The risks of harvest failure were further compounded by low yield ratios. Data from the early 1500s show, for example, that wheat yielded an average of between 1:2 and 1:3.5 (in Estremadura), although plots of land in more

61 See Marques (1987a: 104), Magalhães (1993: 274–275) and Pereira (2003: vol. 1, 328).

62 See Magalhães (1970: 45) and Pereira (2003: vol. 1, 323).

63 See Marques (1978: 89–99) and Magalhães (1993: 244–246).

64 See Reis, Chapter 6 in this book.

fertile areas produced yields of 1:4 to 1:5, as also seen in central Portugal.[65] These yields were far below those in the Low Countries (1:7.4), although lagged less behind the yield ratio estimated for France, Spain and Italy (1:6.3).[66] As a further comparison, it is worth noting the average yields of 1:4 and 1:6 recorded for the Alentejo region in the early 19th century.[67]

Looking beyond mainland Portugal and at the Atlantic islands, agriculture in the Azores presents a different story in terms of land productivity since these islands' high soil fertility allowed intensive exploitation of the land. On the most important islands of the archipelago, São Miguel and Terceira, wheat had been cultivated in continuous crop rotation, without resorting to fallow, since the islands were first settled in the 1440s and 1450s. The excellent volcanic soil and climate conditions produced unusually high average yields of 1:15 (São Miguel island) and 1:10–12 (Terceira island).[68] By the 1530s, however, soil depletion affected most of the land on São Miguel island, with yields dropping to 1:6–7 at a time when the population was increasing and demanding more food. These conditions prompted the introduction of two-field crop rotation and the cultivation of a fodder crop, sweet lupines, that enhanced the fertilization of the soil and resulted in more efficient integration of cereal and livestock farming (Brito 1955: 71–72). These improvements boosted productivity again and, by the 1570s, São Miguel had assumed the role of the main wheat producer of the archipelago, exporting regularly to Madeira island, North African garrisons and even to the mainland (Vieira 1985: 137–138).

The next section aims to place agriculture within a wider context, looking at its interrelations with other economic sectors in order to assess incentives that could have encouraged and obstacles that could have prevented the development of more dynamic agricultural practices.

The Marketplace

It has long been claimed that peasants in 16th-century Portugal continued a trend of crop diversification dating back to the previous century in response to new market opportunities. Such opportunities derived from the growth in urban demand, albeit limited to a few bigger cities such as Lisbon, and also from the Portuguese economy's increasing openness to foreign demand. The

65 See Marques (1978: 49–50) and Pereira (2003: vol. 1, 323).

66 See Braudel (1992: vol. 1, 98) and Malanima (2009: 139).

67 See Silbert (1978: vol. 2, 483).

68 See Vieira (1985: 130–131) and Gregório (2007: 329, 345).

evidence of new plantings of vines and olive trees, intermingled with cereal cultivation, is usually seen as a sign of a slow rise in more market-oriented agriculture. Across the country, institutional landlords such as military orders and monasteries encouraged viticulture, while urban elites in Lisbon and Porto made large-scale investments in vineyards, particularly in the final decades of the 1500s.[69] Although most of the production was consumed in the domestic market, casks of wine, admittedly of negligible total value, were exported to Flanders, Spain, France and England (Braga 1998). Olive oil's export potential was more significant, and this probably accounts for the continued expansion of olive trees in the 1500s. Though far from representative of the structure of Portugal's foreign trade, data from trade carried privately to Antwerp in the years 1535–1537 reveal that olive oil accounted for 88 percent of the total value (8.2 million *réis*) of agricultural exports. Olive oil also represented the main share of agricultural exports in another record of trade carried by a Portuguese merchant to Antwerp in 1552, accounting for 76 percent of the total value of his merchandise (Costa *et al.* 2016: 90). In addition, farmers took advantage of foreign demand for other Mediterranean products such as oranges, grown in Estremadura and exported to London, while figs helped integrate the Algarve region into the world of international trade.[70]

Signs of crop diversification can also be seen in the increase in flax growing, which provides a particularly interesting case of incentives generated not only by urban demand, but also by the shipbuilding industry in response to overseas expansion. Regional studies attest to flax being grown in most provinces – Beira, Estremadura, Trás-os-Montes and even Algarve – wherever the soil and climate conditions were right.[71] It was in Minho, however, that the conditions for the plant to thrive were best and where linen production was most intensively developed as early as the first decades of the 1500s (Rodrigues 1998: 175). Several reasons account for this success. On the one hand, Minho, as the most densely populated *comarca* and struggling with low agricultural productivity, had a sufficient labor supply, along with a strong rural tradition of spinning and weaving flax, and this favoured the development of a domestic-based linen industry. On the other hand, the shipbuilding industry's increasing demand for sailcloth, as well as reasonable communications by sea and river, generated significant stimuli. Rural production was coordinated by regional centres of manufacturing and consumption, such as Braga and Guimarães. The latter alone was able to secure annual production of over 100,000 *varas*, part of

69 See Magalhães (1993: 264) and Pereira (2003: vol. 1, 385–389).

70 See Magalhães (1970: 125–128) and Godinho (1982–1984: vol. 3, 187).

71 See Magalhães (1970: 172–173), Garcia (1986: 331, 334) and Pereira (2003: vol. 1, 338–340).

which was distributed to more distant markets.[72] Sailcloth, for example, found its way to the royal shipyard of *Ribeira das Naus* in Lisbon, as well as to foreign markets, through the north-western seaport of Vila do Conde, while flax fibres for rope and cordage, grown and processed in Trás-os-Montes (Moncorvo) and Estremadura (Santarém), were braided by the *cordoaria* (royal rope factory) in Lisbon.[73]

Although new market opportunities were exerting influence on the primary sector, the benefits of these changes were limited to the few regions where communications allowed access to markets at a reasonable cost. This was because despite the long Atlantic coast enabling maritime transportation, the sparse river network made internal communications difficult and costly.[74] Market integration was also hindered by the prevalence of institutional constraints dating back to medieval times. Municipalities continued to charge tolls and to prohibit exports of local farm produce as part of a strategy to avoid famine, but these measures limited the incentives for agricultural development (Magalhães 1993: 326–328). Not surprisingly, the bulk of the demand for a staple food – grain – continued to be met locally, and this trend was in line with the rest of Western Europe. There were, however, a few exceptions, with Lisbon being the best-studied case.

As its population rose rapidly, Lisbon increasingly demanded more cereals. By 1530 these were arriving by waterborne transportation from all around the country (except from the southernmost region, Algarve), as well as from the Azores (Magalhães 1993: 330–331). The Tejo River provided the capital with links to several *comarcas* of the kingdom, notably Estremadura, Beira and Alentejo. According to a contemporary record, some 1,490 freight boats operated in the numerous river ports in the 1550s, supplying the capital with foodstuffs, including grain, and raw materials (Gaspar 1970: 159–160). The number of boats was downsized to around 900 in the 1620s,[75] with this difference indicating, on the one hand, that the first account was probably overestimated and, on the other hand, that the volume of exchanges within this interregional trade did not increase during the period, probably owing to high transaction costs. By then, Lisbon's population was growing rapidly and the capital was consequently consuming grain from overseas at an ever-increasing rate. To guarantee supply, the crown even waived customs duties on grain imports in 1525, a measure swiftly extended to other port cities, such as Viana do Castelo

72 *Vara*, a variable unit of measure equal to approximately 1 metre (Garcia: 1986, 341–342).

73 See Costa (1993: 302–304 and 1997: 361).

74 See Map A13 in the Appendix.

75 See Frei Nicolau de Oliveira (1620), Tratado 1, Chapter 3.

and Porto that were dependent on foreign cereals (Magalhães 1993: 332). As frequently mentioned in descriptions by foreign travellers, the country had a permanent shortage of grain, and this necessitated regular imports from France, Flanders and Hamburg, and also from England and the Mediterranean region.[76] According to a description by an unknown traveller around 1580, "The kingdom as a whole does not produce enough bread to sustain its people. It is, however, supplied in large quantities, from France, England, Germany, sometimes also from neighbouring Castile (…). The grain that grows in the kingdom is very little but very good".[77]

Although the diffusion of maize in Minho in the late 16th century probably helped to lessen the impact of grain shortages and to sustain population growth, the effects of this remained at best regional (Godinho 1982–1984: vol. 4, 38). The lack of data on cereal production and cereal imports makes it hard to assess the extent of the grain shortages and variations across the 16th century. However, an estimate for the years 1527–1533 allows a few calculations to be made. Over those six years, volumes of imported grain amounted to 100,000 *moios* (1 *moio* = 828 litres) or, in other words, around 828,000 hectolitres. With an estimated population of 1,215,000 and average estimated annual consumption of three hectolitres *per capita*, annual imports of around 138,000 hectolitres would have roughly covered the consumption needs of 46,000 people (3.7 percent of the population).[78] Over this time span, with Lisbon alone having 56,000 to 60,000 inhabitants in the 1520s, the *pão de mar* (grain from overseas) did not exceed the needs of the costal and urban populations for which it was destined. Although Portugal's dependence on imports may have increased in the second half of the 16th century, it is unlikely, therefore, ever to have reached the levels seen in the Burgundian Netherlands, where, as a result of a different path towards agricultural specialization, 15 to 20 percent of the population in the 1560s was fed with grain grown in the Baltic (Vries and Woude 1997: 198).

Conclusion

The literature has focused on an optimistic view of agriculture's performance during most of the 16th century, and one that befits the notion of a Golden Age of Portuguese history coinciding with the establishment of an empire with offshoots in Asia, Africa and South America. The country's population grew

76 See Godinho (1982–1984: vol. 3, 217–223) and Magalhães (1993: 250–252).

77 Referred to by Marques (1987b: 187–189).

78 See Godinho (1982–1984: vol. 4, 21) and Costa *et al.* (2016: 88).

steadily during this period and, as both demand and prices rose, most peasants responded by bringing more land into cultivation through land-clearing and especially by converting pasture, waste and woodland into arable land. With no significant improvements in farming techniques, however, this response was likely to have resulted in only a slow shift from animal to arable husbandry. The upswing, according to the literature, lasted until the late 16th century, when a general agrarian stagnation set in. Within this long cycle of agrarian expansion, and providing a few preconditions such as secure tenure rights and access to urban and overseas markets were met, peasant farming also responded by diversifying crops. In this way, wine, olive oil, flax and fruit helped to infuse dynamism into the primary sector, particularly in the hinterland of the larger cities, while maize spread rapidly across the fields of the north-west, thus helping to sustain regional population growth. However, despite the positive signs conveyed by qualitative sources, several factors hampered innovation and agricultural improvements. For one, the trend towards diversification in response to urban demand should not be overvalued. Urban expansion was limited mainly to Lisbon, and while the populations of smaller towns increased, by the end of the century some of these towns had stopped growing. Market integration was also hindered both by poor and costly land communications and by institutional constraints. Not unlike other countries in Europe, subsistence agriculture continued to prevail, while the correlation between the urbanization rate and agricultural performance seems to have been weak.

Lastly, recent research shows that agriculture did not experience a long and continuous upswing in the 16th century. Instead, its output underwent three shorter phases of growth, of which the second, lasting from the 1520s to the late 1560s, was in fact merely a return to the level attained in the 1500s. As the 16th century drew to a close, agriculture was coming under strain and a general stagnation ensued, for reasons that need to be further researched. In southern Portugal, at least, the combination of harvest failures, famines and soaring death rates points to Malthusian constraints operating in the final decades of this century. Whatever the reasons for this situation, the burden of seigniorial exactions would not seem to be a major factor in the explanation. Seigniorial rights had been crystallized in Portugal in the 1520s and did not increase, contrary to what happened in Spain, where the intensification of these rights, together with fiscal dues, blocked rural initiative from the late 16th century onwards. Regardless of the problems, however, and even though food production *per capita* declined, the major achievement of agriculture between 1500 and 1620 was its ability to sustain an increase in the population by a factor of almost two.

References

Almeida, Luís Ferrand de (1992). "A propósito do milho 'marroco' em Portugal nos séculos XVII e XVIII". *Revista Portuguesa de História*, 27, pp. 103–143.

Álvarez Nogal, Carlos and Prados de la Escosura, Leandro (2007). "The decline of Spain (1500–1850): conjectural estimates". *European Review of Economic History*, 11 (3), pp. 319–366.

Álvarez Nogal, Carlos and Prados de la Escosura, Leandro (2013). "The rise and fall of Spain (1270–1850)". *Economic History Review*, 66 (1), pp. 1–37.

Amorim, Inês (1997). *O Mosteiro de Grijó. Senhorio e Propriedade: 1560–1720 (Formação, Estrutura e Exploração do seu domínio)*. Braga: Barbosa e Xavier.

Barros, João de (1919). *Geografia d'Antre Douro e Minho e Trás-os-Montes*. Porto: Biblioteca Pública Municipal do Porto.

Borges, Emília Salvado (1996). *Crises de Mortalidade no Alentejo Interior – Cuba (1586–1799)*. Lisbon: Edições Colibri.

Braga, Isabel Drummond (1998). "A circulação e a distribuição dos produtos". In João José Alves Dias (Ed.) *Portugal do Renascimento à Crise Dinástica*. In Joel Serrão and A.H. de Oliveira Marques (Eds.) *Nova História de Portugal*. Lisbon: Presença, vol. 5, pp. 195–247.

Braudel, Fernand (1992). *Civilização Material, Economia e Capitalismo (Séculos XV–XVIII)*. Lisbon: Editorial Teorema.

Brito, Raquel Soeiro de (1955). *A Ilha de S. Miguel. Estudo Geográfico*. Lisbon: Instituto de Alta Cultura – Centro de Estudos Geográficos.

Campbell, Bruce M.S. (2010). "Nature as historical protagonist: environment and society in pre-industrial England". *Economic History Review*, 63 (2), pp. 281–314.

Clark, Peter (2009). *European Cities and Towns (400–2000)*. Oxford: Oxford University Press.

Clavero, Bartolomé (1986). "Enfiteusis: que hay en un nombre". *Anuario de Historia del Derecho Español*, 56, pp. 467–519.

Coelho, Maria Helena da Cruz (1989). *O Baixo Mondego nos finais da Idade Média*. Lisbon: Imprensa Nacional – Casa da Moeda.

Congost, Rosa (2003). "Property rights and historical analysis. What rights? What history?". *Past and Present*, 181, pp. 73–106.

Costa, João (2010). *Palmela nos finais da Idade Média. Estudo do Códice da Visitação e Tombo de Propriedades da Ordem de Santiago de 1510*. Lisbon: FCSH-UNL (unpublished master's thesis).

Costa, Leonor Freire (1993). "A construção naval". In Joaquim Romero Magalhães, (Ed.) *No Alvorecer da Modernidade (1480–1620)*. In José Mattoso (Ed.) *História de Portugal*. Lisbon: Círculo de Leitores, vol. 3, pp. 293–310.

Costa, Leonor Freire (1997). *Naus e Galeões na Ribeira de Lisboa. A Construção Naval para a Rota do Cabo no Século XVI*. Cascais: Patrimonia Historica.

Costa, Leonor Freire, Lains, Pedro and Miranda, Susana Münch (2016). *An Economic History of Portugal, 1143–2010*. Cambridge: Cambridge University Press.

Devy-Vareta, Nicole (1986). "Para uma geografia histórica da floresta portuguesa. Do declínio das matas medievais à política florestal do Renascimento (séc. XV e XVI)". *Revista da Faculdade de Letras – Geografia*, 1 (1), pp. 5–37.

Dias, João Alves (1984). "Uma grande obra de engenharia em meados do século XVI. A mudança do curso do Tejo". *Nova História*, 1, pp. 66–82.

Disney, A.R. (2009). *A History of Portugal and the Portuguese Empire*. Cambridge: Cambridge University Press, vol. 1.

Dupâquier, Jacques (1997). "Les vicissitudes du peuplement (XV–XVIII siècles)". In J.P. Bardet and J. Dupâquier (Eds.) *Histoire des populations de l'Europe*. Fayard: Paris, pp. 239–261.

Garcia, João Carlos (1986). "Os têxteis no Portugal dos séculos XV e XVI". *Finisterra Revista Portuguesa de Geografia*, 21 (42), pp. 327–344.

Gaspar, Jorge (1970). "Os portos fluviais do Tejo". *Finisterra. Revista Portuguesa de Geografia*, 5 (10), pp. 153–204.

Gil, Maria Olímpia da Rocha (1965). *Arroteias no Vale do Mondego durante o Século XVI. Ensaio de História Agrária*. Lisbon: Instituto de Alta Cultura.

Godinho, Vitorino Magalhães (1978a). "L'émigration Portugaise (XVe–XX siècles). Une constante structurale et les réponses aux changements du monde". *Revista de História Económica e Social*, 1, pp. 5–32.

Godinho, Vitorino Magalhães (1978b). *Ensaios II. Sobre História de Portugal*, 2nd edn. Lisbon: Sá da Costa.

Godinho, Vitorino Magalhães (1982–1984). *Os Descobrimentos e a Economia Mundial*. Lisbon: Presença.

Gomes, Saul António (1994). *As Cidades Têm Uma História: Caldas da Rainha das Origens ao Século XVIII*. Caldas da Rainha: Património Histórico.

Gonçalves, Iria (1989). *O Património do Mosteiro de Alcobaça*. Lisbon: Universidade Nova de Lisboa.

Gregório, Rute Dias (2007). *Terra e Fortuna: os Primórdios da Humanização da Ilha Terceira (1450?-1550)*. Lisbon: FCSH-UNL.

Guerreiro, Manuel Viegas and Magalhães, Joaquim Romero (1983). "Duas descrições do Algarve do século XVI". *Cadernos da Revista de História Económica e Social*, 3.

Hespanha, António Manuel (1994). *As Vésperas do Leviathan: Instituições e Poder Político, Portugal, Século XVII*. Coimbra: Livraria Almedina.

Johnson, Harold (1998). "A colonização portuguesa do Brasil, 1500–1580". In Leslie Bethell (Ed.) *América Latina Colonial – História da América Latina*. São Paulo: Editora da Universidade de S. Paulo, vol. 1, pp. 241–281.

Kelly, Morgan and Ó Gráda, Cormac (2014). "The waning of the Little Ice Age: climate change in Early Modern Europe". *Journal of Interdisciplinary History,* 44 (3), pp. 301–325.

Langhans, Franz-Paul (1949). *Apontamentos para a História do Azeite em Portugal.* Lisbon: Junta Nacional do Azeite.

Livi-Bacci, Massimo (1999). *La population dans l'histoire de l'Europe.* Paris: Éditions du Seuil.

Lobo, A. de Sousa Silva Costa (1984 [1903]). *História da Sociedade em Portugal no Século XV.* Lisbon: Rolim.

Magalhães, Joaquim Romero (1970). *Para o Estudo do Algarve Económico durante o século XVI.* Lisbon: Edições Cosmos.

Magalhães, Joaquim Romero (1988). *O Algarve Económico. 1600–1773.* Lisbon: Estampa.

Magalhães, Joaquim Romero (1993). "As estruturas da produção agrícola e pastoril". In Joaquim Romero Magalhães (Ed.). *No Alvorecer da Modernidade (1480–1620).* In José Mattoso (Ed.), *História de Portugal.* Lisbon: Círculo de Leitores, vol. 3, pp. 243–281.

Malanima, Paolo (2009). *Pre-modern European Economy. One Thousand Years (10th–19th Centuries).* Leiden-Boston: Brill.

Marques, A.H. de Oliveira (1978). *Introdução à História da Agricultura em Portugal.* Lisbon: Cosmos.

Marques, A.H. de Oliveira (1987a). *Portugal na Crise dos séculos XIV e XV.* In Joel Serrão and A.H. de Oliveira Marques (Eds.). *Nova História de Portugal.* Lisbon: Presença, vol. 4.

Marques, A.H. de Oliveira (1987b). *Portugal Quinhentista. Ensaios.* Lisboa: Quetzal Editores.

Mata, Joel Silva Ferreira (2007). *A Comunidade Feminina da Ordem de Santiago: a Comenda de Santos em Finais do Século XV e no século XVI: um Estudo Religioso, Económico e Social.* Porto: Fundação Eng. António de Almeida.

Monteiro, Nuno Gonçalo (1998). *A Corte, as Províncias e as Conquistas: Centros de Poder e Trajectórias Sociais no Portugal Restaurado (1668–1750).* Lisbon: Colibri.

Monteiro, Nuno Gonçalo (2005). "A ocupação da terra". In Pedro Lains and Álvaro Ferreira da Silva (org.) *História Económica de Portugal (1700–2000).* Lisbon: Imprensa de Ciências Sociais, vol. I, pp. 67–91.

Monteiro, Nuno Gonçalo (2007). "Nobility and aristocracy in Ancien Régime Portugal (seventeenth to nineteenth centuries)". In H.M. Scott (Ed.) *The European Nobilities in the Seventeenth and Eighteenth Centuries. Western and Southern Europe.* London: Palgrave Macmillan, pp. 256–284.

Monteiro, Nuno Gonçalo and Salvado, João Paulo (2014). "La administración de los patrimonios de las grandes casas aristocráticas en el Portugal del Antiguo Régimen". In Raúl Molina Recio (Ed.). *Familia y Economía en los Territorios de la Monarquía Hispánica (ss. XVI–XVIII)*. Badajoz: Mandalay Ediciones, pp. 241–264.

Mota, Salvador Magalhães (2006). "A penetração da cultura da oliveira no Entre Douro e Minho: algumas achegas ao contributo do senhorio cisterciense de Santa Maria de Bouro nos séculos XV e XVI". In *Estudos em Homenagem ao Professor José Marques*. Porto: Faculdade de Letras, vol. 4, pp. 321–328.

Oliveira, António de (1971–1972). *Vida Económica e Social de Coimbra de 1537 a 1640*. Coimbra: Faculdade de Letras da Universidade de Coimbra.

Oliveira, António de (1991). *Poder e Oposição Política em Portugal no Período Filipino, 1580–1640*. Lisbon: Difel.

Oliveira, Aurélio (2005). "Vila Verde. As terras e as gentes: 1500–1850". *Boletim Cultural de Vila Verde*, pp. 27–37.

Oliveira, Fr. Nicolau de (1620). *Livro das Grandezas de Lisboa*. Lisbon: Por Jorge Rodrigues.

Oliveira, João Nunes (1990). *A Produção Agrícola de Viseu entre 1550 e 1700*. Viseu: Câmara Municipal de Viseu.

Palma, Nuno and Reis, Jaime (2016). "From convergence to divergence: Portuguese demography and economic growth, 1500–1850". Groningen Growth and Development Centre. Research Memorandum.

Pereira, António dos Santos (2003). *Portugal, o Império Urgente (1475–1525). Os Espaços, os Homens e os Produto*. Lisbon: Imprensa Nacional-Casa da Moeda, 2 vols.

Phillips, Carla Rahn and Phillips, William D. (1997). *Spain's Golden Fleece. Wool Production and the Wool Trade from Middle Ages to the Nineteenth Century*. Baltimore: Johns Hopkins University Press.

Ramos, Rui, Sousa, Bernardo Vasconcelos and Monteiro, Nuno Gonçalo (2009). *História de Portugal*. Lisbon: Esfera dos Livros.

Rau, Virgínia (1982). *Sesmarias Medievais Portuguesas*. Lisbon, s.n.

Ribeiro, Orlando (1979). "Significado ecológico, expansão e declínio da oliveira em Portugal". *Boletim do Instituto do Azeite e dos Produtos Oleaginosos*, 7 (2), pp. 50–54.

Ribeiro, Orlando (1984). "Milho". In Joel Serrão (Ed.). *Dicionário de História de Portugal*. Porto: Livraria Figueirinhas, vol. 4, pp. 294–300.

Rodrigues, Ana Maria (1995). *Torres Vedras. A vila e o termo nos finais da Idade Média*. Lisbon: Fundação Calouste Gulbenkian.

Rodrigues, Ana Maria S.A. (1998). "A produção agro-pecuária". In João José Alves Dias (Ed.). *Portugal do Renascimento à Crise Dinástica*. In Joel Serrão and A.H. de Oliveira Marques (Ed.) *Nova História de Portugal*. Lisbon: Presença, vol. 5, pp. 165–181.

Rodrigues, Ana Maria S.A. and Duarte, Luís Miguel (1998). "A propriedade". In João José Alves Dias (Ed.). *Portugal do Renascimento à Crise Dinástica*. In Joel Serrão and A.H. de Oliveira Marques (Eds.). *Nova História de Portugal*. Lisbon: Presença, vol. 5, pp. 83–160.

Rodrigues, Lisbeth de Oliveira (2013). *Os Hospitais Portugueses no Renascimento (1480–1580): o Caso de Nossa Senhora do Pópulo das Caldas da Rainha*. Braga: Universidade do Minho (PhD dissertation).

Rodrigues, Teresa Ferreira (1993). "Estruturas populacionais". In Joaquim Romero Magalhães (Ed.). *No Alvorecer da Modernidade (1480–1620)*. In José Mattoso (Ed.). *História de Portugal*. Lisbon: Círculo de Leitores, vol. 3, pp. 197–241.

Rodrigues, Teresa Ferreira (Ed.) (2008). *História da População Portuguesa*. Porto: Edições Afrontamento – Cepese.

Salvado, João Paulo (2009). *Nobreza, Monarquia e Império: A Casa Senhorial dos Almotacés-mores do Reino (Séculos XVI–XVIII)*. Lisbon: Universidade Nova de Lisboa (PhD dissertation).

Salvado, João Paulo (2011). "An aristocratic economy in Portugal in the first half of the seventeenth century: the House of the Marquises of Castelo Rodrigo". *e-Journal of Portuguese History*, 9 (2), pp. 35–67.

Santos, Rui (2003). *Sociogénese do Latifundismo Moderno. Mercados, Crises e Mudança Social na Região de Évora. Séculos XVII a XIX*. Lisbon: Banco de Portugal.

Santos, Rui (2005). "The agrarian economy of the region of Évora in the first half of the 17th century (1595–1660): an exploration of main indicators". *Revista de Historia Económica*, 23, pp. 349–378.

Santos, Rui and Serrão, José Vicente (2011). "Land policies and land markets: Portugal, late eighteenth and early nineteenth centuries". In Gérard Béaur *et al.* (Eds.), *Property Rights, Land Markets and Economic Growth in the European Countryside (13th–19th Centuries)*. Turnhout: Brepols Publishers, pp. 317–341.

Serrão, José Vicente (1993a). "O quadro humano". In António Manuel Hespanha (Ed.). *O Antigo Regime (1620–1807)*. In José Mattoso (Ed.), *História de Portugal*. Lisbon: Círculo de Leitores, pp. 49–69.

Serrão, José Vicente (1993b). "O quadro económico". In António Manuel Hespanha (Ed.) *O Antigo Regime (1620–1807)*. In José Mattoso (Ed.). *História de Portugal*. Lisbon: Círculo de Leitores, pp. 71–88.

Serrão, José Vicente (1996). "População e rede urbana nos séculos XVI–XVIII". In César de Oliveira (Org.). *História dos Municípios e do Poder Local*. Lisbon: Círculo de Leitores, pp. 63–77.

Serrão, José Vicente (2000). *Os Campos da Cidade. Configuração das Estruturas Fundiárias da Região de Lisboa nos Finais do Antigo Regime*. Lisbon: ISCTE (PhD dissertation).

Silbert, Albert (1978). *Le Portugal méditerranéen à la fin de l'Ancien Régime. Contribution à l'histoire agraire comparée*. Lisbon: Instituto Nacional de Investigação Científica, vol. 2.

Silva, Álvaro Ferreira da (1997). "A evolução da rede urbana portuguesa (1801–1940)". *Análise Social*, 143–144, XXXII (4–5), pp. 779–814.

Silva, Teresa Rebelo da (2006). "Drenagem do paul de Lagos (finais do século XV)". *Paisagens Rurais e Urbanas. Fontes, Metodologias, Problemáticas*. Lisbon: Centro de Estudos Históricos – Universidade Nova de Lisboa, pp. 207–212.

Subrahmanyam, Sanjay (1995). *O Império Asiático Português (1500–1700). Uma História Política e Económica*. Lisbon: Difel.

Thompson, I.A.A. and Yun Casalilla, Bartolomé (Ed.) (1994). *The Castilian Crisis of the Seventeenth Century. New Perspectives on the Economic and Social History of Seventeenth-century Spain*. Cambridge: Cambridge University Press.

Trindade, Maria José (1965). "Alguns problemas do pastoreio em Portugal nos séculos XV e XVI". *Do Tempo e da História*, 1, pp. 114–134.

Vieira, Alberto (1985). "A questão cerealífera nos Açores nos séculos XV–XVII. Elementos para o seu estudo". *Arquipélago, Série História e Filosofia*, 7 (1), pp. 123–201.

Vries, Jan de and Woude, Ad van der (1997). *The First Modern Economy: Success, Failure, and Perseverance of the Dutch Economy, 1500–1815*. Cambridge: Cambridge University Press.

White, Sam (2014). "The real Little Ice Age". *Journal of Interdisciplinary History*, 44 (3), pp. 327–352.

CHAPTER 4

Conflict and Decline, 1620–1703

Margarida Sobral Neto

Introduction

This chapter deals with the period from 1620 to 1703, which coincides with the later phase of the displacement of the core of European economic activity from the Mediterranean to northern Europe. The period has long been described, namely by Hobsbawm (1965), as one of crisis, or a succession of crises, but that interpretation has been rejected by subsequent scholarship. Currently there is a more diversified interpretation of the 17th century, according to which growth and depression occurred in different periods and regions across Europe.[1] Indeed, economic conditions differed, and a divergence can be observed between economic development in Western and Mediterranean and Eastern Europe. There were, on the other hand, "opposing conjunctures" between Europe and Iberian America (Romano 1992). It is in the context of these general transformations that the Portuguese economy will be analysed, while providing a perspective for studying the interplay between the European and the Atlantic economies. The chapter starts in 1620, the year traditionally associated with a normalization of the country's economic conditions, and ends in 1703, with the signing of the Anglo-Portuguese commercial treaty in the context of the War of Spanish Succession, which reinforced the alliance between the two countries and consolidated their commercial links.

To understand how this period was lived through in Portugal, we need to analyse the specific national economic and social constraints stemming from the new position occupied by the Iberian Peninsula in the context of the European correlation of forces.[2] The combination of the different factors can be summarized in four points. First, we need to take into account the international competition over the control of the main sources and flows of goods, carried out mainly by the Dutch East India Company (founded in 1602) and the Dutch West India Company (founded in 1621), which directly affected Portuguese interests in Asia, Africa and the Americas (Boxer 1969; Subrahmanyam 1993). Secondly, between 1580 and 1640, the union of the crowns of Portugal

1 See De Vries (1976 and 2009) and Kindleberger (1991). See also Wallerstein (1974).
2 See Boxer (1969), Hanson (1981), Russell-Wood (1992) and Bettencourt and Curto (2006).

 | DOI 10.1163/9789004311527_006

and Castile resulted in the military and financial involvement of the kingdom in wars waged by Castile in Europe and in imperial territories (Boyajian 1983; Álvarez 1997). Thirdly, following the restoration of the independence of the Portuguese kingdom from Spain in 1640, a war lasting nearly 30 years ensued, negatively affecting economic activities in the metropole and the income of different social groups (Valladares 2006). Finally, the several commercial treaties signed with England (1654, 1661 and 1703) allowed English protection to safeguard Portuguese independence. But while these agreements consolidated an Anglo-Portuguese alliance, they also established an asymmetric economic relationship linking the two countries: Portugal secured the export of Douro wines to England, and in return imported British textiles (Sideri 1970; Pedreira 1994).

Bearing in mind these four points, this chapter is organized in the following way. The next section addresses the most relevant population dynamics in the period under analysis, giving particular emphasis to migration to imperial territories. The third section examines the place of Portugal within European and colonial trade flows in conjunction with the domestic economy. The fourth section shows the composition of agricultural production, identifying continuities, innovations and constraints. The fifth section discusses some of the trends in agricultural production articulated with market income, the framework of institutional constraints on agriculture, namely ownership and taxation, as well the development of the manorial system. And a final section presents the main conclusions.

Domestic Settlements and Emigration

There are no reliable population figures for the 17th century and the existing demographic studies for the entire country do not allow us to draw an accurate portrait of population dynamics in the regions of the metropole. By comparing the numbers presented in the first Portuguese population census of 1527 and the information contained in a chorography drawn up in the early 17th century, we can nevertheless conclude that the Portuguese population doubled between 1527 and 1700. If we take as an indicator an average of 4.3 persons per household, the population would be around 1,216,000 souls in 1527, reaching 2,427,000 in 1700 (see Table 4.1). Furthermore, according to estimates, in 1640 around 1.3 or 2 million people lived in the metropolitan territory. It can thus be argued that there had been a proportional increase in population during the 16th century. This growth was hindered by the mortality crises that occurred in the late 16th century and the first decades of the 17th century.

TABLE 4.1 *Population estimates*

Region	Territory (km²)	1527		1700	
		Households	Pop. Total[1]	Households	Pop. Total[1]
Entre-Douro-e-Minho	7252	55016	236569	131183	564087
Trás-os-Montes	11493	35629	153205	59438	255583
Beira	15298	67696	291093	120586	518520
Estremadura	19930	65515	281715	132928	571590
Alentejo	30319	48934	210416	104178	447965
Algarve	4989	9918	42647	16148	69436
Total	89281	282708	1215645	558461	2427181

[1]4,3 inhabitants per household.

SOURCE: RODRIGUES 2008: 177.

There is more reliable data on demographic dynamics in Lisbon. The information gathered from different historical sources helps to establish that between 1620 and 1700 the capital's population stagnated. In 1620, 143,608 individuals were counted, while in 1700 this figure had declined to 140,136. The estimates show that the city's population decreased between 1620 and 1641 (−0.65 percent), recovering very slowly from 1642 onwards. In about 60 years, up to 1700, there was an average population growth of 0.11 percent (Rodrigues 2008: 196). Several factors explain Lisbon's demographic behaviour, especially the difficulties in the supply of grain and meat during many years of the 17th century.[3] From the limited information gathered, historians have argued that the trend in the other regions of the country would have been analogous to that found in Lisbon. Thus, we can conclude that, from 1620 to 1668, in most regions the population stagnated due to a combination of several factors. One of these factors was the occurrence of more than a dozen mortality crises, which were also *crises of poverty*, distributed throughout all decades for 80 years (from 1619/1620 to 1700). Another factor was the *Restauração* wars that consolidated

3 The annual growth of the Porto population is estimated at 0.3 percent during 1622–1687 (Oswald 2008: 316). In turn, the Algarve registered a negative rate of growth during 1591–1631 (−0.44 percent) and a small increase during 1631–1672 (0.15 percent) and 1672–1717 (0.53 percent) (Magalhães 1988: 20). The annual growth rate of Évora between 1593 and 1700 was 0.26 percent (Santos 2003: 216).

the independence from Castile, but led to the loss of human lives and the breakdown of the organization of local families' lives.

Emigration to overseas imperial territories can also be mentioned as one the factors affecting the demographic evolution in Portugal. According to Godinho (1978: 9), the following are estimates for migration to the empire: 280,000 individuals departed between 1500 and 1580; 300,000 to 360,000 between 1580 and 1640; and 150,000 between 1640 and 1700. This means that in 200 years about 800,000 individuals may have left the metropole. Apart from economic reasons, emigration could also have been caused by the persecutions carried out by the Inquisition against the kingdom's inhabitants (Rodrigues 2008: 246). Similarly to Spain's, in Portugal demographic behaviour varied from region to region. The population stagnated in the inland regions and along the borders with Castile (extending from Trás-os-Montes, in the extreme north east, to the Alentejo, in the south east), which were less economically active and more affected by the *Restauração* wars. In contrast, the seashore regions of the north and centre, geographically located between the Lima and Mondego rivers, recorded population increases. This demographic evolution, similar to that of Galicia (Eiras 1998: 119–120), was influenced by the spread of maize and by the socio-economic dynamics linked to the Brazilian trade.

Unsurprisingly, one of the main topics of the economic literature in this period was demographic problems. Indeed, several of the kings' *advisers*, namely the authors that became known as *arbitristas*, mainly concerned with the depressing situations of Portugal and Spain, suggested in their published works several solutions to the economic and financial problems that, in their opinion, blocked their development. *Arbitristas* specifically identified the causes of the "lack of people" and offered "remedies" to stimulate population growth.[4] From the early 16th century, several authors raised doubts about the benefits of Portuguese overseas expansion, especially to India, which they considered a "drain" of people. In 1608, Luis Mendes de Vasconcelos feared that, over time, India would "consume our men, and, possibly money, for it will be necessary to defend it with heavy arms, the two (men and money) being those things which states cannot do without – so I say that it is better not to pursue the conquest of India" (Sérgio 1974: 46). Manuel Severim de Faria, one of the most distinguished members of this *school of thought*, presented a similar opinion in 1655: "The first cause of the lack of people in this kingdom are our conquests [...]. Therefore, even though population naturally increased (as we have proven), our Portuguese Nation, after these conquests, decreased, not for lack of natural multiplication, but because the Portuguese leave their homeland to settle and

4 See Oliveira (1991: 80–82).

found many cities and places, as we have said, in remote lands" (Sérgio 1974: 127–128). During the *Restauração* wars, Portuguese kings actively tried to control migratory flows to Brazil that took men from agriculture, but also from the army, by passing specific legislation (Rodrigues 2008: 242).

Colonial Trade

In the 16th century, the sea route linking the Atlantic and Indian Oceans, the route of spices and drugs, continued to be a Portuguese monopoly (Godinho 1970; Subrahmanyam 1993). In the final decades of that century, English and Dutch merchants however began to penetrate the Eastern markets, gradually building their empires, which Braudel called "an immense spider's web of trafficking and trade" (Braudel 1979: 181). The Dutch East India Company (WOC), established in 1602, intensified competition, reducing to "less than a third of what was the trade of the Portuguese by the sea route from Lisbon to Goa" (Disney 1981). From 1621, navigation between Portugal and Asia, the so-called *Carreira da Índia*, recorded an irreparable reduction (Duncan 1986). In an attempt to cope with the strong competition on the Cape route, Portugal decided to create the *Companhia das Índias* (India Company) (1628–1633), with both public and private capital investment. This project was, however, not successful: in spite of causing an influx of gold and silver to the *Casa da Moeda* (the Mint), it failed to raise the capital needed for the project (Mauro 1989: II: 164–166). Faced with these problems in Eastern trade, the King tried to consolidate the South Atlantic economy, based in Angola, the supplier of slave labor, and in Brazil, where sugar cane plantations developed and led to the growth of sugar manufactures (Schwartz 1985; Bethel 1987). The pace at which sugar mills were set up – they increased from 170 to 528 between 1612 and 1710 – was linked to the intensification of the African slave trade (Mauro 1989: I: 257).

According to Vilar (1974: 279), the fact that in the 17th century the Brazilian and Portuguese economy did not suffer "a continuous and catastrophic decline" was largely due to the success of these businesses. The exploitation of Brazil wood, the slave trade and sugar production sustained the Brazilian economy in the first half of the 17th century. This interpretation agrees with that of Mauro (1989: II: 282), who shows the vitality of the Portuguese Atlantic economy during this century and argues that this "makes Portugal a kind of isolated phenomenon of prosperity in the middle of the stabilization, even stagnation, of Southern Europe in the same period". The Brazilian colonial economy differed from the Eastern one, in terms both of its internal features and its articulations with the outside world. Unlike in the East, where Portugal found an organized economy, in Brazil the Portuguese had to build the entire economic system. This allowed them to control both production and trade. Another key difference was that the Eastern trade was connected to high finance, based in Lisbon, the imperial capital. It was also articulated with the Spanish Empire to obtain the silver required for commercial transactions in the East. Brazilian trade developed without the dependence on Spanish silver, a factor that would make the connection to Spain less necessary (Russel-Wood 1992).

Since it required lower capital spending, trade with Brazil allowed for the participation of the middle bourgeoisie active in the ports spread along the Portuguese coastline.[5] Data shows that in the 17th century not only Lisbon, the largest sea port in the country, but also other port cities recorded an economic growth linked to the Brazilian trade and the set of activities that supported it, such as shipbuilding. This trend is particularly noticeable in Viana do Castelo, located in the extreme north west of the Portuguese territory (Costa 2002: 35–111). From the seashore ports, the economic stimulus spread to other regions of the country, following the journey of goods through inland rivers and roads. The impact of the Brazilian trade is documented in regions as far away as the Algarve, in the south, or the Entre-Douro-e-Minho, in the north (Oliveira 1991: 68).

The flourishing trade with Brazil however experienced problems, as proved by the crises registered during the century. One of these crises occurred in 1621, when the resumption of hostilities between Castile and the Netherlands led to the absence of foreign ships in Portuguese ports. At the same time, the Dutch West India Company (WIC) began operating in the Atlantic, causing great instability to Portuguese and Spanish shipping. The Dutch were not only interested in controlling trade: they also wanted to control production. To that end, they started settling in towns located in the north east of Brazil (like Baía in 1624; Olinda and Recife in 1630; Pernambuco from 1635 to 1645), closer to the source of sugar. At the same time, they seized settlements on the African coast (São Jorge da Mina, 1637 and Arguin, 1638), ensuring the supply of slaves.

Due to all these adverse factors, Portuguese external trade, which had taken an upward curve between 1609 and 1621, reversed the trend in the 20 years that followed. The disruptions of trade with Brazil particularly hindered the activities of northern seaports, like Viana do Castelo, Vila do Conde and Porto, and according to historical sources the revenues of the municipalities of Viana do Castelo, Porto and Funchal (Madeira) were decreasing in the 1630s. The crisis seems to have been felt especially in the port of Viana do Castelo, once the

5 See Map A13 in the Appendix.

home port of a maritime fleet of over 70 vessels, of which only five were left in 1635. The lack of Dutch ships in Setúbal, a seaport near Lisbon in the Estremadura province, also caused significant problems in the local economy, based on salt extraction and trade (Rau 1984).

But the consequences of the end of these trade relations were even greater, as the ships that carried salt were the same that brought the foreign wheat that fed Lisbon. Thus, the kingdom's small grain production during these years, associated to difficulties in importing wheat, meant that the years 1620 and 1621 were *a time of famine and strife* in Lisbon. The commercial crisis, felt between 1621 and 1640, reflected in customs revenues and in every national activity connected with foreign trade: grain imports, salt extraction or wine, olive oil and fruit production. The decrease in revenues from foreign trade led to an increase in fiscal pressure over economic activities, from both the state and noble families. These demands affected different social groups and gave rise to popular riots against the new fiscal policies (Oliveira 1991: 65–80).

In 1640, with the recovery of independence, Portugal was able to reach an agreement with the Dutch covering their mutual interests in Europe. This did not happen immediately for the Brazilian and Eastern imperial territories, though throughout the 1640s the Dutch were eventually removed from the places they occupied in Brazil and the African coast. Nevertheless instability on the seas remained very great. In just two years, 1647 and 1648, the Portuguese fleet circulating in the Atlantic lost more than half of its 259 vessels. To contain the meltdown, in 1649 the *Companhia Geral do Comércio do Brasil* (General Company for Brazilian Trade) was founded, and its fleets were escorted to allow them to travel in the safest conditions (Costa 2002). The regularization of trade relations with Brazil allowed Brazilian sugar to become a major source of financial support for the Portuguese Wars of Independence, which were fought in the Iberian Peninsula until the late 1660s.

When peace with Castile was finally signed, Portugal was faced with a new commercial crisis. In the 1670s, the Dutch and French began to produce sugar and tobacco in the Antilles, removing Brazilian colonial products from European markets, which caused a sharp drop in prices.[6] In the decade of 1670–80, the influx of silver into the country also decreased, due to the reduction in quantities of metal coming to Seville and the collapse of trade with the Dutch. Thus, for Portugal "the crisis involved at once sugar, tobacco and

6 For example, an *arroba* (around 14.688 kg) of sugar, sold in Lisbon at 3,800 *reis* in the mid-17th century, was worth only 2400 *réis,* in 1668, and about 1300 *réis* 20 years later. The price of tobacco has a similar pattern: an *arratel* (around 0,459 kg), sold at 260 *réis* in 1650, dropped to 200 *réis* in 1668 and 70 *réis* in 1688 (Godinho 1980b: 423–448).

silver" (Godinho 1980b: 433). In attempting to contain deficits in the trade balance, it was necessary to reduce imported goods, especially textiles, paid for by colonial trade profits. To overcome the shortage of imported goods, the crown and several economic agents joined forces to reactivate some national manufactures. Their main goal was to promote the manufacture of wool and silk and the production of glass and iron. At the same time, some mercantilist measures were taken to protect domestic production from competition from foreign products. While these investments took place in the metropole, the King promoted the production of oriental spices in Brazil (Almeida 1975) and encouraged a gold and silver rush across Brazilian territories. By the late 17th century, the discovery of Brazilian gold allowed the Portuguese trade balance to reach equilibrium (Costa *et al.* 2013). This period was a turning point in the Portuguese economy. The "cycle of sugar, tobacco and salt", which structured economic activities during the 17th century, came to an end, and "the cycle of Brazilian gold, Port and Madeira wines" (Godinho 1980b: 439), the main products in the 18th century economy, began.

Subsistence Farming and the Market

Continuing the trend noted in the late 16th and early 17th centuries, during the period analysed in this chapter grain kept its leading position in the structure of agricultural output, and the production of maize (*zea mais*), imported from Central America, increased substantially, particularly in the north of Portugal, where it partially replaced other cereals, namely, wheat and rye. The food diet was complemented by a variety of vegetables and fruit consumed both fresh and dried, and the production of wine and olive oil also increased significantly (Ferreira 1983; Oliveira 1986). In turn, the development of shipbuilding led to an increase in the output of flax, hemp and wood. The area under cultivation also increased, particularly in the fertile areas closer to river ports. The overall dynamism in agriculture was met by population growth and a rising demand for land, leading to an increase in prices. The extension of cultivated areas and the intensification of farming affected the relationship, at the same time necessary and conflictive, between farming and animal husbandry. In an attempt to solve the problem, a reorganization of crop rotations, aimed at harmonizing human and animal food production, was implemented, and other changes were introduced in livestock.

The reorganization of livestock activities took two forms. On the one hand, there was increased livestock kept indoors. On the other, an effort was made to remove livestock from the fields in the period between sowing and harvest.

For these months, the animals would graze in the hills or in nearby wastelands. Thus the mobility of animals was promoted within a short radius. Despite these changes, the reduction of available pastureland caused a decline in livestock production and, predictably, a decrease in meat supplies (Tenenti 2012: 186). Herders and flocks nevertheless continued to have other rhythms: transhumance. This was a centuries-old rhythm, inherent to herding "independent of agriculture" (Ribeiro 1941: 41), required by the search for pastures. Transhumance was associated with the raising of sheep and goats for the marketing of wool and meat. In Portugal, the great pendulum movement occurred between the mountainous areas around Serra da Estrela (Beira province) and the flat lands of Campo de Ourique (Alentejo), a distance of over 400 kilometers. Transhumance across long distances interested herders, "industrialists" from Covilhã town, located in the Serra da Estrela mountainous area, wool merchants and a multitude of other agents linked to the sale of pasture for flocks, particularly in the Alentejo and southern Beira areas (Ribeiro 1941).

In Spain, where transhumance had a major economic and social impact, the first decades of the 17th century witnessed a decrease in livestock. The sharp drop occurred as a result of reduced grazing areas, with a consequent increase in the cost of herbage, resulting from the clearing of private and communal lands (Vassberg 1983). In Portugal there was also a decrease in transhumant livestock, particularly from the Serra da Estrela. A document dating from 1605 denounced the existence of a sharp decline in the flocks in this mountainous region. Of an estimated 300,000 sheep, only a third remained at the beginning of the 17th century. The decline in the flocks was reflected in the production of meat and wool, raw materials that fed the industry installed in the neighboring villages of this region. In order to revive the livestock industry, privileges were granted to herders applicable to both Portuguese and Spanish herds (Garcia Sanz 1990).

In Portugal, the increase in cultivated areas did not solve the problem of the urban supply of wheat, which remained dependent on imports and, therefore, vulnerable to disruptions in foreign trade, recurrent throughout this period. Records from the municipality of Lisbon show shortages of wheat from 1621/1622 until 1668, sometimes in successive years (Mauro 1989: II: 56–77). In some of these years, wheat deficiencies were also noticed in other regions of the country. During 1621, a famine occurred in the Alentejo due to poor harvests (Rodrigues 2008: 183). Between 1630 and 1637, the food crisis shaped the conditions for the outbreak of riots in the area of Évora, one of the most important cities of Alentejo province.

The *arbitristas* were also concerned with the supply of goods to the urban centres, particularly those on the coast. Severim de Faria, for example, wrote in

1624: "The places in Portugal which suffer the most with the lack of wheat are coastal towns and cities, such as Lisbon, Porto, Setúbal and others. This is due to the fact that more people flock to these areas than that which the land can support, and thus what is harvested in the fields is not enough, and more has to come from outside" (Godinho 1980a: 108). Assuming a mercantilist perspective, the same author argued that in order to prevent the outflow of currency and ensure regular supply to the cities, dependence on external markets had to be eliminated. A solution for exceptional moments would be to import wheat from the Portuguese Atlantic islands. But if imported wheat arrived on the domestic market at lower prices than the wheat produced in the country, farmers would have little incentive to increase production, nor would merchants have incentive to organize more efficient trade routes. To increase the national production and trade in wheat, the author proposed that each year public authorities would for three months grant national grain the exclusive supply of the domestic market, starting in the beginning of August (harvest month). This is a curious transposition for the wheat market of a manorial privilege called *relego*, which had been applied since the first centuries of the country to the sale of wine.

Severim de Faria identified other problems in Portuguese agriculture that he attempted to solve. His proposals mainly focused on the difficulties he recognized in the Alentejo, then the top producer of wheat in Portugal, and gave particular attention to the conditions of land tenure and land exploitation. In his view, due to the dominance of large estates, there were difficulties in accessing land, which became the main cause for the human depopulation that occurred in the Alentejo plains. Indeed, these farms were subject to extensive agriculture and fallow lands, with long growing vegetation used only for livestock, while property was concentrated in the hands of a small group of landowners, some of which acted as mere administrators of entailed property. To solve this problem, the same *arbitrista* proposed moving populations living in the north-west province of Entre-Douro-e-Minho, the most densely populated area of the country and the most important for maize production, to the Alentejo province. To make access to land easier and to safeguard the inhabitants of this southern and dry region, he argued that large estates should be divided into small and medium-sized farms.

In line with the suggestions made by Severim de Faria, in 1624 the government in Madrid created a board to promote land clearing throughout the territory of the Portuguese kingdom, with a particular focus on the Alentejo and the Algarve. The implementation of this royal order involved conducting an inventory of all uncultivated lands: communally owned lands linked to entailed estates and chapels, crown properties, and farm buildings belonging to manorial houses. At the same time, the royal officials in charge of adapting uncultivated land for crops, gardens and orchards were entrusted with the task of registering all city, town and village dwellers. Each one of them should in principle have been given a portion of land that they were required to cultivate. This directive did not allow for privileges, and excluded only widows, orphans and destitute persons. The main target groups were the holders of capital, including merchants and other businessmen, who were to contribute with the *cabedal* (capital) needed to invest in land and to make it productive. As nobles and aristocrats were unable to cultivate all the lands they held, it was determined that they would allocate them to people who would till them.

The impact of this policy across the Portuguese kingdom is unknown. There is only evidence of the Madrid government's insistence on those objectives in order to enforce land reforms in the Alentejo, but without success (Oliveira 1991: 84). Collective land customs that, in this region, supported the interests of large-scale livestock owners prevented the enclosure of the fields. The enclosures were necessary to introduce technical changes aimed at increasing agricultural production and productivity, as well as improvements in the livestock sector (Silbert 1978). Ultimately, changes in the Alentejo's large estates, which had been formed in the *Reconquista* period, could not be implemented. The inability to change the production of wheat and marketing conditions in the Alentejo contributed to the continuation of the difficulties in supplying the much-appreciated *white bread* to Lisbon. Because the Algarve was another grain-deprived area, in 1623 the Madrid government ordered the region's governor to force producers to bring the *hidden bread* (grain that had been concealed) to the market, and likewise decisions were taken for the clearing of uncultivated land in the mountainous area of Serra de Tavira with the purpose of increasing land under grain cultivation.

The *arbitristas*' proposals also had an echo in the decisions taken by the Madrid government regarding the bogs and swamps that existed on riverbanks, including in the main rivers, the Tejo and the Mondego, viewed as another factor that hindered the increase in cereal production. Yet, the policy of draining marshlands often amounted to no more than good intentions, as it did not receive financial support from the crown or assistance from landlords. As demonstrated by Fernand Braudel, the plains that have a history of agricultural success, such as those of Lombardy (Italy), Lower Languedoc (France) and Andalusia (Spain), were those where drainage infrastructures were deployed and maintained. This continuity was based on constant financial capital investment made by state, mercantile cities, capitalists and nobles (Braudel 1979). In the 17th century the development of Dutch and English agriculture also benefited from land drainage. However, the same did not occur in the plains of major Portuguese rivers, where agricultural activities continued to be hampered by the existence of large permanently flooded areas and the occurrence of cyclical floods.

Another important trend in Portugal's agriculture in the 17th century was the change in the traditional pattern of grain cultivation due to the spread of maize (*zea mais*). The introduction of this crop into the country was part of a broader movement of exchange and acclimatization of plants and animals caused by the discovery of overseas territories by the Portuguese and the Spanish. The incorporation of new territories and the intensification of exchanges, coined as the *Columbian Exchange* (Crosby 1972), improved human diets in Africa, Brazil and Europe (Tenenti 2012: 183–186). According to Almeida (1995: 241), "in the present state of knowledge, it is not possible to determine accurately, or even approximately, the date of the introduction of *zea mais* in Portugal". In any case, he suggested that "It must have occurred, however, during the 16th century, probably in the first half, as a realistic or stylized representation seems to show the cob as a decorative element of Manuelina art". There seem to be no doubt that in Portugal the spread of this new maize variety started in Coimbra's irrigated fields in the Mondego River valley.

To clarify the problem under consideration, it is particularly important to determine the shapes and rhythms of spread, as well as its impact on subsistence economies. According to the evidence of Severim de Faria, from the mid-17th century maize was the most important foodstuff for *ordinary people*, "almost across the border and Entre-Douro-e-Minho maize cultivation resulted in large "benefits" for these provinces because "as the crops are irrigated they are never deficient; and yields, can be maintained in an inexpensive way, so that people can be sustained" (Sérgio 1974: 144–145). In spite of the exaggerated beliefs about the impact of maize in solving the problem of feeding the population, the positive effects brought by the grain to these subsistence economies were undeniable: used for bread-making, maize was also consumed in porridge with vegetables, preventing hunger in the periods that preceded major harvests.

Recent studies suggest a slower diffusion of maize in Portugal. The crises that marked the last years of the 16th century (1598–1599) and first decade of the 17th century (1608–1609) favored the introduction and spread of this American grain (Eiras 1998). Its introduction probably began in the last decades of the 16th century, in the fields of Coimbra, where it was eaten dried. During the first half of the 17th century, the new agricultural product was disseminated through northern coastal fields, where it could benefit from irrigation. By the end of that century and during the 18th century, maize extended, as a rain-fed crop, across suitable countryside regions. The sowing of maize was inserted in the existing triennial and quadrennial crop rotations, replacing millet, an old variety of maize already cultivated in medieval centuries (Gil 1965: 73–74). There are also indications that more land was cleared to accommodate the new variety coming from the New World territories (Neto 2013).

Traditional grain (wheat, rye and barley) was cultivated especially in extensive farming, with long fallow periods, from the northern mountainous regions to the Alentejo plains. The spread of maize caused profound changes in cultivation systems and in the entire organization of the agrarian landscape. Aside from furthering the clearing of uncultivated land and changing existing farming systems, it promoted the reclamation of wetlands and better use of the soil. Along the sandy soils near seashores, maize fields were surrounded by pine forests, which sheltered them from the sands and provided compost for land fertilization. In the province of Entre-Douro-e-Minho, the enclosure of the field-meadow was for fruit trees and vines (to produce white wines), freeing the soil of small plots and making it fresher and suitable for growing maize. In the Beira region, terraces were built on hillsides, creating conditions for growing this rain-fed maize.

The expansion of intensive agriculture required by maize led to the occupation of land previously used for grazing transhumant livestock, which may have contributed to reducing the number of flocks. However, the use of maize as forage permitted the housing of animals employed in farm work and transportation, as well as in milk production. In fact, livestock would establish itself as an important factor of economic dynamism in the regions that hosted this new maize variety (Neto 2013). While it continued to rely on organic fertilization from livestock, the association of maize and leguminous plants, which, in addition to their food uses, promoted the renewal of soil fertility, favored land cultivation. Apart from these changes, maize spurred technological innovations in irrigation, threshing and drying systems, as shown in the case of the Benedictine abbey of Tibães (Braga), in the Entre-Douro-e-Minho province (Oliveira 1979: I: 248). New wells for irrigation were dug and equipped with wheels and other mechanisms driven by animals to bring water to the surface. At the same time, irrigation canals were built to channel water from rivers, streams and fountains. In turn, the monks erected earthen threshing floors for manual threshing and drying of the maize. Finally, in the province of Entre-Douro-e-Minho, granaries – small buildings made of wood and stone – were built to store grain.

Although grain was the main production of Portuguese and European agriculture, other crops that contributed to the subsistence of populations were present in farm fields. Fruit was an important supplement to food diets, and had the advantage of avoiding noblemen's taxation impositions more easily.

Across the country, there was a diversified fruit orchard or, more often, trees were scattered across fields that were cultivated with other crops. In 1610, Duarte Nunes do Leão refers to several of the existing fruit trees, like oranges, ciders, lemons, pears, apple trees, quinces and cherries. In some areas fruit production in fact became particularly important. Trade that reached both domestic and external markets stimulated production in regions where soil and climatic conditions were beneficial for fruit trees. The territory of Algarve was the most relevant of these cases. Fruit production, consisting mainly of oranges, lemons, almonds, carob and especially figs, was the "support" of the regional economy (Magalhães 1988: 161–174).

In the colder lands of Entre-Douro-e-Minho, Trás-os-Montes and Beira provinces, landscapes were filled with chestnut woods. The chestnut tree was extremely important for *Ancien Régime* subsistence economies. The poorest populations mainly ate green chestnuts during harvest time and dried chestnuts throughout the year, using them for making the "bread of the poor" (Oliveira 1990: 38). In regions further south, other fruits could replace the bread. In Alentejo, acorn was mainly consumed (produced by oaks and holm trees), and in Algarve figs, endogenous fruit along the seashores of the Mediterranean Sea, were more used.

Fresh vegetable products also constituted an important component of agriculture for self-consumption geared to the supply of the domestic market. Small gardens were cultivated near dwellings, in both rural and urban areas. For example, in the 17th century gardens of Coimbra, both in the periphery and inside the city, the following vegetables were produced for urban consumption: garlic, onions, sprouts, beans, peas and melons (Oliveira 1972: II: 30–337). An increase in the production of beans and pumpkins, already known in Europe in previous centuries, but enriched with new varieties from the American territories, was linked with the spread of maize (Ferrão 1992: 205). Among this set of farming produce, tomatoes and capsicum peppers were a novelty. They reached Europe in the 16th century, brought by the Spanish from Central America, but spread very slowly (Ferrão 2006). Yet, in Portugal, the first known records, dating from the 16th century, indicate that tomato was already part of the food diet in the city of Évora in the Alentejo province (Braga 2007).

Sweet potato (*ipomoea batatas*) and potato (*solanum tuberosas*) were other novelties that came from America to Europe in the 16th century, and also spread slowly (Eiras 1988). In the 16th and 17th centuries, the cultivation of tubers was confined to gardens and orchards, except in Ireland, where it spread in the 16th century to deal with subsistence crises. In Portugal, several pieces of historical evidence attest that they were particularly appreciated. In 1643, a lawsuit filed by the monastery of Santa Cruz in Coimbra, a large estate in the centre of the country, indicates that in the sandy lands of this region there could be harvests with over a thousand *alqueires* (around 11 kg) of potatoes (probably sweet potatoes). Other registers show that potato growers resisted paying taxes, proving the continuous cultivation in the poor lands of this region. Potato production however became widespread in inland fields only from the late 18th century, and especially in the early years of the 19th century, during the food crisis linked with the French military invasions (1808–1811). At this time, the suspension of tax payments to the manors was one of the stimuli for the expansion of potato cultivation. This circumstance corroborates the thesis that in the 17th century land taxation was an important factor inhibiting agricultural innovation (Neto 1994).

These new products coming from other continents were integrated into the different regional mixed culture systems. In these systems, seasonal products were often combined with long-term crops, such as olives and vines. Though olive oil and wine were already very relevant in the livelihoods of farming families, they became even more attractive as trade opportunities for these products grew. In the 17th century, the trading possibilities of these goods increased with the pace of intercontinental travel and the colonization of other continents.

Olive trees were spreading from south to north, implanting themselves in the sheltered and sunny slopes or in the peripheries of large urban centres. In the 17th century, the Algarve, Alentejo, Estremadura and Beira were the regions with the highest concentration of these trees. In the central region, the monastery of Santa Cruz encouraged cultivation in the outskirts of Coimbra city, where it was based, giving freedom to *emphyteutae* to plant vineyards and olive groves in *bread lands*. However, the need to protect the olives growing on the outskirts of Coimbra caused constant conflict with livestock owners, who did not comply with the legislation that defined the times of grazing in areas with groves (Oliveira 1972: II: 83–93). In the Minho area, as the availability of animal fat decreased due to the expansion of cultivated areas, thanks to the expansion of maize, olive oil grew in importance. In these provinces, like everywhere in Portugal, olive trees ensured vegetable fat necessary for food, lighting and, in some cases, for industrial use (namely the textile industry).

In the Algarve, merchants also boosted the plantation of vineyards and olive groves on the outskirts of major urban centres. As wine and olive oil were mainly produced for export, the evolution of production could reflect the disruption in foreign trade affected by piracy and wars: in the Algarve by the mid-17th century there was an interruption or even a reversal in the expansion of those crops. The expansion occurred again in 1678–80, in connection with the general recovery in the Algarve's economy (Magalhães 1988: 164). Despite the reduction in commercial activities in Algarve's ports, the agricultural sector was critical for regional economic dynamism. Fruit, wine and olive oil were thus exported from the Algarve to both cities of Northern Europe (London, Bristol, Dublin, Amsterdam, Rotterdam, Hamburg, Danzig, and Stockholm), and France and neighboring Castilian cities, particularly to Andalusia (Magalhães 1988: 282).

The first decades of the 17th century also saw an expansion in vine cultivation in the Riba Douro area (Douro River valley). This expansion led to intense commercial activity, involving nobles, clergy and merchants, artisans, farmers and institutions that received seigniorial rents. Religious orders promoted the cultivation of vines in order to increase the ground rent that constituted the economic support of ecclesiastical establishments.[7] These institutions also had the financial capacity to introduce the technical improvements suitable to the production of various types of wine. If the Douro River terraces produced mature wines, in the region of Entre-Douro-e-Minho the Benedictine abbeys built stone and wood pillars that supported the vine trellises to produce white and fresh wines (known as green wines).

These wines were mainly exported to Brazil, Angola and neighboring Galicia. In the first decades of the 17th century, the product occupied a leading role in the economy of the city of Porto, streamlining the entire region by complementary activities related to the dynamics of the wine trade: cooperage, boat building and river transportation. However, the origins of the economic dynamics witnessed in the late 17th century and the first half of the 18th century in the Douro valley and Porto city must be traced back to the 16th century. It was during the period from 1580 to 1640 that the qualities of sweet Port wines were discovered, becoming established in Europe from the second half of the 17th century onwards (Silva 1986: 163–238). The place achieved by viticulture in the Douro throughout these centuries created the conditions for the enactment of legislation in 1756 defining one of the first wine *terroirs* in the world.[8]

The protection of national wine production led King Pedro to ban, in 1676, the entry of wines from Castile, since the import of wine from other countries reportedly hit the national wine industry. The order was renewed in 1683, with the agreement of the Municipality of Lisbon, where the largest commercial

7 Among the religious institutions that sought to breach the slopes of the Douro River banks to plant vines were the Jesuits, the Benedictines of Porto and the Braga region, the Cistercians of Tarouca and the priests of the Chapter of Porto and Braga.

8 About this topic see Reis, Chapter 6 in this book.

movement of the country took place.[9] The intention of the Portuguese government was to stimulate exports. In the late 17th century, English, Dutch and French ships were moving to various seaports in the north and the centre of the country (Porto, Figueira da Foz, Lisbon, Setúbal) to load wines for their countries (Almeida 1981: 97–98). At this time, England had been the main destination of Port wines and the expansion of trade contributed to wine production specialization, evident in the more accessible areas of the Douro Valley (see Table 4.2).[10] In the following centuries, England would continue to be one of the main consumers of fortified wines, which received the designation of Port, but which had always been produced on the slopes of the Douro River valley stretching from Régua town to the inland areas.

TABLE 4.2 *Exports of Portuguese wine (Lisbon, Porto and Madeira) to England*

Years	Tons	Years	Tons
1675	20	1691	2964
1676	83	1692	6052
1677	176	1693	8200
1678	199	1694	9454
1679	1013	1695	3983
1680	1003	1696	6668
1681	1718	1697	4774
1682	13860	1698	4057
1683	16772	1699	8703
1684	1611	1700	7757
1685	12185	1701	7408
1686	289	1702	5924
1687	327	1703	8845
1688	540	1704	9924
1689	579	1705	8449
1690	1115		

SOURCE: MACEDO 1982: 53.

9 For example, in 1685, 77 percent of imports and 75 percent of exports were concentrated in Lisbon (Magalhães 1988: 282).

10 For example, in the period from 1678 to 1691, the average annual export was 1300 barrels, and this average rose to 11,100 barrels annually within four years (Pedreira 1994: 32).

Although foodstuff production was the main purpose of farming, arable land was also used for "industrial" crops. One of them was flax, then cultivated in irrigated lands across the country. The plant supplied the raw material for the manufacture of domestic and industrial textiles intended for clothing and household linen. The production of linen cloths clearly increased during the 17th century. Indeed, according to Duarte Ribeiro de Macedo, in 1680 production satisfied domestic demand and linen fabrics were exported to the colonies and Castile (Pedreira 1994: 26). The production of flax acquired special relevance in the fertile fields of Santarém (on the Tejo River valley), Coimbra (on the Mondego River valley) and Moncorvo (taking advantage of the confluence of several rivers in this region of Trás-os-Montes) and the irrigated lands of Entre-Douro-e-Minho. In this region, endowed with abundant female labor, two important centres of linen textile production developed: the towns of Guimarães and Braga (Oliveira 1985: 87–88).

Another textile fibre cultivated at the time was hemp (*cannabis sativa*) or linen-hemp, as it was then known, used for the manufacture of fishing nets and rigging (sails and rigging used on ships). The need for this raw material increased during the 16th and 17th centuries. In order to stimulate the production of starching necessary for the armies, King Filipe III of Portugal established a trading post on the outskirts of Santarém, granting it the monopoly of production and manufacturing of linen-hemp in this region. At the same time, factories were also created in Moncorvo and Coimbra, which received royal charters in 1656 and 1659, respectively (Costa 1997: 342–359). According to existing regulations, the factory controlled the production of the raw material, making it available to the rope makers, who transformed it in domestic workshops. Thus, factories or manufacturing centres of flax and hemp functioned as "coordinated centres of dispersed production" nuclei (Pedreira 1994: 26).

Mulberry production and silk manufacturing were other activities that showed a strong momentum in the second half of the 17th century. They mainly benefited the district of Bragança (Trás-os-Montes). Since the 16th century, the region had been linked to textile production, to which foreign artisans were dedicated, particularly Jews expelled from Castile (Madureira 2001: I: 9). The decline in trade with Asia, the main supplier of silks, combined with the need to reduce imports, sparked an internal dynamism of the sector in the 17th century. One of the advocates of the silk industry was the French cleric Rafael Bluteau. But the manufacture of silk required expertise, which Portugal lacked at the time. To deal with this situation, in 1677 and 1683 privileges were granted to the French family Duclos to install silk manufactures in Lisbon. The dynamism of the activity is confirmed by the number of looms in operation (50 looms in 1679) and the need to import machines from England.

Output Cycles

In this section we will first be looking at trends in agricultural production, drawing mostly on data regarding the collection of tax, particularly the tithe, and secondly at the development of the manorial system and its effects on agriculture. The values set in the auctioning of lease contracts are also taken as indicators of trends in production. The available data does not allow us to make an overall assessment of the trajectory in the kingdom, but studies in the field of economic history provide some evidence on regional dynamics of agricultural production (Neto 2012). The route of the region of Entre-Douro-e-Minho can be evaluated through the incomes of Benedictine monasteries, particularly the monastery of Tibães (the most important one that the order had in Portugal). There was an income drop in the late 16th century. Nevertheless, in the early 17th century, beginning around 1609, there was a phase of recovery and growth, which continued until the eve of 1640. Thereafter, the war affected economic and demographic stability in neighboring Galicia (Eiras 1986), and, presumably, also on the Portuguese side of the border. The resumption of growth would only occur in the decade of 1670–1680, reaching the highest levels of the century in 1680/82. The final years of the 17th century and the first decade of the 18th century were again marked by declines in the revenues of the Benedictine monasteries (Oliveira 1979: 489–493).

In central Portugal, the changes can be estimated from the assets of the monastery of Santa Cruz and the university, both located in Coimbra, which collected revenues from several areas of the country. The 17th century proved clearly unfavorable for these estates, which suffered a steep decline in income from land in the first four decades of the century, reversed after 1640. The trend may reflect a possible increase in production, which can be added to a more efficient collection of revenues, arising from the use of updated collection methods. For the last three decades of the century, there were more solid signs of economic recovery (Neto 2012).

In turn, the evolution of agricultural production in the Beira Alta province, northern interior of the country, can be gauged from the lists of direct recoveries of tithes for the church Chapter of Viseu city (see Table 4.3). The amounts of tithes collected show that after a period of regression in production (1575–1605), there was a reversal of the trend. In spite of this, the years

1635–1653 and 1659–1670 were marked by great irregularity in the recovery of income and years of steep decline in production (1637, 1643, 1647, 1659–1669). From 1680 onwards a period of "firm" expansion began, lasting until the late 17th century (Oliveira 1990: 39).[11]

The evolution in the cost of property leases belonging to the Misericórdia of Évora reflects the development of production in the Alentejo. According to Rui Santos, during the 17th century, the income of that institution showed a general downward trend that was reversed only in the 1680s. The author distinguishes between "a very sharp decline since the values of the late 16th century, an attempt at recovery between 1610 and 1630, a new violent descent in the 30s only partially recovered, and a further, though softer decline", recorded between 1640 and 1655, and this was followed by a depression, "which hit bottom in the second half of the 60s", and a recovery in the 1680s and 1690s (Santos 2003: 382–383).

The leases of the tithes of the church Chapter of Faro city are good indicators of the evolution of the Algarve's economy (see Table 4.4). In central Algarve, income from tithes reveals a fall, between 1618 and 1650, and a small rise until 1675. From this date on, there was a rise that stretches back to the early 18th century. An improvement in grain production took place in the 1670s, a period that marked "the beginning of the general recovery of the Algarve's economy" (Magalhães 1988: 176). Although the exploration of most farms was oriented towards family subsistence, there was a strong component of commercial agriculture in the subsectors of fruit, wine and olive oil. Thus, the evolution of income reflects the vicissitudes of foreign trade.

Analysing the performance of the various regional incomes during the 17th century, it can be concluded that in all provinces of the country the curves evince a sharp decline in the late 16th century and in the beginning of the 17th century, returning to sustained growth only in its final decades (1670–1680). Another common element to several provinces, with the exception of Entre-Douro-e-Minho, is the breakdown that occurred in the thirties, which, in some areas, interrupted a trend towards growth. By 1640, there was a shift towards moderate income growth, a phenomenon recorded in the Iberian Peninsula, with the exception of Old Castile (Sanz 2004: 392). However, the recovery of the Portuguese agrarian economy was heavily disturbed by war with Castile that, in addition to affecting agricultural and livestock production, disorganized

11 These irregularities are expressed in gaps in the data series, without taxation information for the years of 1636, 1638, 1640, 1642, 1644, 1646 and from 1648 to 1653 and the following years of 1655, 1659, 1661, 1663, 1664, 1666, 1668 and 1669.

TABLE 4.3 *Income of tithes from the Chapter of Viseu (1622–1700)*

Years	Alqueires (bushels)	Years	Alqueires (bushels)	Years	Alqueires (bushels)
1622	11722	1651	–	1680	8733
1623	18448	1652	–	1681	–
1624	15060	1653	–	1682	–
1625	–	1654	15204	1683	13351
1626	16097	1655	11549	1684	–
1627	12476	1656	13475	1685	12142
1628	14320	1657	13981	1686	14711
1629	–	1658	7726	1687	–
1630	13082	1659	–	1688	–
1631	–	1660	9499	1689	15229
1632	–	1661	–	1690	12010
1633	13365	1662	10268	1691	14552
1634	14429	1663	–	1692	14434
1635	13260	1664	–	1693	14552
1636	–	1665	9031	1694	13818
1637	9086	1666	–	1695	–
1638	–	1667	11366	1696	15181
1639	11318	1668	–	1697	7496
1640	–	1669	–	1698	13196
1641	15749	1670	13321	1699	16574
1642	–	1671	13531	1700	17089
1643	9421	1672	–		
1644	–	1673	10518		
1645	12168	1674	–		
1646	–	1675	10236		
1647	9751	1676	11374		
1648	–	1677	11212		–
1649	–	1678	10617		
1650	–	1679	11933		

SOURCE: OLIVEIRA 1990: 134–136.

TABLE 4.4 *Income of tithes from the Chapter of the Algarve*

Years	Income (in reis)
1618	1,323,400
1636	1,234,700
1646	1,191,900
1656	1,595,500
1666	1,666,500
1676	1,715,000
1686	2,034,750
1696	2,134,750
1706	3,002,000

SOURCE: MAGALHÃES 1988: 188.

trade routes. The outbreak of war had a very negative impact on agricultural producers. It caused a rise in wages, due to the need for manual labor for the conflict. There was also more pressure due to the tax burden in the form of indirect taxes on the consumption of meat and wine, and a new direct tax (the *decima*), amounting to 10 percent of all income. Although the Alentejo and Beira Interior were the regions most affected by the war, the unstable environment was experienced throughout the territory, expressing itself in the pronounced cyclical fluctuations of income. In this context, the sustained growth in agricultural production is noticeable only at the end of the century, especially from 1690. This upward trend is visible in the dynamics of prices and in demographic and commercial recovery too.

With regard to the manorial system, it is important to note that from the early 15th century onwards, some nobility groups and the bourgeoisie began to reap earnings and fortunes overseas: in the Atlantic islands (Madeira and the Azores), India and Brazil. Albeit articulated with colonial trade, especially in the coastal belt, economic life within the metropole continued to be primarily based on agriculture and on a set of complementary or interdependent resources in this sector. Among these, first, were hunting, fishing and livestock. But several rural "industries" were also important, like the extraction of timber, minerals, weaving, coal, lime, and tile production, grain milling, and the manufacturing of wine and olive oil. Land, rivers and forests were the main sources of livelihood, income and wealth of almost all social groups and institutions. The land yields were the basic support base of noble and ecclesiastical manors, as well as of cultural, educational and welfare institutions. Land revenue took on a strong cultural and symbolic dimension: land was a symbol of social prestige and power. Land ownership (through entailed estates) was a condition of survival and reproduction of many noble families, including the local oligarchies that ruled urban municipalities. In turn, rural properties were part of wedding dowries and gave access to ecclesiastical careers. From the late 16th century onwards, income paid for masses and pious works absorbed a part of agricultural production and, at the same time, delayed the mobility of land. The possession of rural buildings (e.g. country houses) was also assumed as a condition for the exercise of bourgeois activities (e.g. contracts for the collection of rents for ecclesiastical institutions), as well as a prerequisite for obtaining capital loans. Finally, it is important to note that the monarchs rewarded services rendered to the crown in the metropole and empire (the *Restauração* wars) with the assignment of land revenue.

During the 17th century, a strengthening of the manorial system was felt across Europe, especially in Eastern regions (with the exception of England and the Netherlands), a sort of *refeudalization* (Romano 1992: 4). In Portugal, the manorial regime was deeply rooted in rural areas, limiting uses of cultivated areas and uncultivated lands, or even the exploration of several natural resources, such as water and forests. Manorial territories were organized over time with the integration of land from donations (with royal and private origins), purchased plots, as well as appropriations of uncultivated commons for community use. As in other European regions, landlords did not dominate mountainous areas, where the private use of land was complemented with practices of agro-pastoral communitarianism. A royal taxation reform, enacted in the early 16th century, recognized the domains and privileges of many secular or ecclesiastical entities in Portuguese rural areas. It established the payment of a set of taxes for economic activities, with a particular emphasis on agriculture, fishing and domestic trade (Neto 1993: 165–175).

Both manorial and ecclesiastical entities, much like the crown, lacked instruments that allowed them to efficiently manage their extensive and often scattered lands. Landowners behaved as income receptors, collecting revenue mostly by contractors (one of the modalities that allowed the bourgeoisie to receive land revenue). Because they did not have a direct and close contact with their domains, manors needed periodic reorganization and demarcation of their estates. The clearing of land taking place during the 16th century was done, in many cases, without the supervision of property owners. However, in the late 16th century and early decades of the 17th century, the crown and several manors made inventories of their estates and ecclesiastical institutions organized records of their own assets and rights. This entailed the placement of landmarks that defined the "borders" of manorial estates, the internal reorganization of land (integrating what had been recently sold or cultivated) and the reconfirmation, by local communities and municipal councils, of the manorial rights enshrined in administrative documentation and agrarian contracts.

Yet, the compilation of new information and a review of these registers did not entail an increase in manorial rights. As an example, the monastery of Santa Cruz in Coimbra, one of Portugal's major landowners, maintained manorial rents. In fact, the manorial rights enshrined in registers and in agrarian contracts were already too onerous, which prevented the manors from increasing them. This kind of administrative registration and inquiries nevertheless created conditions for the improvement of the manorial pressure on tenants, especially on those with outdated contracts or who lacked a contract altogether. As a result, many *emphyteutae* were forced to enter into contracts in order to secure the control of lands that they already cultivated. The registrations were also used by the manors to achieve recognition of dominance over vacant areas, thus securing potential sources of income. After these inquiries, the principle of *no land without a lord* would become effective in many manors (Neto 1997: 39–56).

The *emphyteusis* was the most widespread legal institution ruling the use of agricultural land.[12] This contract could be perpetual or last a lifetime, but generally lasted three lifetimes. From the 16th century onwards, the type of contract in lives came to be prescribed in the bishoprics' formal constitutions.[13] In a period of intense demand for land, it was considered the most appropriate form of contract for periodic control of church property, as it was more favorable to updating rents. These bishoprics' documents recognized the desirability of perpetual contracts for tenants, advising the clergy to sign them as an incentive to the clearing of uncultivated lands, drying of bogs, and turning moors into orchards, vineyards or *bread lands*. Yet, when the demand for land surpassed supply, the peasants had no option but to submit to the conditions offered by the landowners. It appears that, from the 16th century onwards, contracts in lives were the preferred contract offered by manorial entities. The decrease in the duration of *emphyteusis* contracts was met with resistance. In a text dating from the early 17th century, an *arbitrista* author, Duarte Nunes Leão, put the contract in lives forward as a cause of the "decadence" of agriculture. In turn, the people's representatives in the royal courts assembly of 1641 asked for the abolition of these contracts, which were considered precarious.

12 See Henriques, Chapter 1 and Rodrigues, Chapter 2, both in this book.

13 The bishoprics were ecclesiastical administrative divisions headed by a bishop.

The agrarian contracts signed in the 17th century testify to the existence of another phenomenon: the creation of income through the system of *censo consignativo*. In the Portuguese case, the extent of this practice is unknown, but studies have shown that it became well established in Spain in that century (Marcos 2000: 506–507). The *censo consignativo* was a contract between an entity owning capital, an individual or an institution, and another possessor of immovable property. Through the loan of a sum of money an annual income in yields, called a *censo*, was established, which was assigned to a rural or urban building. One of the legal requirements of this contract was that it was an "open retroactive pact", allowing the owner of land burdened by a *censo* (quick rent) to be released from this onus. The "capitalist" could still recover the capital by transferring his right to others. There is evidence that this contract was practiced in Coimbra in the 16th century and the first half of the 17th century. It was a credit – granted mostly by priests, ecclesiastical establishments, lawyers, noblemen and citizens – channelled to rural areas to cope with everyday problems: to pay off debt, to buy seeds or food or make improvements (Oliveira 1972: 107).

In a time of instability and scarce investment opportunities, the creation of income in yields was a strategy used by holders of capital, including rent contractors, to apply their capital in a safer way. In this way, they could get an agricultural income, while at the same time benefiting from the economic, social and symbolic rewards of the ownership of land revenue. As stated by Vries (1976), a businessman who accumulated profits had few investment alternatives in addition to buying land, public offices, palaces, and artwork, which were desirable forms of investment. These were also the most honorable forms of capital investment. Investment in agriculture through mortgage loans started in Portugal in the 17th century, much as in Spain in the same period (Garcia 1990: 222–226), though over indebtedness. In the medium and long-term, debts contributed to the integration of small farmers into the assets of creditors. During the 17th century, there was a strengthening of social uses of land. This allows an enlarged group people to benefit from the redistribution of agricultural incomes. As a result, the economic conditions of farmers worsened, and they were living asphyxiated by manorial and royal taxation.

The social uses of land mentioned above, combined with the expansion of the manorial system, configured forms of ownership and land use, improved throughout the 18th century. They constituted a powerful handicap for agricultural development, resulting in shortage of capital required by the introduction of technical innovations. In this, Portugal would be similar to neighboring kingdoms, to the extent that "in Western Europe and the Mediterranean free peasants worked the land whose land taxes, rents and tithes constituted the bulk of state revenues, the landowning nobility and the Church" (Nash 2001).

Conclusion

During the 17th century, Portuguese agriculture reinforced the strong foundation of self-consumption rooted in traditional grains (wheat, rye, barley and millet), to which was added maize, welcomed as a result of Portuguese irrigation since the 16th century. While grains were the basis of the diet, they were complemented by a variety of vegetables and fruit, consumed fresh and dried. If the urban supply of wheat, particularly in Lisbon, remained dependent on imports, the diet of the rural and urban popular classes improved thanks to the expansion of American maize. The introduction of this cereal into the regular diet led to an increase in population, which became patent from the late 17th century onwards. In addition to diversification in grain and food production, Portuguese agriculture experienced a niche specialization aimed at supplying domestic and export markets. These trends were particularly evident in the fruit subsector, as well as olive oil, associated with the intensification of olive trees, and wine. The specialization in wine mainly occurred in the Douro region, where were implanted the support structures for the production of Port wines that would secure their success in the international markets from the 18th century onwards. In a context of little innovative agriculture, the most dynamic spaces were gardens, vineyards and olive groves, on the outskirts of urban centres, and areas that welcomed maize stood out. The spread of this grain was associated with the introduction of technical innovations, including those resulting from some noticeable replacement of extensive livestock production by agriculture intensification.

The analysis of regional income trends leads to the conclusion that in all country provinces there was a sharp decline in the late 16th and the beginning of the 17th centuries, returning to a sustained growth only in the final decades of that century (1670–1680). By 1640, there was a shift towards moderate income growth. However, the recovery of the agricultural economy was heavily interrupted by war with Castile that, in addition to affecting agricultural and livestock production, disorganized marketing channels. The outbreak of war had a very negative impact on farming and farming families. Although the Alentejo and Beira Interior were the regions more affected by the war, the unstable environment was felt across the territory, expressing itself in pronounced cyclical fluctuations in income. In this context, the sustained growth in agricultural production is noticeable only at the end of the 17th century, especially from 1690. This upward trend is visible in the dynamics of prices and in demographic and commercial recovery too.

In the metropolitan area, it was necessary to address a series of adjustments and readjustments in the management of agricultural land, the composition of the product and the institutional framework of the agrarian economy. In this respect, it is important to note that excessive taxation burdened agriculture, preventing Portuguese farmers from investing in improvements, which could lead to techniques aimed at increasing productivity. Actually, in the 17th century the Portuguese economy must be understood in its articulations with the changes in the colonial empire. The progressive decline in trade with the Eastern colonial territories and the strengthening of the Atlantic areas (Africa and Brazil) reconfigured the Portuguese economy and the place held by Portugal in the balance of European political forces.

References

Almeida, Luís Ferrand de (1975). "Aclimatação de plantas do Oriente no Brasil durante os séculos XVII e XVIII". *Revista Portuguesa de História*, 15, pp. 339–481.

Almeida, Luís Ferrand de (1981). "Problemas do comércio luso-espanhol nos meados do séc. XVIII". *Revista de História Económica e Social*, 8, pp. 95–131.

Almeida, Luís Ferrand de (1995). "Sobre a introdução e difusão do milho mais em Portugal". In Luís Ferrand de Almeida (Ed.). *Páginas Dispersas. Estudos de História Moderna de Portugal*. Coimbra: Instituto de História Económica e Social. Faculdade de Letras da Universidade de Coimbra, pp. 229–259.

Álvarez Nogal, Carlos (1997). *El crédito de la monarquía hispánica en el reinado de Felipe IV*. Madrid: Junta de Castilla y León.

Bethel, Leslie (Ed.) (1987). *Colonial Brazil*. Cambridge: Cambridge University Press.

Bettencourt, Francisco and Curto, Diogo Ramada (Eds.) (2006). *Portuguese Oceanic Expansion, 1400–1800*. Cambridge: Cambridge University Press.

Boxer, Charles (1969). *The Portuguese Seaborne Empire, 1415–1825*. New York: Knof.

Boyajian, James G. (1983). *Portuguese Bankers at the Court of Spain 1626–1650*. New Brunswick, New Jersey: Rutgers University Press.

Braga, Isabel Drumond (2007). *A Herança das Américas em Portugal*. Lisbon: CTT.

Braudel, Fernand (1979). *Civilisation matérielle, économie et capitalisme*. Paris: Armand Colin.

Costa, Leonor Freire (1997). *Naus e Galeões na Ribeira de Lisboa. A Construção Naval para a Rota do Cabo no Século XVI*. Cascais: Patrimonia Historica.

Costa, Leonor Freire (2002). *O Transporte no Atlântico e a Companhia Geral do Comércio do Brasil (1580–1663)*. Lisbon: Comissão Nacional dos Descobrimentos Portugueses.

Costa, Leonor Freire, Lains, Pedro and Miranda, Susana M. (2016). *An Economic History of Portugal, 1143–2000*. Cambridge: Cambridge University Press.

Costa, Leonor Freire, Rocha, Maria Manuela and Sousa, Rita Martins de (2013). *O Ouro do Brasil*. Lisbon: Imprensa Nacional-Casa da Moeda.

Crosby, A.W. (1972). *The Columbian Exchange: Biological and Cultural Consequences of 1492*. Westport: Greenwood Press.

Disney, Anthony R. (1981). *A Decadência do Império da Pimenta: Comércio Português na Indía no Início do Século XVII*. Lisbon: Edições 70.

Duncan, T. Bentley (1986). "Navigation between Portugal and Asia in the sixteenth and seventeenth centuries". In E.J. van Kley and C.K. Pullapilly (Eds.). *Asia and the West. Encounters and Exchanges from the Age of Explorations. Essays in Honor of Donald F. Lach*. Baltimore: Notre Dame, pp. 3–25.

Eiras Roel, Antonio (1986). "Agricultura y poblacion en la Galicia moderna (XVI–XVIII)". *III Xornadas de Historia de Galicia*. Ourense: Servicio de Publicacions da Diputation Provincial de Ourens, pp. 14–41.

Eiras Roel, António (1988). "Los productos alimentarios de ultramar. La Europa mediterránea". In Simoneta Cavaciocchi (Ed.). *Prodotti e tecniche d' Oltremare nelle economie europee. Secc. XIII–XVIII*. Prato: Instituto Internazionale di Storia Economica "F. Dantini".

Eiras Roel, Antonio (coord.) (1998). *El reino de Galicia en la monarquía de Felipe II*. Santiago de Compostela: USC.

Ferrão, José Mendes (1992). *A Aventura das Plantas e os Descobrimentos Portugueses*. Lisbon: IICT, CNCDP, FJB.

Ferrão, José Mendes (2006). *A Aventura das Plantas e os Descobrimentos Portugueses*. Lisbon: Chaves Ferreira Publicações.

Ferreira, J.A. Pinto (1983). "A economia do vinho e o crescimento do Porto, nos séculos XVII a XIX". In *O Vinho na História Portuguesa, Séculos XIII–XIX*. Porto: Fundação António de Almeida.

Garcia Sanz, Angel (1990). "El sector agrario durante el siglo XVII: depresión y reajustes". In Jover Zamora (Ed.). *Historia de España Menéndez Pidal*. Madrid: Espasa-Calpe, vol. 23, pp. 159–235.

Gil, Maria Olímpia da Rocha (1965). *Arroteias no Vale do Mondego durante o Século XVI*. Lisbon: Instituto de Alta Cultura.

Godinho, Vitorino Magalhães (1970). "Portugal and her Empire". In *The New Cambridge Modern History*. Cambridge: Cambridge University Press, vol. VI, pp. 509–540.

Godinho, Vitorino Magalhães (1978). "L'émigration Portugaise (XVe-XX siècles). Une constante structurale et les réponses aux changements du monde". *Revista de História Económica e Social*, 1, pp. 5–32.

Godinho, Vitorino Magalhães (1980a). "Alguns problemas da economia portuguesa no século XVII de depressão internacional". *Revista de História Económica e Social*, 5, pp. 105–123.

Godinho, Vitorino Magalhães (1980b). "Portugal, as frotas do açúcar e do ouro (1670–1770)". In *Ensaios*. Lisbon: Sá da Costa, vol. II, pp. 423–448.

Hanson, Carl (1981). *Economy and Society in Baroque Portugal, 1668–1703*. Minneapolis: University of Minnesota Press.

Hobsbawm, Eric (1965). "The crisis of the seventeenth century". In Aston Trevor (Ed.). *Crisis in Europe, 1560–1660: Essays from Past and Present*. London: Routledge, pp. 5–58.

Kindleberger, Charles P. (1991). "The economic crisis of 1619 to 1623". *The Journal of Economic History*, 51 (1), pp. 149–175.

Macedo, Jorge Borges de (1982). *Problemas de História da Indústria Portuguesa no Século XVIII*. Lisboa: Querco.

Madureira, Nuno Luís (Coord.) (2001). *História do Trabalho e das Ocupações*. Oeiras: Celta Editora, vol. 1.

Magalhães, Joaquim Romero (1988). *O Algarve Económico, 1600–1730*. Lisbon: Editorial Estampa.

Marcos Martin, Alberto (2000). *España en los siglos XVI, XVII Y XVIII. Economía y sociedad*. Barcelona: Editorial Crítica.

Mauro, Frédéric (1989). *Portugal, o Brasil e o Atlântico. 1570–1670*. Lisbon: Editorial Estampa.

Nash, R.C. (2001). "Economy". In Bergin Joseph (Ed.). *The Seventeenth Century: Europe 1598–1715*. Oxford: Oxford University Press.

Neto, Margarida Sobral (1993). "A persistência senhorial". In José Mattoso (Eds.), *História de Portugal*. Lisbon: Círculo de Leitores, pp. 165–175.

Neto, Margarida Sobral (1994). "Introdução e expansão da cultura da batata na região de Coimbra (sécs. XVII–XIX)". *Revista Portuguesa de História*, 29, pp. 55–83.

Neto, Margarida Sobral (1997). *Terra e Conflito. Região de Coimbra 1700–1834*.Viseu: Palimage Editores.

Neto, Margarida Sobral (2012). "A crise da agricultura portuguesa no século XVII". In Álvaro Garrido, Leonor Freire Costa and Luís Miguel Duarte (Eds.). *Economia, Instituições e Império*. Coimbra: Almedina, pp. 263–277.

Neto, Margarida Sobral (2013). *Tocha. Uma História com Futuro*. Coimbra: Palimage.

Oliveira, António de (1972). *A Vida Económica e Social de Coimbra de 1537 a 1640*. Coimbra: s.n.

Oliveira, António de (1991). *Poder e Oposição Política em Portugal*. Lisbon: Difel.

Oliveira, Aurélio de (1979). *A Abadia de Tibães, 1630/80-1813. Propriedade, Exploração e Produção Agrícola no Vale do Cávado durante o Antigo Regime*. Porto: s.n.

Oliveira, Aurélio de (1985). "Indústrias em Braga. As primeiras manufacturas de curtumes". *Bracara Augusta*, 39, pp. 87–88.

Oliveira, Aurélio de (1986). *Níveis de Produção Vínicola no Entre Douro e Minho, 1629–1822*. Porto: Centro de Estudos Norte de Portugal-Aquitânia.

Oliveira, João Nunes de (1990). *A Produção Agrícola de Viseu entre 1550 e 1700*. Viseu: Câmara Municipal de Viseu.

Osswald, Maria Helena (2008). *Nascer, Viver e Morrer no Porto de Seiscentos*. Porto: s.n.

Pedreira, Jorge Miguel Viana (1994). *Estrutura Industrial e Mercado Colonial. Portugal e Brasil (1780–1830)*. Lisbon: Difel.

Rau, Vírginia (1984). *Estudos sobre a História do Sal Português*. Lisbon: Editorial Presença.

Ribeiro, Orlando (1941). *Estudo do Pastoreio na Serra da Estrela*. Lisbon: Imprensa Nacional-Casa da Moeda.

Rodrigues, Teresa Ferreira (2008). *História da População Portuguesa*. Porto: Afrontamento.

Romano, Ruggiero (1992). *Conjonctures opposées. La crise du XVII siècle en Europe et en Amérique Iberique*. Genève: Librairie Droz.

Russell-Wood, A.J.R. (1992). *A World on the Move: Portuguese in Africa, Asia and America, 1415–1808*. Manchester: Carcanet.

Santos, Rui (2003). *Sociogénese do Latifundismo Moderno. Mercados, Crises e Mudança Social na Região de Évora, Séculos XVII a XIX*. Lisbon: Banco de Portugal.

Sanz Ayán, Carmen (2004). "La decadencia económica del siglo XVII". In Alfredo Floristan (Ed.). *Historia de España en La Edad Moderna*. Barcelona: Ariel, pp. 391–408.

Schwartz, Stuart B. (1985). *Sugar Plantations in the Formation of Brazilian Society: Bahia 1550–1830*. Cambridge: Cambridge University Press.

Sérgio, António (1974). *Antologia dos Economistas Portugueses (Século XVII)*. Lisbon: Sá da Costa.

Sideri, Sandro (1970). *Trade and Power. Informal Colonialism in Anglo-Portuguese Relations*. Rotterdam: Rotterdam University Press.

Silbert, Albert (1978). *Le Portugal méditerranéen à la fin de l'Ancien Régime. XVIII-début du XIX siècle. Contribution à l'histoire agraire comparée*. Lisbon: Instituto Nacional de Investigação Científica.

Silva, Francisco Ribeiro da (1986). *O Porto e o Seu Termo (1580–1640): os Homens, as Instituições e o Poder*. Porto: s.n.

Subrahmanyam, Sanjay (1993). *The Portuguese Empire in Asia, 1500–1700: a Political and Economic History*. London and New York: Longman.

Tenenti, Alberto (2012). *La Edad Moderna, XVI–XVIII*. Barcelona: Editorial Crítica.

Valladares, Rafael (2006). *A Independência de Portugal. Guerra e Restauração, 1640–1680*. Lisbon: Esfera dos Livros.

Vassberg, David (1983). *La venta de tierras baldías. El comunitarismo agrario y la corona de Castilla durante el siglo XVI*. Madrid: Servicio de Publicaciones Agrarias.

Vilar, Pierre (1974). *Or et monnaie dans l'Histoire*. Paris: Flammarion.
Vries, Jan de (1976). *The Economy of Europe in an Age of Crisis, 1600–1750*. Cambridge: Cambridge University Press.
Vries, Jan de (2009). "The economic crisis of the seventeenth century after fifty years". *Journal of Interdisciplinary History*, 40 (2), pp. 151–194.
Wallerstein, Imannuel (1980), *The Modern World_System* II. *Mercantilism and the Consolidation of the European World-Economy, 1600-1750*. New York: Academic Press.

CHAPTER 5

Extensive Growth and Market Expansion, 1703–1820

José Vicente Serrão

Introduction[1]

The signing of the Methuen Treaty between Portugal and England in 1703 may be taken as the starting point for the last stage of the history of Portuguese agriculture in the Early Modern period. The dramatic juncture of the early 19th century, which began with the French invasions (1807) and ended with the first Liberal Revolution (1820), marks the end of this period. Though our knowledge of Portuguese agriculture during this period is lacking in many respects, this has nevertheless been one of the most analysed periods. For one, we have a number of local and regional studies, largely influenced by French rural historiography, into the structures and conjunctures of prices, rents, production, land ownership and farming systems.[2] On the other hand, 18th century agriculture has also been considered in synthesis studies, mostly chapters of general histories, dealing either specifically with this sector or with its place within the Portuguese economy as a whole.[3] Most of these monographs and synthesis studies date from the 1970s to the early 2000s. More recently, a new wave of studies has been seeking to provide quantified responses to some of the questions left open by previous scholars.[4] Although further analyses are required before the estimates can be fully accepted, they already suggest interesting directions for further study and call for revisiting the qualitative sources and the available literature from a new perspective.

In international scholarship, the agricultural history of this period has been much studied too. This is largely because agriculture has been closely associated with (or even at the centre of) some of the great debates of European economic history, – on the transition from feudalism to capitalism, on

1 This study was partially supported by the FCT Research Project "Lands Over Seas: Property Rights in the Early Modern Portuguese Empire" (PTDC/HIS-HIS/113654/2009).

2 For a comprehensive review of this literature see Neto (2007).

3 See Serrão (1993 and 2005), Pedreira (1994), Meneses (2001), Monteiro (2005) and Costa, Lains and Miranda (2011).

4 See Reis, Martins and Costa (2013), Palma and Reis (2014), Costa, Palma and Reis (2015), and Reis, Chapter 6 in this book.

 | DOI 10.1163/9789004311527_007

the interactions between agriculture, demography and urbanization, on the existence or not of an agricultural revolution and its contribution to the industrial revolution and, more recently, on the role of agriculture in the "great" and "little" divergences regarding living standards and economic development. This does not however mean that we now have a clear and definite picture of European agriculture in the 18th century with which the Portuguese case might be compared. On the contrary, all these topics have undergone successive revisions and remain controversial.

Although we cannot provide here a comprehensive and detailed review of the literature in the field, it must be noted that, in spite of its many achievements, it has been affected by some biases and questionable options.[5] One of those biases is that it is too based on the assumption of the superiority of British (and Dutch) agriculture, thus falling into the common "Northwest and the rest" narrative. Consequently, the diversity of pathways for agricultural change is not explored, and instead the indicators of efficiency and growth for the first group of countries *versus* the causes of "stagnation" (understood as divergence from the core pattern) for the rest are emphasized.[6] Another shortcoming is that 18th century agriculture is frequently considered only in a subsidiary role, which means that its success or failure is assessed in a teleological perspective in accordance with whether or not it contributed to industrialization. Yet another problem is to focus excessively on productivity and output when it comes to assessing the overall performance of European agricultures, which obscures the many other faces of change and success in agriculture and rural society. But the greatest weakness of current mainstream historiography is probably methodological, arising from the excessive attention paid to measuring performance indicators. This is well intentioned and understandable, because it seeks measuring tools that might be comparable. However, as the sources available for the pre-modern period provide little primary statistical evidence, the analytical process is mostly based upon estimations which, in the words of Joyce Burnette (2014: 99), "require heroic assumptions" on the part of historians. The variables that enter into the calculations (e.g. consumption, food baskets, population, urbanization rates, per capita income, prices, wages, working hours) result from complex and creative calculations based on projections, interpolations and extrapolations, which means that we are ultimately elaborating upon successive layers of *estimated* data.

Returning to the Portuguese case, controversy is also one of the dominant characteristics of the literature about this period. There are contrasting

5 See Zanden (2002), Hillbom and Svensson (2013) and Scott (2015).

6 See Dennison and Simpson (2010).

positions, particularly concerning the question whether 18th century agriculture is characterized by backwardness or by progress. For a long time, the prevalent view was very pessimistic, and words such as *backwardness, decadence, stagnation* and *crisis* were used to describe the situation and the long-term performance of Portuguese agriculture in this period.[7] The thesis that will be sustained in this essay is the opposite. I will argue that – while it is true that there were some crises, problems and blockages, and that development was by no means the same across all regions, sub-sectors and periods – Portuguese agriculture as a whole actually performed quite well in the 18th century. Extensive growth and an increasing market dependence were the two dominant aspects of that performance. These results can be explained, firstly, by a significant change in market conditions; secondly, by the ability of economic agents to adapt to new conditions; and thirdly by some effective economic policies. The market, the farmers and the government form the three sides of a triangle that proved to be reasonably virtuous for most of the period considered.

This chapter is organized as follows. The next section offers a broad overview of the main characteristics of the agricultural sector and rural landscape. The third section then identifies the main macro-economic changes that took place in the 18th century, highlighting the new opportunities and challenges that affected the farming sector, and their positive impact. The following sections provide the counterpoint to this, discussing some of the main weaknesses and obstacles to growth, with reference to the diagnoses made by observers of the period: thus, the fourth section focuses upon grain production and the grain market, while the fifth section discusses property rights. Finally, the conclusion sums up the argument developed in the chapter.

The Rural Landscape

As in previous centuries, during the 18th century three crops stand out in Portuguese agriculture: grain, vines and olive trees. However, in this continuity several changes were underway, as indicated by the geography of their distribution and the way they were incorporated into the rural landscape. Three main varieties of grain were cultivated in this period in Portugal: wheat, rye and maize.[8] American maize, which had been imported in the 16th century, flourished in the more temperate and rainy north west, and spread so

7 See Oliveira (1979), Justino (1981) and Pedreira (1994 and 2005).

8 See Neto, Chapter 4 in this book.

successfully throughout this region that by the 17th century it had already become the dominant cereal in that part of the country, replacing the more traditional harvests of millet and rye. Continuing the expansion trend, throughout the 18th century it was becoming dominant in the wetlands of Beira Alta, Beira Litoral and Atlantic Estremadura.[9] By the turn of the 19th century, maize was cultivated across most of the country in almost all places where irrigated agriculture was possible. Rye, though a dry crop like wheat, could adapt better to poor soils or adverse climates, and predominated over a vast inland area that included the mountains and plateaus of Trás-os-Montes and Beiras, as well as in the southernmost province, the Algarve. Wheat, in turn, was mostly produced in the Tejo valley, the Atlantic Estremadura and the Alentejo, the last often considered the granary of Portugal.

As grains were the dominant and most widespread crop in Portugal, they tended to command, or at least strongly influence, the general layout of the rural landscape. Depending on the region and the various combinations of crops, this resulted in a variety of modes of organizing agricultural fields. In accordance with the classification grid proposed by Orlando Ribeiro (1991: 33–47; 2011: 59–67), which is generally valid for the period under study, there were three main types of fields: open, closed and interspersed. The last, found above all in the province of Atlantic Estremadura and in the Algarve, were so called because grains were grown in the spaces between fruit trees, olive trees or vines that constituted the farm's dominant crop. Closed fields were surrounded by some kind of barrier (dry stone walls, ditches, canes, bushes, rows of trees and other types of fence), which dissuaded people and animals from entering, thus protecting crops and also property rights. They were found in various parts of the country, but their most characteristic type was the so-called *campo-prado* (meadows) of the north west, which were the Portuguese equivalent of the *bocage* fields of the Atlantic coasts of France, Spain or Ireland. They were small irregular plots bounded by trees and bushes, sometimes organized into a system of terraces. In the summer, they were occupied by an intensive irrigation polyculture (maize, beans, green vegetables, etc.) and in the winter were left for pasture or for rye production. These closed fields should not be mistaken for the larger and more distinguished walled properties, the so-called *quintas*, which were one of the most distinctive elements of the Portuguese rural landscape, at least from the 16th century. The *quinta*, always comprising a country house with some distinction, was simultaneously a symbol of status and a farming unit, mostly oriented towards cash crops (vines, orchards, olive trees,

9 See Map A19 in the Appendix.

vegetable garden). Because of such features, *quintas* were found mostly on the outskirts of towns and in prosperous agricultural regions, such as the Douro Valley. The tendency, in the period under consideration, was for a proliferation of this type of rural estates.[10]

The open fields, completely unfenced, predominated in most of the country: from the province of Trás-os-Montes in the far north to the southern mountains of the Algarve. There were various types of open fields. In the northern inland region (provinces of Trás-os-Montes and Beira Alta), the fields were practically bare of trees, cultivated using a system of two-yearly crop rotation involving two rotations: one comprising grain (mostly rye, but also some wheat) and the other left fallow for pasture. These open fields were often associated with community practices and collective rights, complementing an individualistic farm structure of closed irrigated properties (composed of gardens, fruit trees and wetlands) around villages. Further south, in Beira Baixa and the Alentejo, treeless fields co-existed alongside fields with oaks, cork oak and olive trees, though they were open. These fields' organization was adapted to differing levels of soil fertility. Thus, in the most fertile regions (Beira Baixa and Alto Alentejo), with more favorable environmental conditions, there were clean open fields, using a cultivation system of a three-yearly crop rotation. In the Baixo Alentejo, low soil fertility required longer fallow periods, upset the three-yearly rotations, and forced the sparse arborization of grain lands. The extensive cultivation of dry wheat, often with a low yield, alongside more extensive animal husbandry, was the most common way of occupying these fields, often in a logic of competition rather than complementarity. Thus, where grain farming was comparatively less profitable, these fields were taken over by plantations of trees without grains, or were simply left for pasture.

In the hierarchy of Portuguese agriculture, wine growing came second, immediately after grain. Vineyards, usually associated with small and medium-sized family farming, were quite widespread around the country, and their cultivation was expanding considerably from the end of the 17th century. But in the second half of the 18th century, following a process of selection and concentration led by market forces and public policies, between 66 and 75 percent of the total wine production came from three regions only – the vast Lisbon hinterland, the Minho province and the Douro Valley.[11] The last was throughout the century the main area of growth and technological innovation, with

---

10 On *quintas* see Link (1803), Fauvrelle (2001) and Serrão (2002).

11 Estimates based on tax on wines (*Subsídio Literário*) for the year 1772. See Archives of the Tribunal de Contas, Lisbon, collection Erário Régio, books 3914–3917.

the development of Port wines. It was there too that one of the world's oldest demarcated wine-producing regions was officially created in 1756 (Pereira 1996, Martins 1998).

Olive trees could be found mostly in the provinces of Estremadura, Beira Baixa, Alto Alentejo, and parts of the Baixo Alentejo. Yet, during this period, continuing the trends noted in the previous century, they stretched as far north as some parts of the Minho and Trás-os-Montes, namely in the Douro Valley, thus reaching the area of maximum implantation of this crop given the country ecological conditions.

In addition to these three main crops, a few others also made their mark on the rural landscape, though they contributed less to the total agricultural output, or were more restricted in their geographical expression. These included different fruit trees, which were responsible for the production of large quantities of figs, almonds, carob, apples, pears, citrus and drupes. There were also countless vegetable plots, which played an important role in the provision of daily foodstuffs. For this reason, they were an indispensable part of the peri-urban (or even intra-urban) landscape. Thus, in addition to their role in subsistence farming, these horticultural crops also formed part of a local market-oriented agricultural specialization. Their marked presence in and around the outskirts of Lisbon is the best example. Another important element of the Portuguese countryside was the forestry species, both spontaneous and planted, namely the groves of oak and cork oak typical of the Alentejo plains, the pine forests of the Atlantic seaboard, and the oak and chestnut groves that covered extensive areas of the Minho, Trás-os-Montes and Beiras provinces.

Another major component of the primary sector of the Portuguese economy was animal husbandry, which both complemented and competed with agriculture. Cattle farming was usually associated with small family operations, predominantly in the north and north west of the country, while sheep and goat husbandry was an important activity in the mountainous areas of the centre and north. Extensive animal husbandry was however at its most expressive in the hinterland of Lisbon, Beira Baixa and the Alentejo, where vast herds of pigs and sheep existed. In the 18th century, this sector grew to such an extent that the balance between arable and pastoral agriculture was clearly tilted in favour of the latter, particularly in the Alentejo.[12]

Finally, a note should be made about the proportions of cultivated and uncultivated land. There is no credible quantitative assessment available, only some estimates. The 19th century historian, Rebelo da Silva (1860: vol. 4,

12 See Silbert (1978), Santos (2004) and Fonseca and Reis (2011).

452–453), calculated the cultivated area in 1636 as a maximum of 850,000 hectares, while Vitorino Magalhães Godinho (1955: 149) suggested 1,500,000 hectares for 1800. These figures, which correspond, respectively, to only 9 and 17 percent of the total surface of the Portuguese mainland, are rough estimates, but may perhaps indicate the order of magnitude. They are in keeping with eyewitness accounts from the period, which insistently complained about the extent of uncultivated lands. However, these testimonies are not entirely reliable. Firstly, because the term "uncultivated" was applied to many different kinds of situations: lands that were completely unproductive, lands that could be farmed but were not, lands left fallow for long periods, sometimes decades, and woodlands. Secondly, many of these so-called "uncultivated" lands were not entirely dissociated from the agricultural economy, as they were often used for pasture, firewood reserves, cork, acorns and other products, or as a support for beekeeping (one important complementary activity for agriculture).[13] Despite these reservations and uncertainties, it is clear that the early modern Portuguese landscape would have been greatly marked by extensive swathes of uncultivated lands.

The Macroeconomic Context

Having offered a static portrait of the Portuguese farming sector in the 18th century, let us now introduce a dynamic perspective so as to attempt to identify what changed during this period. These shifts were essentially triggered by changes in the macroeconomic context, so this should be our starting point. One of the events that most directly affected the farming sector was the Anglo-Portuguese treaty of 1703, named after its English negotiator (John Methuen), which consolidated and simultaneously conditioned economic relations between the two countries. The terms of this commercial treaty were quite simple. It just stated that British woollen goods would be admitted to Portugal duty-free in exchange for Portuguese wines being imported into England subject to a third less duty than wines imported from France. The interpretation of this treaty's nature and long-term consequences has remained controversial since its own time, including in the remarks of economists like Adam Smith and David Ricardo. But, for our purpose, it is undeniable that this treaty secured the British market for some Portuguese agricultural products,

13 For the uncultivated lands, see Silbert (1978: 195–204, 445–475).

particularly wine, and helped a community of British businesspersons and investors to settle in Portugal.[14]

Another factor that greatly influenced the macroeconomic scenario was gold mining in Portuguese America, which began at the end of the 17th century and continued throughout the 18th, with a total estimated yield of 850 to 1,000 tons of gold. Although much of this wealth did not remain in either Portugal or Brazil, it had a tremendous impact upon the Portuguese economy. Amongst its direct and indirect effects were the growth in the money supply, a drop in interest rates and an increase in public and private incomes, variables that positively influenced consumption and investment. With regard to income only, it has been recently argued that it would have risen so fast that in 1755 it was up to more than 170 percent that of 1700 (Palma and Reis 2016). Gold also indirectly stimulated the increase in sea traffic and foreign trade, especially during the first half of the 18th century, when Portugal became one of the most sought-after trading partners in all of Europe. Even if imports had grown much faster than exports, Portugal saw new opportunities open up for the export of products such as wine, salt, fruits, olive oil and wool.

In addition, gold mining stimulated large-scale emigration to Brazil.[15] This emigration tempered the demographic growth of the metropole, but was decisive for the creation of a market with European consumption habits in the Brazilian colony. In fact, by the end of the 18th century, the white population of Brazil alone was around one million, roughly one third of the homeland population. Moreover, it generated a flow of capital remittances from the emigrants to their families at home, thereby contributing to the increase in the levels of income, consumption and investment in Portugal.[16]

Regarding population and other demographic indicators, it should be stressed that mainland Portugal still was, as it had always been, a sparsely populated country.[17] In 1700, at the start of the period covered by this chapter, the average population density was only 25.1 inhabitants per square kilometer, although there were many regional differences, ranging from a maximum of 76.6

14 On the Methuen Treaty, see Francis (1966), Shaw (1998), Cardoso *et al.* (Eds.) (2003), Duguid (2003) and Watson (2015).

15 Other estimates for emigration for 1700 to 1760 range from 3,000 to 10,000 a year (Livi-Bacci 2002).

16 On Brazilian gold see Pinto (1979), Morineau (1985), Sousa (2006 and 2008), Costa, Rocha and Araújo (2011), Costa, Rocha and Sousa (2013), Costa, Rocha and Brito (2014) and Costa, Palma and Reis (2015).

17 For population see Serrão (1993 and 1996). See also Moreira (2008) and Palma and Reis (2016). See also Henriques, Chapter 1 in this book.

in Minho to a minimum of 10.2 in Alentejo. Inverting these values, this gives a land-population ratio of 4 hectares per inhabitant across the country, varying between a minimum of 1.3 in Minho and a maximum of 9.8 in Alentejo. It is important to highlight these numbers as they place Portugal amongst those regions of Europe, such as the rest of the Iberian Peninsula, with a high land-population or land-labor ratio, and therefore with a huge margin for growth of the cultivated area (Álvarez and Prados 2012; Santiago 2013).

Concerning its evolution, the Portuguese population, which had been growing moderately since the mid-17th century, entered a phase of stagnation or slow decline around the turn of the century.[18] This phase lasted until the beginning of the 1730s, when it started to grow vigorously until the end of the 1770s, slowing down again in the following decades. The main period of growth was therefore in the middle decades of the 18th century. Focusing the calculations on round dates, in order to simplify international comparisons, we find that there was an increase in the effective population of only 3 percent between 1700 and 1750, and of 26 percent between 1750 and 1800. From 1700 to 1800 the total population increased by 30 percent, growing from 2.2 to 2.9 million inhabitants. It is thus important to stress that there was a population growth in Portugal during the 18th century, though in comparative terms it was considerably slower than the European average, estimated at around 50–55 percent (Malanima 2010: 257). There are probably two reasons for this 18th century divergence. One is that it reflects an effect of adjustment, given that the Portuguese population had grown much faster than the European average in the 16th and 17th centuries, and thus arrived at the early 18th century with a lower margin for growth. The other is that it directly correlates with the strong emigration registered in the 18th century.

The indicators of urbanization are somewhat more expressive. In 1706 Portugal had an urban population rate of 15.3 percent (towns of more than 5,000 inhabitants), dropping slightly to 14.9 percent in 1801. These figures are quite high for the standards of the time. In Europe, the average urbanization rate was 11.4 percent in 1700 and 11.9 percent in 1800 (Bairoch *et al.* 1988: 259).[19] The number of towns increased throughout the century from 25 to 30, an important achievement in a country the size of Portugal (approximately 90,000 square kilometers). Taking the last number (30), this meant that each urban

18 See Neto, Chapter 4 in this book.

19 If the 10,000 inhabitants threshold is used, the urbanization rate is 10.3 percent and 9.8 percent on the dates mentioned above, compared to the European average of 8.2 percent and 9.0 percent (Malanima 2010: 245). See also Map A11 in the Appendix.

centre had at its disposal an average area of 2,970 square kilometers, equivalent to a circle with a radius of approximately 30 kilometers. This urban profile should naturally exert a significant impact on the rural world and on the farming sector (Epstein 2001), even if geographical distribution of the towns and urban population was uneven across the country. In fact, though no less important, roughly one third of that urban population was concentrated in Lisbon, the capital city, whose growth of 56 percent over the course of the 18th century was much greater than that of the population of the whole country, reaching almost 200,000 inhabitants by 1800.

These features and trends created a new macroeconomic framework for the agricultural sector, mostly by changing market conditions. The effects were felt on both the supply and demand sides, and affected all types of markets – labor land, capital, and agricultural commodities – the last being especially stimulated by the expansion of foreign, domestic and overseas demand. Though these changes also brought some risks and were hindered by several obstacles, the final balance of their effects was generally quite positive. They ended up generating growth and modernization dynamics and even a revaluing of the agrarian sector within the overall Portuguese economy.

Probably the most relevant consequence was that the agricultural sector and Portuguese farmers became increasingly market-oriented. We do not have a quantified notion of the balance between agricultural production for self-consumption and production for the market at the national scale, but there are some suggestive regional indicators. In the Lisbon region, in the 1770s, about two thirds of the agricultural output was marketed (Serrão 2000). In the north west, by the end of the century, a third of agricultural produce in the province of Viana do Castelo was exported through its seaport. Studies on the Alentejo and the Algarve refer abundantly to agriculture's responsiveness to the markets, both internal and external.[20] The active presence of foreign merchants close to farm production in regions as diverse as Minho, Douro, Algarve, Setúbal and Lisbon is another suggestive sign of market dependence. The same can be said about the creation of dozens of new fairs all over the country, adding to the many already existing, as well as the growing interdependence of markets and agricultural prices.[21]

Another interesting response of farmers to market stimuli was their move towards specialization. This should not be mistaken for monoculture. In most

20 See Silbert (1978), Magalhães (1988), Vidigal (1993), Borges (2000) and Fonseca and Reis (2011).

21 See Macedo (1963: 142–144), Godinho (1955) and Justino (1981).

regions, farmers kept land namely for vegetables, giving them the capacity to face market volatility. During the 18th century, Portuguese farming specialized in crops that offered comparative advantages. The most noticeable case was wine growing, which increased extensively during this period, in a process that has been described as a "vine revolution" (Meneses 2001: 242), benefiting from new market conditions, especially the reorientation of the British market for Portuguese wines. Even if the expansion process began earlier, soon enough the Methuen Treaty enshrined the position of Portuguese wines in relation to its direct French, Spanish, Italian and German competitors, and the commitment of Great Britain in maintaining close commercial ties with Portugal in order to get access to the gold of Brazil did the rest.[22] Another impulse came from the development of an important colonial market in Brazil. Although these were the two main destinations, and stimuli, for Portuguese wine production, the domestic consumption and the new markets in the Baltic Sea (Sweden, Hamburg, Russia) that began to be explored in the last quarter of the 18th century should be added. Speaking only for the external demand, wines probably accounted for around 40 percent of total Portuguese exports by the end of the century.[23]

The growth of the wine sector also benefited from some supply-side factors. The main requirements for the expansion of wine-growing – available land and intensive labor – were not very difficult to fulfil at the time. In the first case, this was because the vines adapted well to poorer soils, precisely those that were the easiest to find, and at the best price on the land market. In the second case, the population was growing and providing more labor force. This was not necessarily a paid workforce, for wine growing adapted very well to the small family-based peasant economy. For these farmers, vines were an excellent farming choice, as wine was easy to exchange for money, even when produced in small quantities. Nonetheless, waged labor was also widely employed in the wine industry, mainly during the grape harvesting period, in which seasonal and inter-regional migration of workers played an important role.

All this together meant that vineyards could, and did, spread enormously in the 18th century, especially during its first half, and across the whole country. Two regions however stand out as the most remarkable cases of regional specialization – the Douro Valley and Madeira Island. It was there that the

---

22 See Neto, Chapter 4 in this book.

23 There are no reliable estimates for wine exports during this period. See however Martins (1990 and 1998), Serrão (1993b) and Costa (2005). See also Shillington and Chapman (1907), Tenreiro (1942), Schneider (1970), Sideri (1970), Fisher (1971) and Francis (1972).

expression "vine revolution" gained its full meaning, giving rise to two new wines, Port and Madeira, the result of real technological innovations. These were no longer common table wines, but refined drinks with great added value. Their production required operations such as fortification with grape brandy, the addition of natural colouring, careful blending of different wines, ageing, and, in the case of Madeira wine, artificial ageing through heating and the shaking of barrels.[24] But the development of the wine sector in this century also showed signs of what could sometimes be an unhealthy kind of growth: overreaction to market stimuli (which ended up leading to a crisis of overproduction in the 1750s), excessive dependence on the English market, incidences involving wines that were technically defective or even falsified.

Other market-oriented crops were clearly ascending and expanding, particularly in regions most sensitive to the trends and rhythms of urban and foreign demand. Studies on local and regional agricultural history offer plenty of evidence of new plantations of fruit orchards and olive groves, as well as of the increasing number of vegetable gardens. One of the most interesting cases was the production of citrus fruits. Particularly sweet orange (*citrus sinensis*), an exotic species supposedly brought from Asia to Europe by the Portuguese circa 1630, or possibly earlier in the 16th century (Ferrão 1992), whose cultivation rose in the second half of the 18th century, becoming one of the prime agricultural export items. The main producing areas were the Algarve, Setúbal Peninsula, the region of Lisbon, and the S. Miguel Island of the Azores. It was on this island that actually the most refined specialization in orange production occurred, totally aimed at foreign markets, mostly the British one.[25] Olive oil was another important and expanding item of Portuguese agricultural exports, both to the European and the colonial market. Another case of reaction to the expanding markets was the investment in livestock, especially sheep and pigs in the Alentejo, as farmers sought to respond to the increase in the demand for meat, dairy products and wool (Santos 2003, Santos and Serrão 2013). They did this either by converting the former cereal-growing areas into pastures or by occupying marginal lands.

Contrasting with the expansionist tendencies of market-oriented crops, staple farming tended to behave in a more "defensive" way. This was particularly the case for traditional bread grain crops (wheat and rye), which faced several structural and conjunctural constraints. Their combination meant that the more traditional grain sector production developed modestly over the 18th

24 See Bennett (2005), Hancock (2009) and Serrão (2009).

25 See Miranda (1989), Dias (1995) and Rodrigues (2003: 206–208).

century, even retracting in certain regions. Farmers and consumers took refuge in other grains, particularly maize, which produced a much higher yield per seed, was easy to integrate into intensive mixed farms, and adapted particularly well to demographic growth. Maize was expanding during this period, in terms both of cultivated area and of volume of production, thus continuing its "revolution", which undoubtedly constituted one of the main achievements of Early Modern Portuguese agriculture.[26] Rice production was also a possibility, but had yielded poor results and generated little interest amongst farmers. Conversely, there was less resistance to the potato, which already achieved significant output in the first decades of the 19th century (Oliveira 2002: 116–120). Any deficit in staple crops that was not satisfied by these routes was made up for by imports.

This diversity and, to some extent, divergence in the performance of different crops led to a reconfiguration of agricultural output and of the Portuguese rural landscape over the course of the 18th century. While the production of wheat and rye remained stationary or even regressed, at least in comparative terms, there was a growth in the area of lands allocated to maize, vines, olive and fruit trees, vegetable gardens, pasture and animal husbandry, as well as experiments in new crops such as rice and potatoes. These changes in the hierarchy of priorities of farmers and Portuguese agriculture, even though we cannot fully quantify the degree of their extension nor how they translated in terms of the agricultural output composition, stand out as one of the most extraordinary features of the period under analysis.

In order to increase the output of these expanding crops, farmers basically followed three strategies, depending on the availability of production factors, the configuration of property rights and the natural resources of each region. In some cases, farming became more intensive in lands that were already farmed. In others, they simply changed the land use, for example by converting areas that had previously been used for grain into vineyards and pastures. In most cases, however, there was an expansion of the cultivated areas into marginal lands, which were usually of poorer quality, though cheaper and easier to find on the land market. There is plentiful historical evidence of these processes, namely the building of more vine terraces on the slopes of the Douro Valley and on Madeira Island; the draining of swamps and marshes (for instance in the Tejo valley); the privatization of common lands, previously uncultivated and used as wood and pasture reserves; the clearing of marginal lands by

26 The term "maize revolution" was coined by Ribeiro (1991: 193–206). For a broader European perspective see, *inter alia,* Crosby (1994: 148–166) and Braudel (2000: 158–167).

slash-and-burn processes, as recorded for some parts of the Alentejo and the Algarve. Despite the significant regional differences, Portugal as a whole ultimately had a potentially elastic supply of land, though not necessarily good land, particularly when taking into account the population density figures. The expansion of the cultivated area thus constituted one of the main tendencies of 18th century Portuguese agriculture. We do not have a global quantitative assessment of this process, but it is worth remembering the estimates indicated above in this chapter, which suggested an increase of almost 100 percent in the cultivated area between 1636 and 1800. In the same period, the population increased by only 50 percent, which reinforces the significance of this phenomenon.

Another positive development was the increasing investment in land and farming. The traditional social prestige of land ownership, now combined with the economic valorization of various farming activities, attracted part of the wealth accumulated during the 18th century by some social groups. Among these were, for instance, businesspersons, urban and rural elites, former emigrants, and even many monasteries, with some foreign investors also playing their role. These were responsible for the countless small and large investments that made possible the clearance of land and the liveliness of the land market, the plantation of vineyards, olive and fruit groves, the building of *quintas*, country houses, irrigation systems, wine and olive oil presses, and many other improvements in rural estates.

Some technological developments occurring during this period also need to be mentioned. Rather than radical or widespread technological changes, most of them arose essentially within the parameters of traditional farming, as indeed happened across Europe, with the exception of England and the Netherlands. However, the small technological revolution in wine making that gave rise to the "invention" of Madeira and Port wines cannot be overlooked. Likewise, the so-called "maize revolution" induced a certain technological shock, particularly by promoting new crop rotations and combinations, as well as some advances in the tools and techniques of irrigated farming. The relatively fast and easy incorporation of other new crops of extra-European origin (e.g. oranges and potatoes) into Portuguese farming also shows its receptiveness to innovation, even if it does not classify as a technological change in itself. Another dimension of these changes was the technical and scientific progress brought in by the creation of botanical gardens in Lisbon and Coimbra (1768 and 1772), by the establishment of a chair of agriculture at the University of Coimbra (1791), and especially by the agriculturalist research and literature produced at the Lisbon Academy of Sciences since 1779, even if all this was limited to the enlightened elites' milieu and had few repercussions on the ground.

In a more general vein, the agricultural sector also seems to have improved its relative position in the Portuguese economic structure throughout the 18th century. Against the background of what could be called the unbalanced economic expansionism of the first half of the 18th century, only the agrarian sector, at least some of its subsectors, was able to keep a growing trend and to contribute positively to Portuguese exports. Later on, after the middle of the century, with declining gold flows, a severe scarcity of means of payment, and overall economic recession, agriculture was one of the first and main beneficiaries of public policies encouraging production, and it became more valued. Towards the end of the century, in a general context of sustained growth, though also of great international volatility, it continued to perform well. The agrarian sector accounted then for a significant share of Portuguese exports and started a convincing relationship with industry on the domestic front, particularly in the wool and silk subsectors, following the impulse given to the production of wool and the plantation of thousands of mulberry trees. In the final decade of the period covered by this chapter, marked by exceptional circumstances that provoked a global rupture in the Portuguese economic model, agriculture did no more than keep up with the general trend. Nonetheless, it should be stressed that political leaders made it into one of the priorities of the new economic model to be followed.[27]

Finally, we should ask how Portuguese agriculture fared in comparative terms vis-à-vis other European countries. Unfortunately, there is no simple answer to this question, because there is no consensus regarding the most appropriate performance indicators, and because all comparative figures obviously depend on the type of approach used. If adopting a quantitative one, it is now possible, thanks to the data recently released by Jaime Reis, to compare the Portuguese case with the estimates of gross agricultural product of a group of nine countries offered by Allen (2000), although the data are not strictly comparable due to the different calculation methods used. In accordance with Reis' estimates, the Portuguese agricultural output increased by around 26 percent between 1700 and 1800, a value that may be considered modest compared with the 42 percent that represents the unweighted average of those nine countries (Germany, Belgium, Poland, Austria, England, France, Holland, Spain and Italy).[28] However, what is surprising is the somewhat atypical performance of Portuguese agriculture, presenting a sharp contrast between the first and the second halves of that century. Having grown above the average

27 See Cardoso (2009).

28 See Reis, Chapter 6 in this book.

(42 as opposed to 20 percent) during the first half, and registering the second best score immediately behind Belgium, Portugal radically inverted its position during the following period (1750–1800), presenting the worst performance of the group – a regression of 11 percent against a growth of 20 percent of the remaining countries. Such idiosyncrasy of the Portuguese case thus suggests the need for some caution and further investigation.

It is also legitimate to wonder why European economic historiography in the last two decades or so has chosen agricultural productivity and output as the fundamental indicators of success, developing a narrative according to which the development of national agricultures should be measured solely or primarily by their performance in these indicators.[29] As already mentioned in the introduction to this chapter, it is unclear why productivity should be a better indicator of agricultural progress than, for example, the introduction of new crops, the capacity to adapt to the market or to feed the population, among other achievements described in the previous pages. Hence, by adopting a broader and less strictly econometric perspective, 18th century Portuguese agriculture compares very favorably with other countries.

However, this "optimistic" view of 18th century Portuguese agricultural performance does have its limits too. Indeed, agriculture did not grow constantly during the whole century, nor did all its sectors grow or, if they did, they did not necessarily grow in the same way; not all regions participated in the same way in this growth process, and not all this growth was favorable. Thus, in underlining the dynamism of the agrarian sector and the generally upward trend, I am not meaning to suggest that the picture was all rosy, or that this compensated for its inadequacies, deficiencies and inefficiency. Besides, the contemporary discourse actually highlighted the difficulties and obstacles to growth, presenting a view of the sector that was generally rather negative. The question, then, is how to evaluate the real dimension of such drawbacks and the extent to which they could outweigh the evidences and factors of growth described above. This is what we shall see in the following sections.

Grain Issues

One of the main structuring features of the criticizing and negativist discourse about agriculture produced by the contemporary *opinion-makers* concerned

29 See *inter alia* Campbell and Overton (1991), Federico and Malanima (2004), Dennison and Simpson (2010) and Olsson and Svensson (2011).

what was then called the "bread shortage". These words operated as a kind of shorthand for a cluster of problems that were triggered in variable causal combinations: the deficit of grain production relative to domestic consumption; the huge amount of uncultivated land; the misuse of soils; and the negative impact of grain imports.[30] These concerns, which were ultimately sustained by the mythic goal of national food self-sufficiency, were not particularly new. Indeed they had been around since the Middle Ages. Nevertheless, the fact that they continued to attract the attention of intellectuals and politicians in the 18th and early 19th centuries makes us wonder if it was simply discursive inertia, if it was a poorly grounded diagnosis or if the problem was truly worsening during this period. In this section, we shall try to understand whether this discourse corresponded to reality or not. First, we shall see how grain production indicators performed. Then we will look at the problem of grain shortage and at some characteristics of the grain market.

Trends in Grain Production

Although there is a great deal of information about grain production, provided by a number of studies focused on the rural history of many local and regional settings, by no means do they cover the entire country, and the diversity of chronologies, methodologies and sources used makes it impossible to create homogeneous and continuous series. Yet, we have tried to systematize the available information in order to offer the best possible review of the trends in grain production during the 18th century.[31] The first conclusion is that in most cases where a comparison is possible, grain output was higher in the 18th than in the 17th century. The second conclusion, no less important, is that there was no single dominant trend for the whole country or for the whole period. There were periods of output expansion, of decline and of true crisis. Broadly speaking, the available data show that the greatest impulse for expansion took place between the 1720s and the 1740s, extending in some cases into the 1750s. From then on, until the end of the 1780s, there was greater instability in the short term, and some divergence in long-term movements – there are cases of a

30 See the most representative pieces of this sort of discourse in the transactions of the Royal Academy of Sciences of Lisbon, published in *Memorias Económicas* (1990–1991) and *Memórias Económicas Inéditas* (1987). Their interpretation in Pedreira (1988), Serrão (1988), Cardoso (1989), Santos (1993) and Almodovar and Cardoso (1998).

31 The main regional studies used in this exercise are the following (by region). *Minho*: Oliveira (1979, 1982, 1984, 1985), Maia (1991), Silva (1994), Amorim (1997) and Mota (2000), *Beira*: Fonseca (1981), Oliveira (1990 and 2002) and Amorim (1996), *Alentejo*: Silbert (1978), Justino (1981 and 1986), Santos (1995) and Fonseca and Santos (2001), *Algarve*: Magalhães (1985 and 1988) and Vidigal (1993). See for more details Serrão (2005: 155–160).

slight upward trend, others where there is maintenance of levels of production already achieved, and yet others that seem to be marked by a declining trend. Meanwhile, most of the country went through a general drop in the 1790s. In the first years of the 19th century, there were some new signs of production recovery everywhere, though the Peninsular War (1807–1814) once again worsened the situation. This is the general picture, marked by a sequence of ups and downs, from which it is not possible to safely extract a clear upward or downward secular trend.

Regional specificities are deliberately omitted from this short description, which seeks to represent the national picture as clearly as possible. We might merely note that the south of the country seems to have undergone a more problematic cyclical behaviour than the north, especially in the second half of the 18th century. It is also worth mentioning that in their lines and general trends, these fluctuations in grain output show a reasonable resemblance to those that are known about the evolution of the population. However, judging by some crosschecking data, done for specific regions and periods, it is possible that *per capita* production underwent some decrease in the long term, especially in the southern regions (Fonseca and Reis 2011; Santos 2012). This topic warrants special attention in future studies to assess the extent to which productivity and output indicators have followed divergent paths between northern and southern Portugal, which is a plausible hypothesis, given the country's geography of grain production explained in the second section of this chapter.

In addition to these main trends, most of the studies available also acknowledge the existence of a few short-term crises. These usually had their origin in adverse climatic conditions, sometimes worsened by political or military events. The main critical periods were: the years 1708–12, the second half of the 1730s (particularly the years 1738–40), part of the 60s, the turn of the 80s to the 90s, and, in the early 19th century, the French Invasions years (1807 to 1811).[32] Of all these crises, the one in the 1760s is considered to be the most important, an assessment which, to some extent, is justified. Furthermore, it is framed by a wave of disturbances that affected several areas of Europe, and which was expressed in many cases by severe food and grain riots.[33] In Portugal, the main

32 A complete portrait of the climatic indicators for this period is not yet available, but new data has been released in recent years. See Taborda *et al.* (2004), Camuffo *et al.* (2010 and 2013) and Alcoforado *et al.* (2012). See also Figures A1 and A2 in the Appendix.

33 The following cases stand out: Spain: Rodriguez (1973) and Stein and Stein (2003: 81–115), *England:* Thompson (1971) and Bohstedt (2010: 103–164), Czech Lands: Pfister and Brázdil (2006). For a broader perspective see Rudé (1972: 192–206).

evidence of this crisis consisted of poor harvests, price volatility and difficulties with grain imports and payment for them.

This crisis in the grain sector gained a broader scope because it was intertwined with a variety of problems affecting other agricultural and economic sectors. This was the case of the wine-producing sector, affected by a kind of growth crisis: in the 1750s it began to show signs of overproduction, price downfall, contracting exports, overlapping and saturation of the domestic markets. Portugal's involvement in the final stage of the Seven Years War (1756–1763) also had direct implications on the farming sector, due to the conscription of peasants and the compulsory supply of armies. On a broader level, the intense population growth of the previous decades was probably destabilizing the balance between supply and demand, both on the level of consumption of agricultural products and on the use of production factors (land and labor). The income of farmers and other social groups living off the land would have reacted negatively to the crisis, worsened even further by the increase in fiscal pressure arising from the restoration of an income tax, called *décima*, in 1762. The crisis of foreign and overseas trade may also have reinforced investment in land by the merchant elites as an alternative, confronting them with some of the structural barriers to this investment. The scarcity of means of payment abroad, arising from the reduction in the flow of gold and a slowing down of exports, compromised grain imports, which laid bare the shortcomings of national production and of the functioning of the domestic market, as well as the inability of supply systems to withstand crisis situations. Finally, we cannot overlook the effects of the great earthquake of 1755, which destroyed Lisbon and a large number of houses and other buildings, in both urban and rural areas. As a result, a significant part of the domestic income, received by the crown, religious institutions and thousands of families, had to be channeled into reconstruction, making the 1760s appear even bleaker.[34] There were just too many problems in a short period, which explains the severity of this crisis.

This may also help explain the reversal of the agriculture growth curve that Jaime Reis detected in his quantitative study for around this period.[35] However, this crisis does not seem to have had a long-lasting impact, nor does it seem to have constituted a turning point for Portuguese agriculture as a whole, marking the start of a long depression that would extend into the beginning of the 19th century, as suggested *inter alia* by Oliveira (1980) and Justino (1981). In the following years, there were several signs of growth in the agricultural

34 On the economic impact of the earthquake see Cardoso (2007), Serrão (2007) and Pereira (2009).

35 See Reis, Chapter 6 in this book.

sector. In many places, grain production indicators soon began to recover, resuming or even exceeding the indices achieved before the crisis; the wine sector was reorganized; many exports resumed their normal course, benefiting from new, more favorable conjunctures. There is also historical evidence of more land clearing, the planting of orchards, olive groves and vegetable plots, and investment in the construction of terraces and irrigation systems. These improvements, which were carried out in the 1760s and throughout the second half of the century, definitely do not fit in a scenario of general and continued agricultural recession. Therefore, this subject is open to debate and to the results of future research.

As important as the "real" effects of this crisis were, the fact remains that it laid bare some of the weaknesses and blockages of the agrarian sector and placed it on the public and political agenda. This was also stimulated by the agriculturalist vogue (Physiocrats and others) that was then expanding in European economic thought. It was around these years that the criticism and the pessimistic view of contemporary authors began to consolidate. It was also by this time that some important decisions about agrarian policy were taken by Pombal's government. Among these are the legislation concerning the compulsory replacement of vineyards by cereal-growing, the regulation of the wine market and its regional allocation, the removal of institutional barriers to the domestic circulation of grain and other agricultural products, along with other rules more directed towards property rights or to the improvement of transport and storage infrastructures (Serrão 1987). The full and detailed impact of these policies is still unknown, and though they may have produced some contradictory effects, they also corrected some imbalances, and at least started a precedent of fighting some of the structural hindrances to agricultural growth.

Domestic and Foreign Supply of Grain

At the time, the consideration that Portuguese agriculture was unable to feed part of the population was interpreted as a proof of its backwardness or decline. According to this view, domestic production was not sufficient to satisfy consumption, which meant that large quantities of grain had to be imported, creating a situation of national dependence on foreign imports. However, the few statistical data available, which date back to the end of the 18th and beginning of the 19th centuries, provide a different and more realistic perspective. Used with due caution and submitted to some refinement and reconstitution procedures, they can shed some light upon the Portuguese grain market and its degree of dependence upon foreign supply.[36]

36 See, for sources, methodology and further results, Serrão (2005: 164–168).

In short, our results show that Portugal was indeed an importer of foreign grains (wheat, barley, maize, rye), and that these imports were annually worth around 532,000 hl in 1776–1777, rising to 684,000 hl in the period 1778–1787, to 686,000 hl in 1788–1795, and to 841,000 hl in 1796–1810.[37] Those grain imports came from around 20 suppliers, among which, depending on the year, England, Holland, Naples, Genoa, France, Russia and Prussia stood out. If we group these into broad regions, we see that most imports came from northern Europe (60 percent), while southern Europe and the Mediterranean supplied 37.2 percent. There was also a small but noteworthy amount of imports (2 percent) from the newly created United States of America. Another important indication is that most imports entered the country through Lisbon. In 1776–1777, Lisbon absorbed 88.5 percent of the total wheat imports, in marked contrast to the seaports of Porto (3.9 percent), Viana do Castelo (1 percent) and Algarve (0.9 percent) or the border with Spain (5.7 percent). These figures refer only to wheat, though this was clearly the main grain imported. Data is not available for all periods, but in 1778–1795 wheat accounted for 71.1 percent of foreign grain imported and sold in the public granary of Lisbon, followed by barley (17.8 percent), maize (9.4 percent) and rye (1.7 percent). Moreover, sources concerning the period 1778–1787 allow comparing the sales of imported grains with those of domestic origin in the public cellar of the capital. The annual average was 238,878 hl for the former and 601,022 hl for the latter, which means that in those years, domestic supply provided only 28 percent of the Lisbon market's demand for grain, while the remaining 72 percent depended on foreign supply.

There were pronounced oscillations in the annual proportions (varying between 23 and 93 percent for foreign grain), but the series are very short (only 10 years) and the only term of comparison is a 1729 estimate of the entry of domestic and foreign wheat into the Lisbon granary.[38] These correspond to 135,610 hl and 167,894 hl respectively, which would mean a 45 percent share of the market for the former and 55 percent for the latter (compared with the figures of 28 against 72 percent mentioned above). Notwithstanding, these numbers clearly show that the grain market in the capital was largely dependent on foreign supply, and that that dependency probably worsened throughout the 18th century, though there is not enough data available to reach an

37 The 1796–1810 figures also include flour imports, which slightly distorts the comparison with previous periods.

38 Source published by Oliveira (1885: vol. 12, 255).

unequivocal conclusion. Moreover, these numbers alone do not allow us to conclude whether this was a situation generalized to the entire country or only a problem affecting the Lisbon grain market, made less accessible or less attractive for domestic supply, due to an ineffective system of internal commercialization and transportation.

Trying to get beyond this data, referring only to the consumption orientation of the Lisbon market, let us reconstruct grain consumption throughout the country (Table 5.1). Assuming an annual *per capita* consumption of 350 litres (including human and animal needs and expenditure in seeds), and comparing this indicator with data about the Portuguese population, the total expenditure in the country grew from 9.6 to 9.9 million hectolitres per year throughout the period at stake (column 2, Table 5.1).[39] Given these numbers, grain imports, which can be taken as representing the deficit between domestic consumption and production, did not constitute more than 5.5 percent of the aggregated national expenditure during the years of 1776 and 1777. There was a small jump to 7 percent in the following decade, though the figure again fell slightly to 6.9 percent in 1788–95. Unfortunately, the available data about imports in the following years are not directly comparable. However, an author of that time, though based on different calculations, also estimated a 7 percent deficit for the period 1796–1819 (Balbi 1822: 147).

Columns 5 and 6 (Table 5.1) offer another perspective, no less interesting, on the relative significance of imports. Column 6 shows that imports were necessary to satisfy consumption of only 20 to 25 days per year, which domestic production could not ensure. Alternatively, column 5 shows that imports were needed to feed a population of 152 to 196 thousand inhabitants during one year, the equivalent of roughly the population of Lisbon. Even recalculating the data from a *per capita* consumption of only 2.75 hl (the lowest value, and the least plausible, of those supplied by the sources and authors surveyed), the maximum value of the deficit would rise to only 8.9 percent (in 1778–87), equivalent to 32.5 days' consumption or to that of a population of less than 248,000 consumers.

39 *Per capita* consumption established from estimates proposed by various authors is usually between 275 and 350 litres. See Neves (1814: 234), Franzini (1843: 294), Silva (1860–1871: vol. 4, 453), *Boletim do Ministério das Obras Públicas, Comércio e Indústria* (1863: vol. 5, 32), Godinho (1955: 148–149), Silbert (1977: 39), Oliveira (1980: 19–20), Magalhães (1985: 7), Justino (1988: vol. 1, 149–152) and Lains (1995: 171–174, 195). Population figures were obtained by projections calculated on the basis of Serrão (1993a).

TABLE 5.1 *Grain imports and consumption in Portugal (yearly average values)*

	Population (inhabit.)	Consumption (hl)	Imports (hl)	Imports as share of consumption		
				(%)	(inhabit.)	(days)
1776–1777	2743513	9602296	532263	5.5	152075	20.2
1778–1787	2779383	9727841	683946	7.0	195413	25.7
1788–1795	2833187	9916155	686416	6.9	196119	25.3

SOURCE: SEE TEXT NOTES 36 AND 39.

In short, these figures show that there undoubtedly was grain dependence on foreign imports. Yet, overall, on the national scale, this dependence was not dramatic, and the much-blamed deficit was ultimately less than 10 percent. Besides, this was commonplace in most European countries, which often underwent situations of annual grain deficit, plugged by the international grain trade (Persson 1999; Roy 2013). Regarding the Portuguese case, only in what concerns the supply of Lisbon was there a clear dependence on imported grain, a problem that possibly worsened during the 18th century as its population grew. This issue was somewhat overstated by the intellectual and political elites of the second half of the 18th century. More familiar with the situation of Lisbon, the one they knew best, and more affected by political and moral preconceptions (defence of the "public interest", the "common good" and sovereignty), these elites ultimately took the part for the whole picture. Exaggerating the dimension of the country's grain dependence, they took it as a sign of a deep crisis in cereal production, or even as an indication that Portuguese agriculture was experiencing a deep crisis at the end of the 18th century. It was this accumulation of misconceptions that largely gave rise to the "pessimistic" view of 18th century agriculture that even Portuguese historiography has found so difficult to leave behind.[40]

Grain production should not, however, be mistaken for the agricultural economy as a whole, nor can it be reduced to the simple cultivation of wheat, which indeed may not have developed enough, or at least may not have accompanied the growth in population and demand, as concluded by some regional studies, and as can be deduced from the simple fact that imports subsisted

40 See, for example, Justino (1988: vol. 2) and Pedreira (1994 and 2005).

during the 18th and next the centuries.[41] This should not make us overlook the fact that there was actually some growth in grain production during the 18th century due to the expansion of maize cultivation. These are of course diverging trends, but looking at the entire country, they may not have been incompatible, and may have occurred simultaneously, though affecting different regions. The hypothesis that I would like to raise here is that grain imports existed and may even have increased. However, they were needed basically to satisfy Lisbon and the southern areas of the country, whose consumers were less disposed to change their bread diets, and whose producers were more interested in developing other crops and other more profitable activities like wine, fruit, vegetables or livestock, rather than cultivating wheat. Furthermore, Portugal had an economy open enough to be easily supplied from the outside.

Property Rights

Other problems frequently mentioned by the authors of the time were related to land ownership, its social distribution and the mechanisms for the appropriation of land and agrarian revenue. This is what we might generally call the property rights issue. In this respect, Portugal, like most of the rest of Europe, was dominated by the so-called "imperfect property", framed within a complex legal system, which translated into two main features: a plurality of overlapping rights over the same asset, and a widespread dissociation between the nominal or formal ownership of land and its actual possession.

The main criticisms addressed at the system referred to the very complexity and lack of transparency of the legal framework in force, the excessive taxation and rights to which farmers were subject, the insecurity of agricultural contracts, the various forms of property immobilization (entails, ecclesiastic mortmain, and commons) and the concentration of property in the hands of large seigniorial entities.[42] These criticisms, similarly to those regarding the grain and food shortage discussed in the previous section, were only partially correct, and, more importantly, most of the criticized aspects barely affected the actual functioning of the agricultural sector. More than anything else, they were part of the reformist rhetoric of the enlightened and liberal thinkers, in their ideological and political attack upon the *ancien régime*, and did little

41 See Reis, Chapter 6 and Branco and Silva, Chapter 7 in this book.

42 See note 30 above.

more than reproduce many of the topics and criticisms of agrarianism that were then in vogue in Europe.

Admittedly, formal ownership of land had been, since the territorial formation of the country during the 12th and 13th centuries, highly concentrated on a small number of large seigniorial entities: the crown, separate households of the Royal Family, Military orders, big church institutions and the main houses of the Portuguese aristocracy.[43] However, over the centuries, there were significant transfers between these large landowning entities, for example through the donation of crown assets to private landlords, the distribution of the orders' properties through the institution of commanderies, and the transfer of properties to the church. Also, and more often, those large estates were systematically divided and subdivided into smaller ones; this was indeed the only way of bringing the lands into cultivation, and ultimately of valuing them.

By the 18th century, as a result of these centuries-long processes, the much-criticized concentration of property was therefore more "theoretical" than factual. In practice, those large seigniorial landlords did not really own the land itself and had little control over it. They held only the formal deeds to the property (a kind of primary or supreme ownership) as well as the right to collect a share of the rent, characteristics that were reproduced downwards at as many levels as those arising from the successive duplication of property rights. In fact, they owned rights and rents rather than tangible property. The problem was therefore not the concentration of property, but the multiplication of rights in it. Of a different nature were the many duties and seigniorial rights stipulated in the municipal charters (*forais*) to which all residents and landholders of each county were subject, and which applied even to lands held under full ownership.[44] All this was supplemented for everyone by the payment of ecclesiastical tithes (Alves 2012). The reformist discourse was very critical, particularly of seigniorial rights, considering them excessive and anachronistic, but the situations were actually very variable – there were regions where these rights used up a considerable slice of agricultural revenue, while in others it was much smaller (Monteiro 1989).

Meanwhile, seemingly in contrast with the above mentioned processes of dismemberment and multiplication of property rights concerning large holdings, the system had also developed various forms of land immobilization by means of legal restrictions on its transaction in the purchase and sale market. The institution of entails, legal mechanisms that prevented the sale and

43 See Henriques, Chapter 1 in this book.

44 See also Rodrigues, Chapter 2 and Miranda, Chapter 3 in this book.

division of family property, was widespread, even amongst families without special attributes of nobility. The ecclesiastical corporations, despite legal restrictions, continued to accumulate property, which was subject through canon law to the principle of inalienable and perpetual ownership (mortmain). To this were added the large portion of lands known as *baldios* and *maninhos* (the Portuguese words for commons), which were used collectively by the populations and administered by the municipal councils. All these forms of land immobilization necessarily conditioned the investment and the functioning of the land market. However, it is now accepted that they were far from freezing it completely. For instance, the commons, despite frequent resistance from populations, could be allocated to private individuals through emphyteutic contracts, when they were not simply occupied *de facto* without any title. Church institutions resorted copiously to emphyteusis, thus fostering a lively market of property (useful) rights. Also the administrators of entailed properties, upon authorization from the crown, could ultimately alienate part of them by making use of either emphyteusis or the legal institute of subrogation (exchange for other properties or debt securities, for example) (Monteiro 2005).

Thus, emphyteusis arises as the most important agrarian contract within a system that relied for the most part on the contractualization of property and use rights. Emphyteusis, originally a contract for land exploitation, had become over time primarily a way, probably the most widespread one, of transferring property rights, whatever the final use.[45] It was a long-term contract, a kind of hereditary tenure that could extend for two or more generations, by which the full ownership domain was split in two: the eminent and the useful domain. The emphyteut was obliged to pay an annual quit-rent to the landlord, but he would acquire secure tenure for lives and some quasi-ownership rights, including the right to sell or transfer his property, and the ability to take economic decisions. Emphyteusis was thus a legal mechanism that duplicated ownership rights and that actually enlarged, rather than restricted, the land market. Nevertheless, it of course contributed to the proliferation of rights over the same property. Emphyteusis was also of a very different nature than lease tenancy contracts, which were directly aimed at the agricultural exploitation of the land, and were short-term contracts (up to 9 years) that did not involve any transfer of ownership rights, only the (conditioned) rights of use.

These features thus show that the apparent rigidity of ownership regimes was far from absolute, and their "imperfection" did not prevent the system from functioning. The legal framework contained elements of flexibility that

45 For a description of emphyteusis in its early stages see Henriques, Chapter 1 in this book.

enabled its appropriation by social and economic agents in less adverse conditions than might be assumed at first sight (Santos and Serrão 2013). Proof of this is that the land was not immobilized. On the contrary, it was leased, contracted through emphyteusis, bought and sold within and in spite of that framework (Serrão and Santos 2013). Looking, for instance, at the real estate market in the hinterland of Lisbon, even if it cannot be taken as an example for the entire country, we could estimate for the period 1760–1810 an annual turnover as high as 3.6 percent of the total of existing properties, which is quite a high indicator compared to other European cases (Bavel and Hoppenbrouwers 2004: 29).

Another indication that the system had the necessary flexibility was that it still offered various possibilities for responding to the demands of agricultural growth. A good example is what happened in the Alentejo, where an enormous process of social and economic change was under way throughout the 18th century. The result was the consolidation of both the *latifundium* (large agricultural units of hundreds of hectares) and of a wealthy class of capitalist farmers (Santos 2004). On this subject, it is interesting to note the similarities with what *mutatis mutandis* was occurring in England at the same time, generally understood as one of the components of its agricultural revolution. In both cases, there was a similar social pyramid: large absentee landlords, capitalist tenant farmers, and poor landless workers. The difference was that in the Alentejo the leading agents of land concentration were the farmers (either leaseholders or emphyteuts of their lands), while in England the large landowners were the ones who first concentrated the lands and then rented them out to capitalist farmers. Another difference is that in England this occurred alongside the enclosures, whereas in the Alentejo there was no fencing-off of land. Regardless of comparisons, what needs to be stressed is that the rural economy and society of the Alentejo were being renewed within the existing property rights framework.

A different course was followed in the Lisbon region. Here there was also a large class of agricultural entrepreneurs, developing a form of agriculture that was definitely oriented towards the urban and/or foreign market. What is particularly interesting in this case is that amongst these "agricultural entrepreneurs" there were not only farmers proper, but also many noblemen and other members of the government, church and business elites directly involved in farming (Serrão 2000). In northern areas like Minho, in turn, agricultural development occurred mostly within a context of small landownership and small farming, the usually called "family farming" (Brandão 1988; Durães 2009). A different model existed in the Douro region, whereby the agents of growth were a mix of small local nobility (owners of *quintas*) and small peasant farmers, all of them, after 1756, integrated in or subordinated to a powerful joint-stock company under state initiative (Pereira 1986 and 1996; Bennett 1992). On Madeira Island, another case of specialization in the wine sector, land ownership was concentrated in the hands of a small group of large absentee landlords, who were adverse to risk, investment and direct farming. There, the solution consisted of transferring a small cluster of property rights to a mass of poor workers, without land or capital, through the adoption of a new type of long-term tenancy contract, the *colonia,* a particular form of sharecropping (Câmara 2006).

These different examples thus show that the existing system of property rights was not only compatible to some extent with agricultural progress and social change, but also allowed for various kinds of response.[46] Nevertheless, that system also had some negative effects on the agricultural economy that should not be overlooked. The first, and most obvious, was that the large number of ownership and contractual rights meant that land revenue was distributed among multiple beneficiaries, removing capital from agricultural activity and those that were directly involved in it. Secondly, the large amount of land that was entailed, immobilized in mortmain by ecclesiastical corporations or subject to collective rights necessarily constrained to some extent the investment in land and the full functioning of the land market. Thirdly, those restrictions caused reactions to market stimuli to be slower and more expensive, as happened to many farmers that wished to change their crops but were confronted with restrictive clauses imposed by their contracts, by the municipal charters or by the rentiers of seigniorial rents. Moreover, the shift to crops benefiting from the best market opportunities during this period required not only heavy initial investment in land clearing and plantation, but also various additional investments, such as the construction of terraces, wine cellars, irrigation systems, fences, etc. All these required property rights that were relatively stable and could guarantee the tenure of land for a period long enough to allow for the amortization of the investment, which was not the case with short-term contracts.

During the period under analysis, the phases of greater vitality of the farming sector and those that operated in the opposite direction both generated (for different reasons) increased pressure on the land. In those circumstances, the apparent rigidity of the institutional framework that regulated property rights

46 This was the case in Portugal and in many other European regions, as has been highlighted by the literature which is critical of the notions of "immobility" or "imperfection" of the pre-modern property systems. See, for example, Cerman (2008), Béaur and Chevet (2013) and Jones (2013).

may have hindered agricultural development and investment in the land. At least that is how it was perceived. Partly because of this perception, and partly for reasons of a purely political nature, the enlightened government of the Marquis of Pombal implemented reforms in almost all of the aspects mentioned above. For example, new legislation was passed to regulate emphyteusis and other agrarian contracts, and thousands of entails (the least valuable) were abolished. The Church's landed property and income were especially targeted, preventing the incorporation of new assets and even releasing for sale some of their properties. The reparceling of unprofitable smallholdings was made compulsory, favoring the consolidation of larger estates. In the Alentejo the tendency was in the opposite direction, with efforts to protect tenant farmers as they acquired economic possession of their leased lands. Some of the heaviest rights payable to big landlords and seigniorial entities were reviewed in favour of peasants, especially in the Algarve and Madeira. Other legal initiatives targeted the occupation of the commons, the inheritance system, or the credit and mortgages on landed assets.[47] Subsequent governments also promulgated similar reformist measures, intended to change or improve the institutional framework. The full reach of these land policies is yet to be assessed by historians. They did not really foster a revolution in property rights (that would occur only after the 19th century liberal revolutions) and, in some cases, their effects were actually perverse or the opposite of what had been intended (Santos and Serrão 2013). However, in general, they had, or at least intended to have, a positive impact on Portuguese agricultural development.

Conclusion

If asked to describe in a one-sentence summary the 18th century performance of Portuguese agriculture, we might simply say that it succeeded fairly well. This conclusion intentionally emphasizes the more positive aspects of that performance in the long term, which were various and should now be recapitulated. In brief, there was a diversification and reconfiguration of the rural landscape and the agricultural output, a reallocation of land, labor and capital to the emerging crops, an expansion of the cultivated area, a moderate growth in the gross agricultural output, some innovation in farming methods, an increasing market dependence, a good responsiveness to market stimuli,

47 See also Serrão (1987) for an account of Pombal's land policies.

and more investment in land and farming. Furthermore, there was a positive revaluation of the agricultural sector in the general framework of the Portuguese economy, while internationally Portugal consolidated its position in the European market as an exporter of certain agrarian products (wine, olive oil, fruit, and wool).

We could therefore talk of change, modernization, progress and growth regarding 18th century Portuguese agriculture, although all this occurred essentially within traditional boundaries. That is to say, there were no radical breaks in any aspect – for instance, no technological shocks or any revolution in land tenure regimes, or in the labor market. Instead, there was an accumulation of small improvements, which, in the long term, added up to a moderately successful trend.

This performance was due to three main forces: markets, farmers and policies. First and foremost, it was made possible by a favorable development of market conditions, namely the opening of foreign markets to some Portuguese agricultural commodities, mostly wines, the consolidation of a colonial market in Brazil, and the strengthening of the domestic market, stimulated by a growing population and particularly by rising income. The second explanation for the good results lies in the behavior of farmers and all those economic agents that were responsible for the concrete agricultural development, who acted in accordance with their own interests and reacted flexibly to the stimuli and obstacles created by market changes. The third pillar of success was government intervention, intended to remove several institutional obstacles to agricultural growth and to investment in land, and particularly prone to resolving the most critical issues, such as the situation of the wine sector, the supply of grain, the regulation of property rights, the quality of infrastructures, access to foreign and colonial markets, and so on. We cannot say that these three main drivers of change always acted in a coordinated or positive fashion, but each one played its part, contributing together to the generally positive development of Portuguese agriculture in the period at stake. To simplify considerably, we might say that the first half of this long period was the time of the markets and farmers, while the second was the time of policies.

It is also important to note that the 18th century agricultural growth was not a linear or homogeneous process. There were many differences and contradictions across time and space. Geographically speaking, some regions were clearly more dynamic and successful than others. A simplified assessment might assert that the regions that prospered the most from the agricultural perspective were those that had easier access to urban and foreign markets, namely the Douro Valley, the vast hinterlands of Lisbon and Setúbal, the Minho, the coastal Algarve, and the Atlantic islands of Madeira and S. Miguel.

Conversely, some inland regions – especially Trás-os-Montes, the Beiras, parts of Estremadura and the Algarve mountains – remained on the margins of agricultural progress, penalized as they were by difficult access to the markets, low urbanization rates, shortage of capital, and social and institutional settings that were backward and unequal in nature. The Alentejo was a unique case, with both positive and negative signs. On the one hand, it underwent one of the most interesting processes of social change, and its farmers were well integrated into market logics; on the other, the province failed in its role as the granary of Portugal, which many had wanted it to play, thereby demonstrating perhaps more than any other region the limits imposed upon agricultural growth by unfavorable natural conditions.

As regards the chronology of the process, it was characterized by a fluctuation between ups and downs, though it is not yet possible with the knowledge currently available to rigorously establish the periodization. What we can say is that, in general, the first half of this long period was better than the second. This was marked by an interruption in growth, between the late 1750s and middle of the 1770s, but then it underwent a recovery and reorganization until the early years of the 19th century when, due to particularly adverse conditions, the agricultural sector was ultimately dragged into the general turmoil of the crisis of the *ancien régime*. Data recently released by Jaime Reis and his co-workers also suggests that agricultural output probably grew faster than population during the first half of the 18th century, while it was the other way round in the rest of this period, leading to a drop in productivity.

However, assessing its performance solely or predominantly by means of output and productivity data may underestimate the extent of changes in Portuguese agricultural economy and rural society. This leads us to one last observation, concerning the way the Portuguese case may be placed in a comparative perspective. As a simple definitive answer is out of the question, we might say only that Portuguese agriculture performed "decently" and generally within European standards. For what really distinguishes the 18th century in the history of European agriculture, all over the continent though following a variety of ways, was an acceleration of economic, technical, institutional and social changes and improvements: the pre-modern "agricultural revolution", to use an old term. However, apart from England and Holland, which took the lead in that process and whose achievements were exceptions rather than the rule, some countries simply succeeded better than others. Portugal seems to fit in that group. Portuguese agriculture followed its own course towards change, modernization and development. It may not have made exceptional gains in productivity, it may not have evolved with the main "purpose" of contributing to an industrial revolution, and it was clearly far from eliminating all its problems. Nevertheless, as I have sought to show in this chapter, it did bring about its own kind of agricultural revolution – a silent or smooth one – in the 18th century.

References

Alcoforado, Maria João *et al.* (2012). "Early portuguese meteorological measurements (18th century)". *Climate of the Past*, 8, pp. 353–371.

Allen, R.C. (2000). "Economic structure and agricultural productivity in Europe, 1300–1800". *European Review of Economic History*, 4 (1), pp. 1–25.

Almodovar, A. and Cardoso, José Luís (1998). *A History of Portuguese Economic Thought*. London: Routledge.

Álvarez Nogal, C. and Prados de La Escosura, L. (2012). "The rise and fall of Spain (1270–1850)". *Economic History Review*, 66, pp. 1–37.

Alves, Daniel Ribeiro (2012). *Os Dízimos no Final do Antigo Regime: Aspectos Económicos e Sociais (Minho, 1820–1834)*. Lisbon: CEHR-UCP.

Amorim, Inês (1996). *Aveiro e sua Provedoria no Século XVIII (1690–1814): Estudo Económico de um Espaço Histórico*. Porto: Universidade do Porto.

Amorim, Inês (1997). *O Mosteiro de Grijó. Senhorio e Propriedade, 1560–1720 (Formação, Estrutura e Exploração do seu Domínio)*. Braga: Barbosa e Xavier.

Bairoch, Paul, Batou, J. and Chèvre, P. (1988). *La Population des villes européennes de 800 à 1850*. Geneve: Droz.

Balbi, Adrien (1822). *Essai statistique sur le royaume de Portugal et d'Algarve*. Paris: s.n.

Bavel, Bas van, and Hoppenbrouwers, Peter (Eds.) (2004). *Landholding and Land Transfer in the North Sea Area (Late Middle Ages-19th century)*. Turnhout: Brepols.

Béaur, Gérard and Chevet, J-M. (2013). "Institutional changes and agricultural growth". In G. Béaur *et al.* (Ed.). *Property Rights, Land Markets and Economic Growth in the European Countryside (13th–19th Centuries)*. Turnhout: Brepols, pp. 19–68.

Bennett, Norman R. (1992). "The wine growers of the upper Douro, 1780–1800". *Portuguese Studies Review*, 2 (1), pp. 28–45.

Bennett, Norman R. (2005). *That Indispensable Article: Brandy and Port Wine, c.1650–1908*. Porto: GEHVID.

Bohstedt, John (2010). *The Politics of Provisions: Food Riots, Moral Economy, and Market Transition in England, c. 1550–1850*. Farnham: Ashgate.

Borges, Maria E. (2000). *Homens, Fazenda e Poder no Alentejo de Setecentos*. Lisbon: Colibri.

Brandão, Maria F. (1988). *Land, Inheritance and Family in Northwestern Portugal: The Case of Mosteiro in the Nineteenth Century*. University of East Anglia (PhD dissertation).

Braudel, Fernand (2000). *Civilization and Capitalism, 15th–18th Century: the Structure of Everyday Life*. Berkeley and Los Angeles: University of California Press.

Burnette, Joyce (2014). "Agriculture, 1700–1870". In R. Floud, J. Humphries and P. Johnson (Eds.). *The Cambridge Economic History of Modern Britain (1700–1870)*. Cambridge: CUP, pp. 89–117.

Câmara, Benedita (2006). "The Portuguese Civil Code and the colonia tenancy contract in Madeira (1867–1967)". *Continuity and Change*, 21 (2), pp. 235–259.

Campbell, B.M.S. and Overton, Mark (1991). *Agricultural Productivity in the European Past*. Manchester: M.U. Press.

Camuffo, Dario *et al.* (2010). "500-year temperature reconstruction in the Mediterranean basin by means of documentary data and instrumental observations". *Climatic Change*, 101 (1–2), pp. 169–199.

Camuffo, Dario *et al.* (2013). "Western Mediterranean precipitation over the last 300 years from instrumental observations". *Climatic Change*, 117, pp. 85–101.

Cardoso, José L. (1989). *O Pensamento Económico em Portugal nos finais do Século XVIII (1780–1808)*. Lisbon: Estampa.

Cardoso, José L. (2007). "Pombal, o terramoto e a política de regulação económica". In A.C. Araújo *et al.* (Ed.). *O Terramoto de 1755: Impactos Históricos*. Lisbon: Livros Horizonte, pp. 165–182.

Cardoso, José L. (2009). "Free trade, political economy and the birth of a new economic nation: Brazil, 1808–1810". *Revista de Historia Económica – Journal of Iberian and Latin American Economic History*, 27 (2), pp. 183–204.

Cardoso, José L. *et al.* (2003). *O Tratado de Methuen: Diplomacia, Guerra, Política e Economia*. Lisbon: Livros Horizonte.

Cerman, Markus (2008). "Rural economy and society". In Peter H. Wilson (Ed.). *A Companion to Eighteenth-Century Europe*. Malden/Oxford: Blackwell Publishers, pp. 47–65.

Costa, Leonor Freire (2005). "Relações económicas com o exterior". In Pedro Lains e Álvaro F. Silva (Eds.). *História Económica de Portugal, 1700–2000*. Lisbon: Imprensa das Ciências Sociais, vol. 1, pp. 263–298.

Costa, Leonor Freire, Lains, Pedro and Miranda, Susana Munch (2011). *História Económica de Portugal, 1143–2010*. Lisbon: Esfera dos Livros.

Costa, Leonor Freire, Nuno Palma, and Reis, Jaime (2015). "The great escape? The contribution of the empire to Portugal's economic growth, 1500–1800". *European Review of Economic History*, 19 (1), pp. 1–22.

Costa, Leonor Freire, Rocha, Maria M. and Araújo, Tanya (2011). "Social capital and economic performance: trust and distrust in eighteenth-century gold shipments from Brazil". *European Review of Economic History*, 15 (1), pp. 1–27.

Costa, Leonor Freire, Rocha, Maria M. and Brito, Paulo (2014). *Money Supply and the Credit Market in Early Modern Economies: the Case of Eighteenth-Century Lisbon*. Lisbon: GHES (Working Paper 52).

Costa, Leonor Freire, Rocha, Maria M. and Sousa, Rita M. de (2013). *O Ouro do Brasil*. Lisbon: Imprensa Nacional-Casa da Moeda.

Crosby, Alfred (1994). *Germs, Seeds and Animals: Studies in Ecological History*. New York: M.E. Sharpe.

Dennison, Tracy and Simpson, James (2010). "Agriculture". In Stephen Broadberry and Kevin H. O'Rourke (eds.). *The Cambridge Economic History of Modern Europe, 1700–1870*. Cambridge: Cambridge University Press, pp. 147–163.

Dias, Fátima Sequeira (1995). "A importância da economia da laranja no arquipélago dos Açores durante o século XIX". *Arquipélago-História*, 1, pp. 189–240.

Duguid, Paul (2003). "The making of Methuen: the commercial treaty in the English immagination". *Revista da Faculdade de Letras (Porto) – História*, 4, pp. 9–36.

Durães, Margarida (2009). "Providing well-being to women through inheritance and succession: Portugal in the 18th and 19th century". In M. Durães *et al.* (Eds.). *The Transmission of Well-being: Gendered Marriage Strategies and Inheritance Systems in Europe (17th–20th Centuries)*. Bern: Peter Lang, pp. 201–228.

Epstein, S.R. (Ed.) (2001). *Town and Country in Europe, 1300–1800*. Cambridge: Cambridge University Press.

Fauvrelle, Natália (2001). *Quintas do Douro. As Arquitecturas do Vinho do Porto*. Porto: GEHVID.

Federico, G. and Malanima, P. (2004). "Progress, decline, growth: product and productivity in Italian agriculture, 1000–2000". *Economic History Review*, 57 (3), pp. 437–464.

Ferrão, José Mendes (1992). *A Aventura das Plantas e os Descobrimentos Portugueses*. Lisbon: CNCDP.

Fisher, H.E.S. (1971). *The Portugal Trade: A Study of the Anglo-Portuguese Commerce 1700–1770*. London: Routledge.

Fonseca, Helder A. (1981). "O senhorio de São Marcos e a comunidade rural de Vale de Azares no século XVIII (contributo para o estudo da história rural no Antigo Regime em Portugal)". *Revista Portuguesa de História*, 19, pp. 183–270.

Fonseca, Helder and Reis, Jaime (2011). "The limits of agricultural growth in a fragile eco-system: total factor productivity in Alentejo, 1750–1850". In M. Olsson & P. Svensson (Eds.). *Growth and Stagnation in European Historical Agriculture*. Turnhout: Brepols, pp. 37–66.

Fonseca, Hélder and Santos, Rui (2001). "Três séculos de mudanças no sector agrário alentejano: a região de Évora nos séculos XVII a XIX". *Ler História*, 40, pp. 167–178.

Francis, A.D. (1966). *The Methuens and Portugal, 1691–1708*. Cambridge: Cambridge University Press.

Francis, Alan D. (1972). *The Wine Trade*. London: A&C Black Ltd.

Franzini, Miguel Marino (1843). "Considerações acerca da renda total da Nação Portuguesa e sua distribuição por classes, com algumas reflexöes sobre o imposto da décima". *Revista Universal Lisbonense*, 24, pp. 293–297.

Godinho, V.M. (1955). *Prix et monnaies au Portugal, 1750–1850*. Paris: A. Collin.

Hancock, David (2009). *Oceans of Wine: Madeira and the Emergence of American Trade and Taste*. New Haven: Yale University Press.

Hillbom, Ellen and Svensson, Patrick (Eds) (2013). *Agricultural Transformation in a Global History Perspective*. New York: Routledge.

Jones, Eric (2013). "Economics without history: objections to the rights hypothesis". *Continuity and Change*, 28 (3), pp. 323–346.

Justino, David (1981). "Crises e decadência da economia cerealífera alentejana no século XVIII: contribuição para o seu estudo a partir da análise das séries dos preços regionais do trigo e da cevada (1684–1820)". *Revista de História Económica e Social*, 7, pp. 29–80.

Justino, David (1986). *Conjuntura Económica e Relações Sociais numa Aldeia do Alentejo: Monte de Trigo, 1690–1854*. Lisbon (unpublished dissertation).

Justino, David (1988). *A Formação do Espaço Económico Nacional: Portugal, 1810–1913*. Lisbon: Vega.

Lains, Pedro (1995). *A Economia Portuguesa no Século XIX: Crescimento Económico e Comércio Externo, 1851–1913*. Lisbon: Imprensa Nacional-Casa da Moeda.

Link, Heinrich Friedrich (1803). *Voyage en Portugal depuis 1797 jusqu'en 1799*. Paris: Levrault, Schoell.

Livi-Bacci, Massimo (2002). "500 anos de demografia brasileira: uma resenha". *Revista Brasileira de Estudos de População*, 19, pp. 141–159.

Macedo, J. Borges de (1963). *Problemas de História da Indústria Portuguesa no Século XVIII*. Lisbon: Associação Industrial Portuguesa.

Magalhães, Joaquim Romero (1985). "Alguns aspectos da produção agrícola no Algarve: fins do século XVIII – princípios do século XIX". *Revista Portuguesa de História*, 22, pp. 1–29.

Magalhães, Joaquim Romero (1988). *O Algarve Económico, 1600–1773*. Lisbon: Estampa.

Maia, Fernanda de Sousa (1991). *O Mosteiro de Bustelo: Propriedade e Produção Agrícola no Antigo Regime (1638–1670 e 1710–1821)*. Porto: Universidade Portucalense.

Malanima, Paolo (2010). "Urbanization". In Stephen Broadberry, Kevin H. O'Rourke (Eds.). *The Cambridge Economic History of Modern Europe, 1700–1870*. Cambridge: C.U. Press, pp. 235–263.

Martins, Conceição A. (1990). *Memória do Vinho do Porto*. Lisbon: Imprensa de Ciências Sociais.

Martins, Conceição A. (1998). *Vinha, Vinho e Política Vinícola em Portugal: do Pombalismo à Regeneração*. Évora: Universidade de Évora (PhD Dissertation).

Memórias Económicas da Academia Real das Ciências de Lisboa para o Adiantamento da Agricultura, das Artes e da Indústria em Portugal e suas Conquistas, 1789–1815 (1990–1991). Lisbon: Banco de Portugal.

Memórias Económicas Inéditas, 1780–1808 (1987). Lisbon: Academia das Ciências.

Meneses, Avelino de Freitas de (2001). "A exploração da terra". In A.F. Meneses (Ed.). *Portugal, da Paz da Restauração ao Ouro do Brasil*. In Joel Serrão e Oliveira Marques (Eds.). *Nova História de Portugal*. Lisbon: Presença, vol. 7, pp. 226–263.

Miranda, Sacuntala (1989). *O Ciclo da Laranja e os 'Gentlemen Farmers' da Ilha de S. Miguel (1780–1880)*. Ponta Delgada: Instituto Cultural de Ponta Delgada.

Monteiro, Nuno Gonçalo (1989). "Geografia e tipologia dos direitos de foral". In F.M. Costa *et al.* (Ed.). *Do Antigo Regime ao Liberalismo: Perspectivas de Síntese (1750–1850)*. Lisbon: Edições Vega, pp. 259–271.

Monteiro, Nuno Gonçalo (2005). "A ocupação da terra". In Pedro Lains e Álvaro F. Silva (Ed.). *História Económica de Portugal, 1700–2000. O Século XVIII*. Lisbon: Imprensa das Ciências Sociais, vol.1 pp. 67–91.

Moreira, Maria Guardado (2008). "O século XVIII". In Teresa Rodrigues (Ed.). *História da População Portuguesa: das Longas Permanências à Conquista da Modernidade*. Porto: CEPESE/Afrontamento, pp. 247–287.

Morineau, Michel (1985). *Incroyables gazettes et fabuleux métaux*. Paris/Cambridge: Maison des Sciences de l'Homme/Cambridge U. P.

Mota, Salvador (2000). *O Senhorio Cisterciense de Santa Maria de Bouro: Património, Propriedade, Exploração e Produção Agrícola (1570–1834)*. Porto: Universidade do Porto.

Neto, Margarida Sobral (2007). "A historiografia rural portuguesa". *Studia Historica, Historia moderna*, 29, pp. 251–275.

Neves, José Acúrsio das (1814). *Variedades sobre Objectos Relativos às Artes, Comércio e Manufacturas, Consideradas segundo os Princípios da Economia Política*. Porto: Afrontamento.

Oliveira, Aurélio de (1979). *A Abadia de Tibães, 1630/1680–1813. Propriedade, Exploração e Produção Agrícolas no Vale do Cávado durante o Antigo Regime*. Porto: Universidade do Porto.

Oliveira, Aurélio de (1980). "A renda agrícola em Portugal durante o Antigo Regime". *Revista de História Económica e Social*, 6, pp. 1–56.

Oliveira, Aurélio de (1982). *Contabilidades Monásticas e Produção Agrícola durante o Antigo Regime: os Dízimos do Mosteiro de Santo Tirso, 1626–1821*. Santo Tirso: Câmara Municipal de Santo Tirso.

Oliveira, Aurélio de (1984). "Níveis de produção vinícola no Entre Douro e Minho, 1629–1822". *Primeiras Jornadas de Estudo Norte de Portugal/Aquitânia*. Porto: Universidade do Porto, pp. 325–341.

Oliveira, Aurélio de (1985). "Mercados a norte do Douro – algumas considerações sobre a história dos preços em Portugal e a importância dos mercados regionais (séculos XVII–XVIII)". *Revista da Faculdade de Letras (do Porto) – História*, 2, pp. 97–160.

Oliveira, Eduardo Freire de (1885–1911). *Elementos para a História do Município de Lisboa*. Lisbon: Imprensa da Casa Real.

Oliveira, João Nunes de (1990). *A Produção Agrícola de Viseu entre 1550 e 1700*. Viseu: Câmara Municipal de Viseu.

Oliveira, João Nunes de (2002). *A Beira Alta de 1700 a 1840: Gentes e Subsistências*. Viseu: Palimage.

Olsson, Mats and Svensson, Patrick (Eds.) (2011). *Growth and Stagnation in European Historical Agriculture*. Turnhout: Brepols.

Palma, Nuno and Reis, Jaime (2016). "From convergence to divergence: Portuguese Demography and Economic Growth". Groningen Growth and Development Centre Research Memorandum, 161.

Pedreira, Jorge (1988). "Agrarismo, industrialismo, liberalismo: algumas notas sobre o pensamento económico português (1780–1820)". In J.L. Cardoso (Ed.). *Contribuições para a História do Pensamento Económico em Portugal*. Lisbon: D. Quixote, pp. 63–83.

Pedreira, Jorge (1994). *Estrutura Industrial e Mercado Colonial. Portugal e Brasil (1780–1830)*. Lisbon: Difel.

Pedreira, Jorge (2005). "A indústria". In Pedro Lains and Álvaro F. Silva (Ed.). *História Económica de Portugal, 1700–2000. O Século XVIII*. Lisbon: Imprensa das Ciências Sociais, vol. 1, pp. 177–208.

Pereira, Álvaro S. (2009). "The opportunity of a disaster: the economic impact of the 1755 Lisbon earthquake". *The Journal of Economic History*, 69, pp. 466–499.

Pereira, Gaspar M. (1986). "Aspectos sociais da viticultura duriense nos fins do século XVIII". *I Jornadas do CENPA – Actas*. Porto: CENPA, pp. 93–118.

Pereira, Gaspar M. (1996). "A região do vinho do Porto: origem e evolução de uma demarcação pioneira". *Douro: Estudos & Documentos*, 1, pp. 177–194.

Persson, Karl G. (1999). *Grain Markets in Europe, 1500–1900: Integration and Deregulation*. Cambridge: Cambridge University Press.

Pfister, C. and Brázdil, R. (2006). "Social vulnerability to climate in the 'Little Ice Age': an example from Central Europe in the early 1770s". *Climate of the Past*, 2, pp. 115–129.

Pinto, Virgílio Noya (1979). *O Ouro Brasileiro e o Comércio Anglo-português*. São Paulo: Editora Nacional.

Reis, Jaime, Martins, Conceição A. and Costa, Leonor Freire (2013). "From major power to economic backwater: Portugal's GDP 1500–1850". Paper presented at the CAGE conference on Accounting for the Great Divergence. University of Warwick, 28 May 2013.

Ribeiro, Orlando (1991). *Opúsculos Geográficos IV: O Mundo Rural*. Lisbon: Fundação Calouste Gulbenkian.

Ribeiro, Orlando (2011). *Portugal, o Mediterrâneo e o Atlântico*. Lisbon: Letra Livre.

Rodrigues, José Damião (2003). *São Miguel no Século XVIII: Casa, Elites e Poder*. Ponta Delgada: Instituto Cultural de Ponta Delgada.

Rodriguez, Laura (1973). "The Spanish riots of 1766". *Past and Present*, 59, pp. 117–146.

Roy, Jean-Michel (2013). "Food". In Michel Delon (Ed.). *Encyclopedia of the Enlightenment*. New York: Routledge, pp. 532–535.

Rudé, George (1972). *Europe in the Eighteenth Century: Aristocracy and the Bourgeois Challenge*. Cambridge: Harvard University Press.

Santiago-Caballero, Carlos (2013). "The rain in Spain? Climate versus urban demand as causes of agricultural stagnation in eighteenth-century Spain". *European Review of Economic History*, 17 (4), pp. 452–470.

Santos, Rui (1993). "A nuvem por Juno? O tema da fisiocracia na historiografia do pensamento económico português". *Análise Social*, 121, pp. 423–443.

Santos, Rui (1995). *Celeiro de Portugal Algum Dia: Crescimento e Crises Agrárias na Região de Évora, 1595–1850*. Lisbon: FCSH/UNL.

Santos, Rui (2003). *Sociogénese do Latifundismo Moderno: Mercados, Crises e Mudança Social na Região de Évora, Séculos XVII a XIX*. Lisbon: Banco de Portugal.

Santos, Rui (2004). "Economic sociology of the modern latifundium: economic institutions and social change in Southern Portugal, 17th–19th centuries". *Sociologia, Problemas e Práticas*, 45, pp. 23–52.

Santos, Rui (2012). "Food dearth and demographic behaviour in Southern Portugal. The region of Évora, 17th–19th centuries". Unpublished paper presented at the CRICEC Workshop on Harvest Fluctuations and Food Shortages in European Rural Societies. Lisbon: FCSH-UNL.

Santos, Rui and Serrão, José Vicente (2013). "Property rights, social appropriations and economic outcomes: agrarian contracts in Southern Portugal in the late-eighteenth century". In G. Béaur *et al.* (Eds.). *Property Rights, Land Markets and Economic Growth in the European Countryside (13th–19th Centuries)*. Turnhout: Brepols, pp. 475–494.

Schneider, Susan (1970). *The General Company of the Cultivation of the Vine of the Upper Douro, 1756–1777: a Case Study of the Marquis of Pombal's Economic Reform*. University of Texas (PhD dissertation).

Scott, Tom (2015). "The agrarian west". In Hamish Scott (Ed.). *The Oxford Handbook of Early Modern European History, 1350–1750*. Oxford: Oxford University Press, vol. 1, pp. 398–427.

Serrão, José Vicente (1987). *O Pombalismo e a Agricultura*. Lisbon: ISCTE.

Serrão, José Vicente (1988). "O pensamento agrário setecentista (pré-'fisiocrático'): diagnósticos e soluções propostas". In J.L. Cardoso (Ed.). *Contribuições para a História do Pensamento Económico em Portugal*. Lisbon: D. Quixote, pp. 23–50.

Serrão, José Vicente (1993a). "O quadro humano". In A.M. Hespanha (Ed.). *O Antigo Regime (1620–1807)*. In José Mattoso (Ed.). *História de Portugal*. Lisbon: Círculo de Leitores, vol. 4, pp. 49–69.

Serrão, José Vicente (1993b). "O quadro económico: configurações estruturais e tendências de evolução". In A.M. Hespanha (Ed.). *O Antigo Regime (1620–1807)*. In José Mattoso (Ed.). *História de Portugal*. Lisbon: Círculo de Leitores, vol. 4, pp. 71–117.

Serrão, José Vicente (1996). "População e rede urbana nos séculos XVI–XVIII". In César Oliveira (Ed.). *História dos Municípios e do Poder Local (dos Finais da Idade Média à União Europeia)*. Lisbon: Círculo de Leitores, pp. 63–77.

Serrão, José Vicente (2000). *Os Campos da Cidade. Configuração das Estruturas Fundiárias da Região de Lisboa nos Finais do Antigo Regime*. Lisbon: ISCTE.

Serrão, José Vicente (2002). "Os nomes da terra: sobre o padrão predial dos campos de Lisboa no século XVIII". *Ler História*, 43, pp. 85–124.

Serrão, José Vicente (2005). "A agricultura". In Pedro Lains and Álvaro F. Silva (Eds.). *História Económica de Portugal, 1700–2000. O Século XVIII*. Lisbon: Imprensa das Ciências Sociais, vol. 1, pp. 145–175.

Serrão, José Vicente (2007). "Os impactos económicos do terramoto". In A.C. Araújo *et al.* (Ed.). *O Terramoto de 1755: Impactos Históricos*. Lisbon: Livros Horizonte, pp. 141–163.

Serrão, José Vicente (2009). "Land management responses to market changes. Portugal, seventeenth-nineteenth centuries". In V. Pinilla (Ed.). *Markets and Agricultural Change in Europe from the Thirteenth to the Twentieth Centuries*. Turnhout: Brepols, pp. 47–73.

Serrão, José Vicente and Santos, Rui (2013). "Land policies and land markets: Portugal, late eighteenth and early nineteenth centuries". In G. Béaur *et al.* (Ed.). *Property Rights, Land Markets and Economic Growth in the European Countryside (13th–19th Centuries)*. Turnhout: Brepols, pp. 317–341.

Shaw, L.M.E. (1998). *The Anglo-Portuguese Alliance and the English Merchants in Portugal, 1654–1810*. Aldershot/Brookfield: Ashgate.

Shillington, V.M. and Chapman, A.B.W. (1907). *The Commercial Relations of England and Portugal*. London: George Routledge & Sons.

Sideri, Sandro (1970). *Trade and Power: Informal Colonialism in Anglo-Portuguese Relations*. Rotterdam: Rotterdam University Press.

Silbert, Albert (1977). *Do Portugal do Antigo Regime ao Portugal Oitocentista*. Lisbon: Livros Horizonte.

Silbert, Albert (1978). *Le Portugal méditerranéen à la fin de l'Ancien Régime*. Lisbon: Instituto Nacional de Investigação Científica.

Silva, Célia Taborda da (1994). *O Mosteiro de Ganfei: Propriedade, Produção e Rendas no Antigo Regime (1629–1683 e 1716–1822)*. Lisbon: Fragmentos.

Silva, Luiz Augusto Rebelo da (1860–1871). *História de Portugal nos séculos XVII e XVIII*. Lisbon: Imprensa Nacional-Casa da Moeda.

Sousa, Rita Martins de (2006). *Moeda e Metais Preciosos no Portugal Setecentista, 1688–1797*. Lisbon: Imprensa Nacional-Casa da Moeda.

Sousa, Rita Martins de (2008). "Brazilian gold and the Lisbon mint house (1720–1807)". *e-Journal of Portuguese Studies*, 6 (1), pp. 1–22.

Stein, Stanley and Stein, Barbara (2003). *Apogee of Empire: Spain and New Spain in the Age of Charles III, 1759–1789*. Baltimore: Johns Hopkins U. Press.

Taborda, J.P., Alcoforado, M.-J., and Garcia, J.C. (2004). *O Clima no Sul de Portugal no Século XVIII. Reconstituição a Partir de Fontes Descritivas e Instrumentais*. Lisbon: Centro de Estudos Geográficos.

Tenreiro, A. Guerra (1942). *Douro: Esboços para a sua História Económica*. Porto: Tip. Gonçalves e Nogueira.

Thompson, E.P. (1971). "The moral economy of the English crowd in the eighteenth century". *Past and Present*, 50, pp. 76–136.

Vidigal, Luís (1993). *Câmara, Nobreza e Povo: Poder e Sociedade em Vila Nova de Portimão (1755–1834)*. Portimão: Câmara Municipal.

Watson, Matthew (2015). "Following in John Methuen's early eighteenth-century footsteps: Ricardo's comparative advantage theory and the false foundations of the competitiveness of nations". Paper presented at the Research Workshop *Should Nation States Compete?*. City University London, 25th June 2015.

Zanden, Jan Luiten van (2002). "The 'revolt of the early modernists' and the 'first modern economy': an assessment". *Economic History Review*, 55 (4), pp. 619–641.

CHAPTER 6

Gross Agricultural Output: A Quantitative, Unified Perspective, 1500–1850

Jaime Reis

L'agriculture n'a jamais été bien fleurissante au Portugal.

BALBI 1822: VOL. 1, 143

Introduction[1]

In the current state of knowledge, a quantitative approach to the long-run performance of Portuguese agriculture taken as a whole is possible only from the mid-19th century onward. Thanks to recent research mostly based on published sources, we now have at our disposal continuous time series which span the period from 1850 to the present and cover the main dimensions of this sector. Sufficient evidence has also now been assembled to make possible the estimation, at regular intervals, of the efficiency of factors of production at the aggregate, sector and factor levels.[2] For earlier times, however, there is an almost complete dearth in this respect, notwithstanding the considerable volume of micro-oriented scholarship on the medieval and early modern periods, often producing copious and valuable statistical results.

The present chapter has two aims. One is to put together, for the years from 1500 to 1850, a similar though less ambitious collection of macro time series, while naturally using different sources and methodologies. The second is to employ this information in order to establish a broad quantitative view of how

1 This study has received valuable financial support from the Fundação para a Ciência e Tecnologia (projects PTDC/HAH/70938/2006) and (PTDC/HIS-HIS/123046/2010). It has also benefited greatly from comments and discussion with António Castro Henriques, Conceição Andrade Martins, Leonor Freire Costa, Carlos Faísca, Nuno Palma and the editors of this volume.

2 For the period 1846–1992 see Lains (2003). The picture of the 20th century can be completed from Soares (2005) and Chapters 7 and 8.

 | DOI 10.1163/9789004311527_008

the agrarian economy evolved from the end of the Middle Ages to the onset of a modern agricultural sector in the 19th century.

Pursuing these goals has several benefits. Firstly, it enables us to establish, for the first time, a unified long-run perspective of the dynamics of the entire period, thanks to a new set of consistent quantitative indicators and to the inter-temporal comparisons which they allow. Moreover, it does so over an exceptionally long period of three and a half centuries. Secondly, it delivers a macroeconomic framework into which the findings of the many micro studies, which are the foundation of Early Modern agrarian history, can be incorporated. Since this allows a higher degree of interaction between these explorations, it also considerably raises their respective analytical potential. Thirdly, it offers a number of new quantitative instruments which enable us to sharpen traditional and oftentimes unclear images of both the aggregate performance and the role of agriculture in the overall economy. Finally, it makes greater accuracy possible in assessing derived variables such as per capita food consumption, which are at the heart of current long-term macro debates, in particular those concerning the Malthusian regime. Without such a critical link a model of long-run population size and dynamics in the pre-industrial era cannot be properly established.[3]

Much has had to be left out of the present analysis. In particular, this is not an exhaustive compilation of all the measurable aspects of the agrarian past. It does not consider subjects like land tenure and access to land, labor relations, crop acreage, or taxation and institutions, which are considered extensively throughout the chapters of this book but at present are still hard to organize into a global vision. Inevitably, the focus here is on agriculture as a whole. Some attention is paid to sub-sectors and products, but the burden of our exercise is for the most part the macroeconomic picture. In addition, it suffers from the limitation of not embracing quantitative trends prior to the early 1500s, a consequence of the still fragmentary and deficient quality of the data available for this period.

To complete this overview of the Early Modern period, two further aspects should be taken into account. One is to consider what it was that Portuguese farmers produced during those years. The other is an outline of how agriculture may have evolved during those three and a half centuries, given the constraints and the stimuli to which it was subjected.

Natural conditions, in particular agrological and climatic constraints, cultural persistence and the structure of the economy determined a pattern of productive specialization in Portugal which was typical of southern Europe at

3 For a discussion of this problematic in the Portuguese context see Palma and Reis (2016).

TABLE 6.1 *Structure by value of agricultural output (percent)*

	Grain	Wine	Oil	Meat	Total
1515	38	15	3	44	100
1838	53	36	17	n.a.	100
1850	45	19	5	31	100

SOURCE: 1515, GODINHO 1968: VOL. 2, 20; 1838, FRANZINI 1843; 1850, REIS 2000: 39.

the time. Output consisted mainly of grain – mostly wheat, maize, rye, millet and barley – animal products, wine, and olive oil, and was almost entirely directed towards feeding the population. Exportables and raw materials (hides, linen, wool) for domestic industry also played a role, albeit a lesser one. It is likely, as far as we know, that early in our period of study (see Table 6.1) the livestock subsector predominated, followed by grain, wine and olive oil in diminishing degrees of importance, a distribution which is hardly surprising. The lingering effects of the *Reconquista* and of the Black Death left Portugal at the start of the 16th century a thinly populated country (13 to 15 inhabitants per square kilometer in 1527–32) with abundant natural resources. Labor was therefore scarce and the standard of living of this frontier economy was relatively high (Medeiros 1993; Henriques 2015).

An analogous agrarian regime has been observed for the same period in the rest of the Iberian Peninsula by Álvarez and Prados (2013). They noted, however, that during the 16th century a sustained increase in population brought on a switch to a new agrarian regime characterized by falling land-labor ratios, a retreat of pastoralism, and an expansion of more intensive crop husbandry. Lower real wages and a reduction in per capita food consumption were the consequences and, as a result, from the 1600s to the 1810s Spain moved towards "a lower path" of growth (Álvarez and Prados 2013: 2).

Did Portuguese agriculture follow a comparable course? It seems likely that it did. Table 6.1 certainly reveals a long-run shift in productive specialization between 1515 and 1850 which is commensurate with this. In a period when the population was growing significantly in the face of a by now fixed stock of land, as might be expected grain, wine and oil, all of them labor-intensive products, became much more important (56 to 69 percent of total output), and left behind land-intensive animal husbandry, which fell from 44 to 31 percent.[4] There are thus signs that Portugal may have experienced the transition from a more

4 Between 1500 and 1800 in Portugal the total land-labor ratio fell by a factor of 2.4, while in Spain it fell by a factor of 2.1. See Malanima (2009: 16).

extensive to a more intensive agrarian regime which occurred at roughly the same time in Spain, with possibly parallel repercussions in the long-run evolution of its primary sector's output.

The present study depends on the guidelines and benchmarks provided by this "Iberian model", which it uses to analyze some major questions pertaining to Portugal's Early Modern agriculture. Chief among them is whether a similar long-run regime shift had the same negative consequences as were experienced over the border. The first section following this introduction shows, in terms of method and data, how the country's gross annual agricultural output at constant prices may be estimated. The second presents the main results and reveals an overall significant increase in the size of this sector. It also introduces a somewhat revised description of the customary main phases of this expansion. The third discusses the robustness of these findings, and the fourth section draws conclusions.

Method and Data

Method

There are essentially two ways of quantifying the gross output of a nation's agriculture during its pre-statistical period. In the case of countries which have a strong background in quantitative history, this can be done directly by gauging the yearly output of its components at current prices and aggregating them. The real value of this estimate is then obtained by deflating this result with the help of an appropriately weighted index of agricultural prices. These procedures are feasible when sufficient reliable evidence exists for the purpose. To date, this treatment has been possible in three economies – Britain, the Netherlands and Sweden.[5]

The alternative, indirect approach, which we will follow here, has been used in countries like Italy, Spain and Germany, where quantitative history has not yet matured enough to sustain an effort of direct measurement on such a scale (Álvarez and Prados 2007 and 2013; Pfister 2009; Malanima 2011). An indirect indicator is therefore used, namely the total domestic consumption of agricultural products, which is presumed equal to gross agricultural production.

5 For Sweden, Edvinsson (2011) and Krantz and Schön (2012); for the Netherlands, van Zanden and van Leeuwen (2012); and for Britain, Broadberry *et al.*, (2015). It should be noted, however, that in the last of these cases, despite the abundance of good data and the use of sophisticated estimation techniques, considerable discrepancies exist between the results obtained. See the discussion and references in Kelly and O'Grada (2013).

This outcome is deflated, as above, to obtain values for this variable at constant prices.

Since the second approach is well known, it is presented here only briefly. It is inspired by Wrigley (1985) and has since been significantly revised by Allen (2000) and others. It rests on three main assumptions. One is that the balance of exports and imports of agricultural produce is negligible, a condition for accepting the principle of equality between home production and consumption. Another is that food and raw material consumption can be estimated by means of a demand function for total agricultural production (Qa) given by

$$Qa = I^a P^b M^c N \quad (1)$$

in which I is a proxy for real income per capita, P is the real price index for agricultural products, M is the real price index for other consumer goods and N is total population. The coefficients a, b and c are the respective income, own price, and cross elasticities of demand. In accordance with the Slutsky-Schultz relation, their sum must be zero. All the prices and wages used to calculate these variables are indicated in grams of silver to ensure comparability. The third assumption is that markets for outputs and for factors of production function well enough to avoid serious allocative distortions. The following comments serve to establish how well these three postulates are met in the Portuguese case.

Regarding the first one, it should be noted that Early Modern countries usually conducted a certain amount of foreign trade, much of it in agricultural produce, and cannot therefore be termed "closed economies". Their exports and imports, however, were seldom large relative to GDP. Exceptionally, in Britain and the Netherlands the difference between the consumption and production of agricultural commodities might at times account for as much as one tenth of national output. Elsewhere, this difference has been deemed barely significant (Allen 2000).

Was Portugal also an exception or part of the norm? Traditionally, the view has been that the country suffered chronically from food shortages and had to import large quantities of grain in order to meet the population's nutritional requirements. According to Godinho (1955: 147), this particular deficit was "one of the most important structural features of Portuguese economic history".[6] This view has been recently tested using data for Portuguese foreign trade for

6 Author's own translation. For a recent and rare dissenting voice, which concerns the 18th century only, see Serrão (2005: vol. 1, 172) and Serrão, Chapter 5 in this book.

several benchmarks between 1550 and 1850 (Costa and Reis 2015). Grain imports and wine and olive oil exports, the main items, were considered for this, the conclusion being that the foreign food balance by value during this period was generally negligible and exceeded the 3 percent mark in only one benchmark, in 1850. It thus seems fair to accept that agricultural consumption is a reasonable proxy for this sector's output.[7]

Our second supposition is that the aggregate consumption of the fruits of the land was mainly determined by two factors: the number of consumers (i.e. the population) and their purchasing power. The latter was in turn influenced by their real income, the real costs of food and non-food items, and by the respective propensities to consume them reflected in their demand elasticities. Estimating the various dimensions of this function raises several problems.

Possibly the most important of these caveats concerns the calculation of nominal income, from which the real income variable is ultimately derived. Ideally, it should encompass the remuneration, throughout each year, of all productive factors – labor, land and capital. Unfortunately, data for them all are not always easy to come by. In the case of the first, time series for daily rates of pay are readily available for many classes of occupation. On the other hand, average days or hours of work per annum, which undoubtedly varied a good deal over time, as well as across employment categories, are not, and this introduces a significant element of uncertainty in the results. When it comes to land, time series exist too, but covering all forms of land access can be problematic. In the case of income from capital, a lesser share of total income (Malanima 2011), evidence at present is practically unobtainable. Therefore, in keeping with the current practice in this literature, it is left out of our calculation.

The solution adopted for estimating nominal income is that suggested by Álvarez and Prados (2013: 9), with some modifications. It involves adding yearly total values of annual wage income and total annual land rent income in 1850, at current prices and in grams of silver. At this date, we assume their respective shares in national income were 75 and 25 percent.[8] The two series are calculated from this terminal point, used as a base year, all the way back to 1500 and are combined to obtain the yearly current value estimate for total factor income.

The common practice in constructing the wage component of this exercise has been to multiply the daily wage rate (either of skilled or unskilled labor)

7 For an estimation for the 18th century see Serrão, Chapter 5 in this book.

8 Malanima (2011: 178) uses similar proportions for Italy between the 16th and the 18th centuries. Clark (2007: 138) proposes for pre-industrial England 80 and 20 percent.

by the working population and by a fixed number of working days in the year and assume as constant the number of hours in a working day.[9] Our approach is akin to this, but more realistic in that it takes into account the simultaneous existence of different labor categories, with different rates of remuneration and labor intensities. In practice, this entails aggregating three classes of earnings: of unskilled laborers, earning the corresponding wage and working an assumed 120 days a year; of skilled urban workers, who received skilled wages and were occupied 250 days a year; and of rural non-agricultural labor (proto-industry, services), whom we presume received an intermediate wage and were active 180 days a year.[10] The shares of these categories in the overall labor force are drawn mainly from fiscal and demographic data for the different benchmarks and converted by linear interpolation into annual series.[11]

Though an improvement, this reformulation of the standard method for deriving labor income is still subject to a degree of bias. In presuming that the annual number of days of labor per worker remained fixed for 350 years, it ignores the possible occurrence of an "industrious revolution" such as has been claimed for many parts of the Early Modern European economy (Angeles 2008; Vries 2008). Moreover, in Portugal a long-term rise in the supply of labor per worker does not seem implausible during this period, given the sustained spread of labor-intensive products like maize (Ribeiro 1986) and wine (Martins 1998), and the productive opportunities offered by imperial expansion (Costa *et al.* 2015). This intuition can be corroborated with some precision. It is done here by comparing the gross silver values of agricultural output estimated from, respectively, the demand and the supply sides at two distant points in time for which data exist, namely 1515 and 1850.[12]

The result is that to produce the agricultural output actually obtained in 1850 would have meant increasing the average labor effort per worker by about 150 percent relative to 1515. Clearly a fixed labor input-based methodology would

9 While Allen (2001) uses skilled wages, Pfister (2012) and Álvarez and Prados (2013) employ unskilled wages, and Malanima (2011) a combination of the two. The total year's work effort may comprise an average of 250 (Allen 2001; Pfister 2012), 168 (Álvarez and Prados 2013) or 165 days (Malanima 2011).

10 The lengths of these three types of working years are taken from Spain's (1750) Catasto de Ensenada. See Álvarez and Prados (2013: 7).

11 Details on the estimation of these shares and the respective sources are in Section A.1.3 of the online appendix to Costa *et al.* (2015).

12 The supply side benchmarks are derived from Godinho (1968) and Reis (2000). The demand side ones are estimated using the demand-for-food function described above.

entail a substantial under-estimate of actual agricultural output in Portugal for later periods and require a sliding adjustment in the work effort variable. In the absence of direct evidence to enable us to distribute this differential over time, we correct the bias by using an indicator of labor intensity which reflects the annual share of maize, the most diffused labor-intensive crop, in total grain production.[13]

Putting together an indicator of aggregate rental income is not without its problems either. At this time, most of this revenue came from land, little of which was farmed directly by its lords. Possibly a little less than half of this factor of production was under a regime of commercial tenancy (Monteiro 2005). Typically, leases in this case ran for from three to ten years and provide dependable market-based evidence of the value of the services produced by this resource. The remainder was held under long-term or perpetual emphytheutic agreements, whereby the lord received a fixed fee and the tenant enjoyed an assignable right to the exclusive enjoyment of the fruits of the land (Fonseca and Reis 2011; Costa *et al.* 2016).[14] Although the latter leases were more numerous than the former, the corresponding records are far harder to extract from the sources. We have thus assumed that the rent of the first category of contracts provides a reliable guide to the long-run market price for agricultural land as a whole. This finds justification in the fact that tenant turnover in the second, more rigid category of leases might not be infrequent, given the high mortality among tenants and their successors. Since this could lead to changes in fees at every turn, it is reasonable to suppose that emphytheutic charges were thus likely to vary over time and therefore be quite responsive to market forces, and similarly to commercial ones.[15]

A second difficulty posed by the implementation of the agricultural output demand function has to do with the value of the elasticities which the model requires. In the absence of historical estimates for them, the literature has drawn inspiration from present day cases of third world economies where

13 This proxy is based on data for grain production in the district of Viseu from the 16th to the 19th centuries (Oliveira 1990 and 2002). Comprehensive, quantitative knowledge about other sectors is patchier, and we assume that, during these three and a half centuries, the rise of "industriousness" was of the same order of magnitude as this one.

14 While the first of these arrangements was employed mostly for larger units of production, the latter corresponded to small or minuscule farms. Contractual forms for accessing land are discussed at length in the first five chapters of this book.

15 For an illustration of how this could happen with similar leases in a 16th-century Spanish context see Drelichman and Gonzalez-Agudo (2014). For a similar point about the flexibility of emphyteutic leases in 17th century Portugal see Amorim (1997) and Neto (2012).

such parameters have been directly estimated and which have similar traits to those of Early Modern economies. There is a fairly wide range from which to choose and which go, for own-price elasticity, from −0.4 to −0.7, and, for income elasticity, from 0.3 to 0.6. Although we find the arguments on this matter by Álvarez and Prados (2013: 3) convincing, we have decided to employ those from Allen (2000), so that we can compare our results with his panel of estimates for agricultural output in ten European countries. We have opted consequently for the set the latter proposes, namely a = 0.5, b = −0.6 and c = 0.1.

In implementing this demand model, caution is also needed to make sure that the consumer price index (CPI) employed in it is suited to the context under examination. Allen's (2001 and 2009) "respectability" basket of goods consumed by a hypothetical mid-18th century worker's household in Strasbourg is widely used by the current literature on Early Modern European price history.[16] In order to adapt it to Portuguese conditions, we have altered the original list of items, as well as some of their quantities, whilst bearing in mind that the global nutritional content is not significantly altered. The Portuguese consumer basket thus defined comprises six food items – bread (made of either wheat or maize), meat, wine (instead of beer), olive oil (instead of butter), eggs, and hens (instead of cheese) – and four non-food items – charcoal, linen cloth, soap and lamp oil.[17]

16 We have not adopted the recently devised basket for 17th century Madrid, by Andrés and Lanza (2014), because of what seemed remarkably high figures for meat and wine per capita consumption, respectively 42 kilos and 110 liters a year.

17 The quantities in question correspond to a yearly consumption per adult and are the same as in Allen (2001: 421 and 2009: 36), except in the following cases in which we follow his specific alterations for southern Europe. Grain comprises both wheat and maize, in varying proportions according to their respective changing yearly shares in national output. For this sliding scale see footnote 13. Grain is transformed into bread using the formula calculated by Allen (2001: 418–419). Pulses are not included due to a lack of price information. They are replaced by the equivalent amount of grain (52 liters), which provides a similar quantity of calories. Beer (182 liters) is substituted by 68.25 liters of wine, and 5.2 kg of butter by 5.2 liters of olive oil. Five hens take the place of 5.2 kg of cheese, for which Portuguese price evidence is scant. Given the warmer and longer days characteristic of the European south, annual fuel consumption is reduced from 5 to 2 million BTUs (British Thermal Units) produced from 90 kg per (for the energy equivalent of charcoal see http://www.extension.iastate.edu/agdm/wholefarm/pdf/c6-88.pdf). We use for other non-agricultural goods the same weights as in Allen (2001), and replace lamp oil with an equal amount (5.2 liters) of olive oil, which performed the same function. The price of soap is proxyed by that of olive oil, which was the principal raw material in the manufacture of this item.

The last of the three assumptions on which our estimate of agricultural product rests has to do with the extent of the rural population's engagement in markets. As regards the Early Modern Period, the traditional view in European (and Portuguese) historiography has been that peasants were risk-averse and tended to stay away from the imperfect and risky markets of that era. This made them inclined to self-sufficiency and, consequently, impervious, in their production and consumption decisions, to the stimulus of market forces. If this were so, attempting to model their food consumption patterns on the basis of their reactions to market signals, as we are doing here, would be contradictory and would undermine the aims of this endeavor. A growing current in economic history has argued, however, that such a belief misrepresents the situation. Peasants were not as risk-averse as has been claimed and were often able to work their way round hazards, even using markets as part of their strategy for this. One simple reason is that they lacked sufficient land to sustain their families and meet their obligations to pay dues and taxes in cash. They were thus compelled to enter markets in order to sell at least some of their farm produce and even some of their labor.[18]

Although the traditional view has enjoyed a strong influence in Portugal too (Serrão 1993: 88), the available evidence increasingly suggests an important economic role for rural markets from 1500, if not earlier. Marques, cited by Santos (1998: 149), has claimed that already in the 16th century a majority of the rural population regularly worked to some extent in return for cash payments, and this continued to prevail at the end of the 18th century. Both in the most and in the least densely populated provinces at this time at least 60 percent of families were dependent for their survival to a significant extent on the sale of their members' labor (Sousa and Alves 1997; Fonseca and Reis 2011). Revealingly, both Oliveira (1982) and Santos (1998) lay the blame for this at the door of the land scarcity which plagued more than three fourths of rural households and made it unlikely that a regime of economic self-sufficiency might have dominated the Portuguese countryside during those centuries.[19] The great abundance of price and wage data which research has recently unveiled

18 For a useful discussion of these issues and evidence on the market involvement of the French peasantry before the 19th century see Hoffman (1996: Chapter 3). Morrison and Snyder (2000: 64) claim that in 18th century France servants and the lower ranks of those in agriculture received half of their income in cash, and that further up the social ladder this share would have been much greater.

19 This share, in the 17th century, was 80–95 percent in the province of Entre-Douro-e-Minho where a quarter of the Portuguese population lived. See Maia (1991), Oliveira (2002) and Mota (2006).

validates the notion that, throughout the period, the majority of the population was connected to some extent with rural or urban markets.

Data

The long-term series employed here to estimate Portuguese gross agricultural production have been gathered under the auspices of the PWR-Portugal project and can be consulted at its website.[20] The sources for both wages and prices refer entirely to Lisbon and its rural hinterland, the latter defined as the territory administered by the episcopal see of the capital city. We thus assume the validity of the principle of "national representativeness" of the data pertaining to the country's major urban centre and surroundings, a procedure generally adopted by the Early Modern wages and prices literature.[21] Several reasons justify this choice here.

Portugal is a small country (89,000 square kilometers) and Lisbon is centrally located within it. Although land communications in the Early Modern period were poor, the capital had reasonable links by sea, river and land with many of the country's regions and their markets (Macedo 1951: 105).[22] This would permit the circulation of commodities on a not inconsiderable scale. At least by the late 17th and 18th centuries, the integration of markets for food products was hardly insignificant. Undoubtedly, there was a significant transport and transaction cost premium between Lisbon and its more distant markets, but since improvements in transportation during this period were limited, inter-regional price differentials on this account are likely to have remained more or less constant. In the case of labor mobility, formal and practical restrictions were non-existent and qualitative evidence regarding internal labor movements during this period is plentiful, suggesting the existence of well-integrated markets for this factor (Silbert 1966; Reis 2005).

In common with similar projects, our data come chiefly from the accounts of religious foundations, charitable institutions, royal palaces and municipalities. In the case of commodity prices, they refer almost always to market transactions. In the case of wages, they correspond to employment in either agriculture or the building industry and exclude situations in which there were non-monetary complementary remunerations. In order to standardize results,

20 This project is entitled PWR-Portugal: Prices, Wages and Rents in Portugal, 1300–1910 and can be visited at http://pwr-portugal.ics.ul.pt/.

21 For a recent discussion of the pitfalls involved in the cases of Britain and Italy see Malanima (2013).

22 See maps A13, A14 and A15 in the Appendix.

TABLE 6.2 *Years in each century covered by data (percent of total)*

	Wheat bread	Maize bread	Meat	Eggs	Hens	Wine	Olive Oil	Wages skilled	Wages unskilled
16th c.	94	na	78	33	60	62	65	66	75
17th c.	92	84	98	98	100	83	100	42	40
18th c.	98	94	99	97	100	100	100	79	65
1800–1850	100	100	100	100	100	78	100	98	100

SOURCE: SEE TEXT.

all monetary values have been converted into grams of silver and prices have been normalized by the metric system. The coverage afforded by our data compilation displays some variance, with the lowest density of observations occurring in the 16th century, especially during its first half.[23] Some linear interpolation has therefore been necessary in order to fill empty cells in the time series. In Table 6.2, we display the percentage of years for which there are data in each century, for the most important variables in use here. As might be expected, coverage rises over time and varies considerably across series. Since workers' daily rates of pay commonly stayed constant for much longer stretches than prices, interpolation in their case is less problematic.

Land rents have attracted comparatively little attention in Portugal from historians of this period, and no doubt for this reason attempts at quantifying them are hard to come by. In the present study, our proxy for the value of the services of land is the total rent generated every year by an invariant set of 32 large commercial estates leased out by a charitable institution in the region of Alentejo between 1595 and 1850 (Santos 2003). We retropolate this index back from 1595 to 1565, using an analogous index for a similar set of estates owned by the hospital of Todos os Santos (All Saints) and located north of Lisbon; and, from then until 1500, employing the series for Spain published by Álvarez and Prados (2013).[24]

23 The problem of inadequate coverage and the solutions adopted for it – inter- and extrapolation – are common in the historical construction of Early Modern price, wage and rent series. The implication is that the results obtained are more suitable for analyzing the long-term picture than the short-term one. See Zanden and Leeuwen (2012).

24 Between 1595 and 1850, the main Portuguese series, in grams of silver, is consistent not only with the Spanish one by Álvarez and Prados (2013), but also with several other

Until recently, data on the year-to-year variation of the total population of Portugal have been unavailable. The only countrywide demographic information consisted of a few household counts at irregular intervals prior to the end of the 18th century. Thereafter, the two proper population censuses were carried out in 1801, 1841 and 1864 (Rodrigues 2008). This has serious negative implications for the present evaluation, since a linear interpolation between population figures based on such far flung benchmarks is bound to smooth excessively and unrealistically the annual movements of Portugal's aggregate agricultural output per capita. A recent annual series for the country's population during the years 1500–1850 (Palma and Reis 2016) is now available, however, and enables us to overcome this difficulty.[25] It follows the methodology pioneered by Wrigley and Schofield (1989) and combines information on national population stocks, taken from reliable benchmarks, with yearly flow statistics drawn from a sample of regionally representative parish-level demographic studies.

Trend, Cycles and Short Term Fluctuations

For a long time, the tendency of Portuguese historiography has been to regard pessimistically the performance of Early Modern agriculture.[26] This consensus has been constructed on an empirical foundation of impressionistic and occasionally quantitative elements, but has rarely aimed at achieving findings that could be expressed numerically or comparatively. Indeed, until the 1970s, attempts at measuring total production accurately were not a part of the discipline's program.[27] The few historians who contemplated this possibility recognized that the lack of adequate sources rendered this goal unattainable. Writing in the 1960s about agrarian Portugal at the demise of the Ancien

shorter Portuguese series from northern parts of the country collected by Salvado (2010). Moreover, it satisfies the condition that to be accurate it should reflect the principal outputs of the sector (Clark 2002). Grain, olives and sheep are represented but wine is missing from the picture. Nevertheless, its price series tracks that of grain quite closely and our proxy can thus be deemed nationally representative from this point of view.

25 For details of this estimation see the on-line appendix to Palma and Reis (2016).

26 See, however, Serrão, Chapter 5 in this book.

27 In 1951, Macedo (1951) made a list of the time series that would be needed as a foundation for "a clear and precise" study of the agrarian history of Portugal in the 18th century: prices, wages, rents, production areas, yields, taxes, etc. Some of these were published a few years later in Godinho (1955), but even so the kind of agrarian history that Macedo had in mind never took off.

Régime, Silbert warned that "the application of the rigorous present day methods of economic, social and geographic history [to Portugal] seems to us impossible. To change this would require a miracle" (Silbert 1966: vol. 1, 10).[28]

At about this time, however, an important change occurred. As a result of the introduction of the study of tithes into this field, an exciting new tool of analysis of the agrarian past. Thanks to the influence of the French *Annales* School, a flow of new research based on this ecclesiastical source spread rapidly. The result was a significant increase in the quantitative component of the study of agricultural history, as well as a drive in new directions of interpretation.[29] A more precise breakdown of the ebb and flow of agricultural activity into phases of expansion, contraction and stagnation was one important contribution of this advance. The possibility of measuring the different rates of change involved in this sub-periodization and of assessing the relative importance of different crops in its long- and medium-term development was another. A third was the potential for aggregating production data into a single representative variable, given the fact that, with this method, agricultural output could be linked to a due which was charged not only on all the fruits of the earth without exception, but also at a universal, easy-to-calculate uniform rate of one tenth.

Despite all its advantages, not all that was expected of this methodology was attained. The pitfalls proved to be many. The pronounced local character of this type of research made it hard to construct, with tithes, an empirical foundation for the desired national scope of this exercise. One reason is that combining tithe-based micro indicators to reach a higher level of generality, whether regional or national, was hampered by the lack of adequate benchmarks from which to derive the weights for aggregating them. Secondly, even if nominal output values could have been generated from the merger of disparate tithe series, deflating the result always remained a problem because such studies were often not accompanied by the production of broadly based consumer price indices (CPIs), the only ones suited to this purpose. Finally, the frequent patchiness of the sources inhibited the construction of reliable,

28 Similar concerns can be found in Oliveira (1982), Magalhães (1993), Serrão (1993) and Amorim (1997). A recent economic history of Portugal by Costa *et al.* (2016) reveals the persistence of these concerns as evidenced by the caution with which the authors handle in their narrative the macro-economic indicators required for the Early Modern period.

29 The pioneer of this diffusion was Oliveira (1979). His work has been followed, among others, by Campos (1989), Oliveira (1990 and 2002), Maia (1991), Silva (1994), Amorim (1997), Neto (1997) and Mota (2006). For a summary of tithe-based agrarian studies in Europe see Dodds (2007).

long-term, homogeneous time-series such as one needs in order to analyze agrarian history quantitatively over such long and complex stretches of time.[30]

Our present effort circumvents these problems by means of the new metric as described above. Its advantage lies in being strictly macroeconomic and grounded in a substantially broader database. The result, in the form of an annual estimate of gross agricultural product at constant prices, is displayed in Figure 6.1.[31] To assist the reader in making use of it, three main features should be noted. Firstly, it is expressed in abstract numbers, a consequence of the way in which it is calculated. The implication is that, for each year, its value has no concrete significance in itself and can only serve as a relative magnitude, to be compared with that of any other year of the series. For the base year of 1500 we have chosen a value of 100, all other observations being therefore a multiple of this reference point. In the second place, one should note that it is a consistent evaluation. This means that the values presented throughout the period are the result of applying the same methodology systematically to a homogeneous database. All observations are thus strictly comparable with each other, the same being true for the growth rate of any sub-period within the time span considered. Finally, one must remember that we are dealing here with the aggregate output of the sector, not with its per capita value, two very different concepts which have all too often been used interchangeably and confusingly in the literature.

The main significance of this time-series lies in its potential for solving quantitatively some of the key issues and dissipating some of the contradictions which have thrived in the analysis of Portugal's agrarian past. It allows us to deepen and clarify existing debates, but also to assist in raising hitherto unasked questions and probing new avenues of research. The development of appropriate measures of economic efficiency employing these results is an example of a possible area of innovation in the study of Portuguese agricultural history.

An examination of Figure 6.1 brings to light its three main temporal dimensions: the long-run performance over the entire Early Modern period; the cycles of growth, recession and stagnation into which the latter can be broken down; and the short but sharp fluctuations which are characteristic of the economies of this era. The first of these concerns the most critical question in this field: how well did the sector perform between the early 1500s and the

30 Chapter 4 by Neto illustrates the analytical difficulties associated with the use of tithes when it comes to estimating national agricultural output.

31 The data from which this figure is derived are represented in tabular form in the statistical appendix to this chapter.

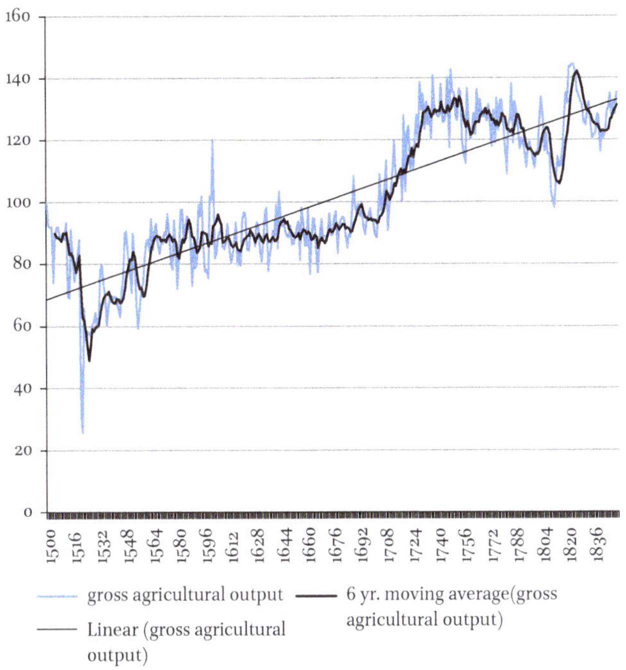

FIGURE 6.1 *Gross agricultural output (1500 = 100) (1500–1850)*

mid-19th century onset of a modern industrial economy? The answer can now be easily determined. Over the 350 years in question, the country's agricultural effort measured in real terms roughly doubled, displaying on trend a cumulative annual growth rate of 0.20 percent.[32] Although seemingly modest, it is hardly a trivial figure bearing in mind the length of the time span involved. On the other hand, as might be expected it stands in sharp contrast ro the 1.01 percent achieved during the next 100 years (Lains 2003: 254–257), a plausible difference bearing in mind that we are comparing a pre-capitalist with a modern capitalist economy (Cabral 1976). Most important of all, however, is that it settles once and for all one of the major issues in Portugal's agrarian history, that of the persistent "decadence" or "stagnation" of this sector. Any claim

32 This growth rate is measured between any pair of points representing centred averages on the linear trend line and not between the end points on the curve. If the latter were used instead, this would yield the much smaller rate of 0.09 percent. An alternative, also used in the growth literature, is to measure this rate between two sufficiently distant peaks or troughs of the curve. Given the difficulty in determining the exact turning points of our curve, we do not use it here.

which depicts the country's agriculture as structurally incapable of sustained long-term expansion is thus finally dispelled.

It is evident from Figure 6.1 that the profile of Portuguese gross agricultural output is far from a straight line. In fact, it can easily be divided into a set of coherent episodes of expansion, contraction or immobility, lasting roughly from a half to a whole century at a time. In the more structured historical narratives currently available, these phases are the backbone of the analytical effort whereby the shifting fortunes of primary production are understood as part of a process. It is therefore important to identify and re-appraise them with care. To do so, we have superimposed on the original time-series a seven-year moving average, which suggests, through a myriad of ups and downs of the output curve, the turning points at which these sub-periods may be held to have started and ended. Obviously, the phases which emerge from this procedure may contain some elements of arbitrariness since their definition depends on the kind of "smoothing" we impose on the data. On the other hand, if other criteria were used, the result would hardly have differed, and thus leaves our conclusions intact.[33]

With the help of this device, we divide the three and a half centuries laid out before us into six meaningful sub-periods. Although some coincidences can be detected with the current conventional historical wisdom in such matters, some interesting differences come to light. They may imply more than minor historiographical corrections. The 16th century illustrates this well. Generally, it has been perceived as a long and unbroken stretch of moderate growth, described by historians as an era of "agrarian expansion" (Costa *et al.* 2016: 94), which for some began already in the 1450s (Santos 1998: 154–156), and from which "crises" and setbacks tended to be absent (Gil 1965: 34). On the basis of our new estimate, however, reality appears to have been somewhat different. The long but moderate secular upswing is replaced by three distinct shorter phases, the first of which was a sharp decline of about 40 percent in gross agricultural product occurring between 1500 and the mid-1520s. This was a quite remarkable episode, which has received hardly any attention in the literature.[34] The ensuing recovery, which lasted until the late 1560s and returned

33 Tests were carried out with 5-, 10- and 15-year moving averages and at most altered the turning points by two years.

34 One factor which helps account for this downswing is the frequency of bad harvests: one in every decade of this period. The most dramatic of these was the harvest failure of 1521–22 caused by a severe drought, which afflicted Spain, Portugal and Morocco. Agricultural output fell by a third of the level of the previous decade (Godinho 1963–1965: vol. 2, 278).

agricultural output to where it had been in 1500, is the second stage. The third carries through from here for the next 100 years as an era of oscillations around a horizontal trend line, which ended in the late 1660s. By this time, agricultural output had still not progressed in any enduring way beyond the level at which our study begins.[35]

Turning to the 17th century, the standard narrative has long perceived this as a classic age of economic torpor after the good fortune which allegedly stamped the 1500s. As regards agriculture, several recent analyses based on micro-evidence from tithe records have concurred that gross agricultural product in 1700 probably still had not exceeded the level attained in 1600 (Oliveira 1990: 55; Neto 2012: 266 and Costa *et al.* 2016: 177). Yet, they are also unanimous in asserting that by the last two or three decades of that century the sector's productive rhythm may have been rising again.[36] Divergence exists mainly regarding when protracted stagnation began. Was it in 1600 or 1620?[37]

The information in Figure 6.1 confirms most of this assessment, but with some adjustment.[38] On the one hand, it corroborates two cherished notions: that the "crisis" (as periods without growth have often been designated by Portuguese historians) lasted for approximately a century; and that it came to an end sometime in the 1670s or 1680s. On the other, it shows that this long absence of growth started much earlier – either in the 1570s or the 1580s – some thirty to fifty years sooner than has been thought.

In the Early Modern era, the 18th century is undoubtedly the sub-period that has enjoyed the least consensus concerning the performance of agriculture. It

For a compilation of "calamities" during the Early Modern period see the specific section in PWR-Portugal entitled "Chronology of calamities".

35 See Miranda, Chapter 3 in this book.

36 While in an earlier work Serrão (1993) declared it impossible to pronounce on the national agricultural trend of the 17th century, he has recently stated (2005: 47–53) that, at least from the 1670s on, it is clear that output was on the increase.

37 In the case of studies which use similar evidence but have a regional rather than a national focus, patterns diverge somewhat from these national perspectives. In the Entre-Douro-e-Minho province, Capela (1987) has found that local agricultural output was taking off already from 1650. In Beira Litoral, there was a similarly precocious economic recovery, according to Amorim (1997), and the same may have happened in Algarve (Magalhães 2008).

38 The narrative of Chapter 4 in this book, which covers basically the same time span, is somewhat at variance with the present analysis. The principal difference is whether there was a sharp decline in the late 16th and early 17th centuries (our view), and whether 1620–1670 was a period of stagnation (our view) or one of oscillation around a horizontal trend line.

is also the epoch which has probably been the most thoroughly scrutinized and which has attracted the largest amount of research. Disagreement has arisen mainly over two issues. One of them is whether growth was sustained over the whole century, or was concentrated entirely in the first half of it, with the second witnessing stagnation or even a reversal. The former view has been espoused by Costa *et al.* (2016: 210) with some caution, by Marques (1973: vol. 1, 519), Serrão (2005: 148–149 and 2009: 48), Oliveira (2007: 256) and Lains (2009: 3, 48). The first has been defended by Justino (1988: 28–30), Pedreira (1994: 386) and Monteiro (1998: 317), mainly on the grounds that the decades after 1750 witnessed a pronounced decline in the grain sub-sector.[39]

The second problem concerns the net variation in output that may have occurred over the course of the century as a whole. For those who have defended the existence of a continuously rising agricultural cycle, the answer to this, obviously, has to be a secular increase. From the opposing point of view, of those who endorse a downswing after mid-century, a net expansion over these 100 years is not a foregone conclusion. The outcome depends on whether the slow-down in the last fifty years annulled the gains of the preceding half-century. Only one author, Pedreira (1994: 389), has so far attempted to quantify this outcome by describing this century as a time of "stagnation".

Our new estimate clarifies these doubts. A glance at Figure 6.1 reveals a very marked contrast between the two halves of this century, with an unmistakable downturn in the early 1750s.[40] Moreover, it shows that growth did not start abruptly *circa* 1700 but constituted, as noted above, a longer trend coming from the 1660s. It accelerated around the turn of the century and ended in the 1750s at a historic peak 50 percent above the figure for 1500. At an annual compound rate of 0.72 percent (measured from 1665 to 1754), this was the longest continuous sweep of agricultural growth in Portugal's Early Modern period and the best result, in terms of levels, ever achieved prior to 1825.

39 This point is developed by Lains (2009: 34) on the basis of the real income of fifty families of the high nobility between 1750 and 1820. A similar decline has been detected by Ribeiro (2012) lower down the social scale, for the provincial nobility of the region of Coimbra. Regionally-based studies are present once again in the debate, this time on the side of the retreat of agricultural output after 1750. See Silbert (1966), Magalhães (1988) and Santos (2003).

40 Chapter 5 by Serrão above endorses this outline of the agricultural output performance of the 18th century, but only in terms of the macroeconomic approach we follow here. It claims, on the other hand, that a more broadly based assessment, which also takes into account variables such as crop innovation, adaptability to markets and capacity to feed the population will yield a "better indicator of agricultural progress" (p. 147).

Figure 6.1 helps illuminate a further aspect of 18th century agricultural history. It shows that the output decline following the early 1750s was significant enough to justify the pessimism of enlightened contemporary opinion, but not sufficient to cancel earlier advances. Although by the next trough, around 1811, gross agricultural product had suffered an overall decline of 40 percent relative to its mid 18th century peak, it was still 20 percent above the previous 17th century low, around 1665. As Serrão (2009 and Chapter 5 above) has suggested, it is as if, during the 1700s, an epochal change had occurred in the nature of Portuguese agriculture, and permanently raised its productive capability to a new plane.

The last of the six stages into which Portuguese Early Modern agrarian history can be divided runs from the trough in 1811 until 1850. In contrast with the sub-periods which made up the preceding 300-odd years, it is probably the least known and understood. The chief virtue of our gross agricultural product estimate is that it thus allows us to fill an empty space in the overall trend of agricultural performance between the demise of the Ancien Regime and the middle of the 19th century. For a period generally held to have been economically adverse owing to the regression in manufacturing and trade and to the prevailing climate of social and political turmoil, the achievement of an annual growth rate of 0.7 percent – an overall increase of 30 percent – may come as a surprise. This presents us with the puzzle of what could have caused such a re-awakening of primary production, bringing it back to the historic peak of the 1750s. Possible explanations are: the Liberal implementation of a partial land reform in the 1830s; a hypothetical flight of capital and enterprise from the less promising performance of the trade and industry sectors; and the establishment during the 1810s of a protectionist regime for beleaguered grain producers (Ferreira 1995). The respective impacts of these factors remain, however, to be tested.

Over the centuries, rapid and often quite significant oscillations in agricultural production have been a salient feature of pre-industrial economies. These successions of "good" and "bad" harvest years have been a major attraction for agrarian historians and are amply documented in their writings. Portugal, in this respect, is no outlier. Indeed, the record of these fluctuations and of their proximate causes constitutes a substantial part of its Early Modern economic historiography. In these analyses, however, agricultural prices, and not the output variations themselves, have played the leading role in the study of the problematic of short-term output variations.[41]

41 For examples see, amongst a great many others, Godinho (1955), Oliveira (2002) and Santos (2003).

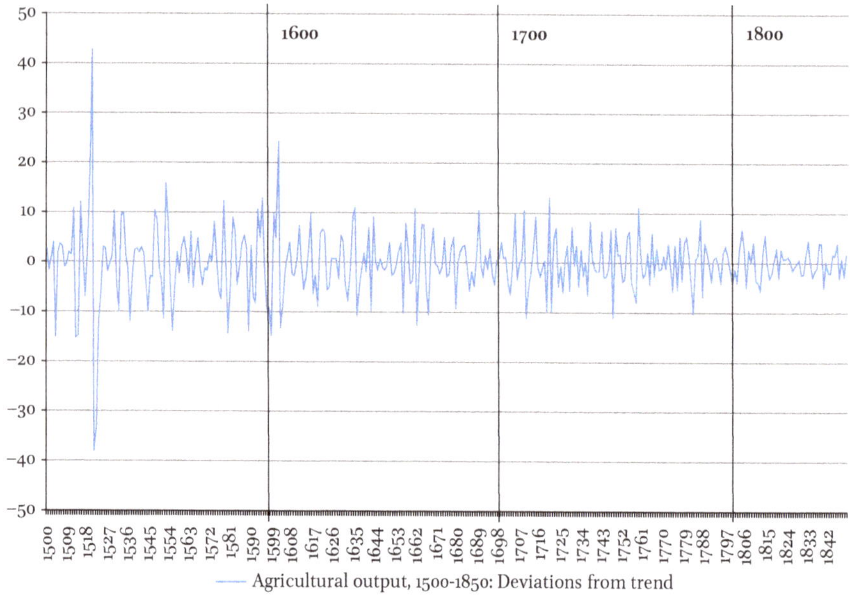

FIGURE 6.2 *Agricultural output, 1500–1850: Deviations from trend (percentage)*

The third major contribution of the newly estimated index of gross agricultural output is the possibility of deriving a better indicator from it for making cross-temporal comparisons of this sector's variability. This is done here by taking annual deviations from the long-term trend of gross agricultural product and dividing them by the respective trend values in those years.[42] Relative to the conventional price-based approach, the advantage of this is, first of all, that it is based on a direct indicator, rather than a proxy, as has happened until now. The second is that, while our indicator reflects the reaction of the entire sector to shocks, in most studies this is not the case. Usually, only one price or a small handful of disaggregated prices have been employed, and in the latter case these may well follow contradictory paths and be difficult to reconcile. A third is that in price-based evaluations a given year is compared with the preceding one or with the mean of a number of preceding years, a standard of comparison which has quite a different meaning from that provided by a trend.

Figure 6.2 displays the deviations of annual agricultural output from trend measured in percentage points.[43] It allows us to observe the frequency and

42 A Hodrick-Prescott filter for annual data is used for the purpose of calculating these trend values.

43 A similar exercise has been carried out for England before the Industrial Revolution by Kelly and O'Grada (2013). Unfortunately, owing to differences in the methodology used

intensity of these episodes and brings to light the archetypal unevenness in the performance of this sector. If we assume a drop in output of more than 10 percent as the threshold for "bad" agricultural years, the total of such occurrences for the whole of the Early Modern period comes to a total of twenty-one such events. This suggests that serious food scarcity was taking place on average once every seventeen years, that is, more than once in a generation. It also corroborates the traditional view that this was an epoch in which exposure to hunger and its trail of misfortunes was common. On the other hand, a less obvious fact and one that is hardly remarked upon by the literature is the decline over time of the incidence of bad years, which these data reveal. In the course of the 1500s, fourteen episodes may be classified as "lean", or worse: 1504, 1513–4, 1521–3, 1532, 1537, 1545, 1552, 1556, 1580, 1589 and 1597. It was also the century in which the only truly catastrophic shock of this kind – a fall in output of more than 30 percent – took place, in 1521 and again 1522. In contrast, during the 17th century bouts of serious hunger were endured only five times – 1603, 1636, 1656, 1662 and 1667 – and in the next century, Portugal was visited only three times by this scourge: in 1710, 1748 and 1783.[44] During the first half of the 19th century, there was not a single calamitous agricultural year.

One may speculate whether it was the colder and wetter climate of the 1500s which determined this "hunger cycle", and then the climatic improvement, from about the 1630s, which later had the opposite effect on harvests.[45] One may also suppose that it was the secular change in the product mix towards greater diversification which rendered agriculture less susceptible to adverse natural conditions and flattened the peaks and troughs of agriculture's productive profile.[46] Or it could be a combination of both and possibly other factors as well. Whichever it is, the main point to be made here is, once again, that shifting from other tools of analysis to output estimates has two important advantages. It enlarges the information pool at our disposal, thus providing a richer and clearer picture of the past. And it takes us further into our exploration of

to estimate gross agricultural product, their results cannot be compared with those presented here for Portugal.

44 Interestingly, the short-term agricultural crises identified in Chapter 5 of this book match only one of these three episodes of "serious hunger" – the one centred on the year 1710.

45 See figures A1 and A2 in the Appendix.

46 The rising integration of regional and local markets and the gradual reduction of institutional barriers detected by Serrão (2009) for the 17th and 18th centuries may have lessened the oscillations of agricultural prices. We cannot infer from this however that it would have reduced variability in output as well.

the past by raising new questions and problems and creating additional demand for new data and for the methodology to produce them.

Are the Results Consistent?

The importance of the issues raised by the construction, for the first time, of a pre-1850 measure of gross agricultural production is sufficient to justify questions regarding the methodology and sources used for this purpose. Some of them have been pondered already in the methods and data section above, and hopefully dispelled. Others, concerning the historical consistency of the results, have not and need now to be discussed. In this section, we try to verify the credibility of our findings by examining them in the light of the broader picture of European agriculture, as well as by checking their plausibility in terms of relevant domestic indicators.

The first step is to establish that our long-term profile of Portuguese agriculture does not imply an unusual growth rate. Allen (2000) provides a convenient standard of comparison in the form of a sample of nine West European countries for which gross agricultural output at constant prices during the period 1500–1800 has been calculated. Table 6.3 displays the percentage increase in their respective agricultural outputs over these three centuries, as well as for Portugal. These data are calculated as the variation between the benchmark values at the extreme points of the distribution, rather than along the trend

TABLE 6.3 *Gross agricultural output increase, 1500–1800*

	Increase (%)
Italy	22
Spain	31
France	33
Portugal	38
Austria	43
Germany	53
Belgium	60
Poland	95
Netherlands	119
England	147

SOURCE: ALLEN 2000: 17 AND TEXT

line, for reasons of comparability with the standard we have adopted here. The conclusion is that Portugal is hardly out of line with the rest, since it belonged to the majority of nations which managed to achieve only relatively modest secular agricultural growth in this period. This contrasts with England and the Netherlands, which reached the exceptional rates of increase of respectively 95 and 147 percent in the same interval.[47]

Although we may not have erred as regards the rate of change, a second concern here is whether we may not have over- or undervalued the actual level of output instead. The current dearth of country estimates of agricultural output at current prices for this period makes it difficult to reach comparative conclusions regarding levels, as opposed to rates, of change. An indirect way of approaching this problem is from a Malthusian point of view. In an era such as this one, of scarce food supplies, each national agricultural system had to deliver a minimum per capita nutritional requirement in order to sustain its population. Anything less than this "subsistence" food consumption over a significant stretch of time would have implied an untenable demographic situation, as preventive or positive checks would sooner or later have kicked in. Observed population movement can thus be a way of determining whether our calculation of agricultural output is realistic, not by direct comparison with other national outputs but by analyzing whether historical demographic dynamics corresponded to estimated food availability.

Figure 6.3 graphs food supply per capita (drawn from our estimate of agricultural output) in Early Modern Portugal and reveals a significant fall in this variable between 1500 and 1850. A long-term decline was not uncommon in Europe at that time, as Allen (2000: 18) has pointed out: "on most of the continent, [agricultural] output ... failed to keep pace with population growth", particularly during the 16th century.[48] Indeed, all nine countries in his sample experienced decreases in food consumption in the range of 10 to 33 percent during these three centuries. In Portugal, however, the drop was considerably greater – 60 percent – yet this happened at a time when population was rising impressively, and by 1800 had reached a level almost three times higher than in 1500. The joint performance of these two variables over such a long time span

47 We are unable here to compare agricultural output levels at different points in time because Allen's (2000) methodology for estimating this variable differs in important respects from ours. In our case, we assume variability in the annual number of days each worker worked per year, in contrast with his preference for assuming a fixed labor input per capita. This biases upwardly the Portuguese figures for levels.

48 More up-to-date figures for Italy and Spain corroborate this finding. See, respectively, Federico and Malanima (2004) and Álvarez and Prados (2013).

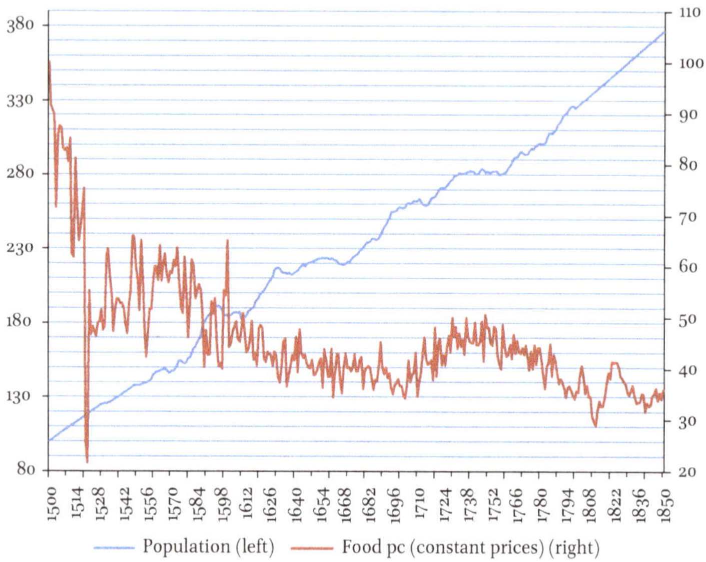

FIGURE 6.3 *Per capita food supply and population in Portugal (1500 = 100) (1500–1850)*
SOURCE: PALMA AND REIS 2016; AND TEXT. NOTES: FOOD PC = RATIO OF GROSS AGRICULTURAL PRODUCT TO POPULATION; FOOD SUPPLY IS ASSUMED AS EQUAL TO AGRICULTURAL OUTPUT.

seems like a flagrant contradiction of the workings of the Malthusian model. To solve this puzzle we consider two possibilities. One is that we have underestimated the food supply and therefore the diet consumed per capita, which may have been after all a lot closer to "subsistence". This would explain the absence of any significant demographic reversion. Alternatively, our calculations might be correct and yet so high initially that even at the lowest subsequent levels of decline it never moved below "subsistence".

We consider two food baskets at current prices and valued in grams of silver, which are accepted as standards of "subsistence" for the Eary Modern period. One is the "barebones" diet proposed by Allen *et al.* (2012: 873) and inspired by the survival efforts of poorly paid labor during this period.[49] The other is Federico and Malanima's (2004: 445) "minimal subsistence basket for Italy".[50] Both

49 It comprises 165 kilos of maize/wheat, 20 kilos of beans/peas (wheat/maize), 5 kilos of meat and 3 kilos of butter (olive oil) for a year. Items in brackets are the replacements in Portugal for those in the original Northern European list.

50 It includes 50 kilos of wheat, 170 kilos of other cereals (maize), 10 kilos of meat, 3 kilos of olive oil and 80 liters of wine. Firewood is in the original but for the sake of uniformity is ignored here.

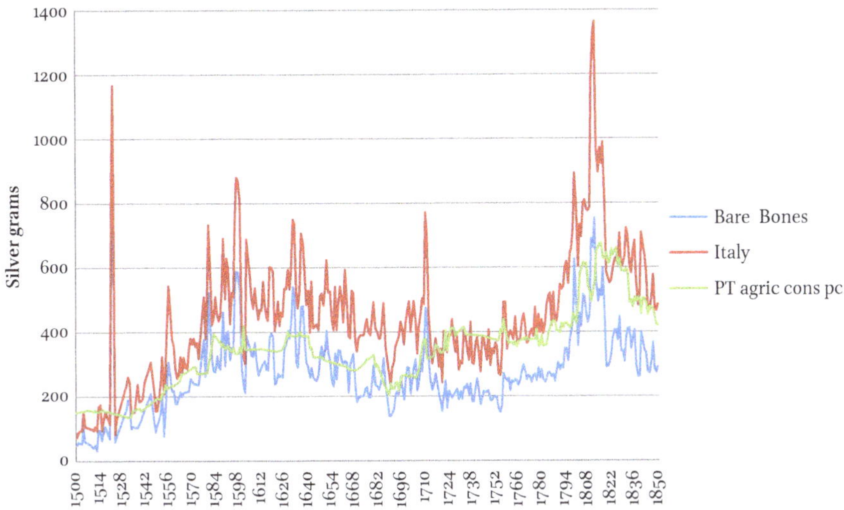

FIGURE 6.4 *Subsistence baskets (Italy and "barebones") and Portuguese per capita food consumption at Portuguese market prices (in silver grams)*
SOURCE: ITALIAN BASKET, FEDERICO AND MALANIMA 2004: 445; "BAREBONES" BASKET, ALLEN ET AL., 2012: 873; PORTUGUESE BASKET, SEE TEXT.

correspond to the needs of an adult and are converted here using Portuguese prices so as to be comparable with Portuguese per capita food consumption estimates. We cannot use here our earlier quantification of national agricultural consumption (see Figure 6.3) since it is in constant, not nominal silver prices. We obviate this problem by employing a short-cut estimation method proposed by Malanima (2011: 179). This yields an acceptable current price estimate for this variable. It involves multiplying total labor income by a factor of 1.4 to obtain the total income of all production factors.[51] Since it is presumed that only a third of the population is economically active (Álvarez and Prados 2007: 326), we multiply this figure by a factor of 0.33 and again by one of 0.6. The latter takes into account Malanima's additional assumption that food consumption is equal to 60 percent of the value of total income.

Figure 6.4 allows us to compare the cost, at Portuguese current prices, of two distinct patterns of "subsistence" food consumption with the value, also at current prices, of the food that, at different points in time, was actually available to the Portuguese population. We conclude that Portugal's agriculture

51 The total income from labor in grams of silver is obtained as described above. Malanima (2011: 178) considers that labor income equals 70 percent of total income.

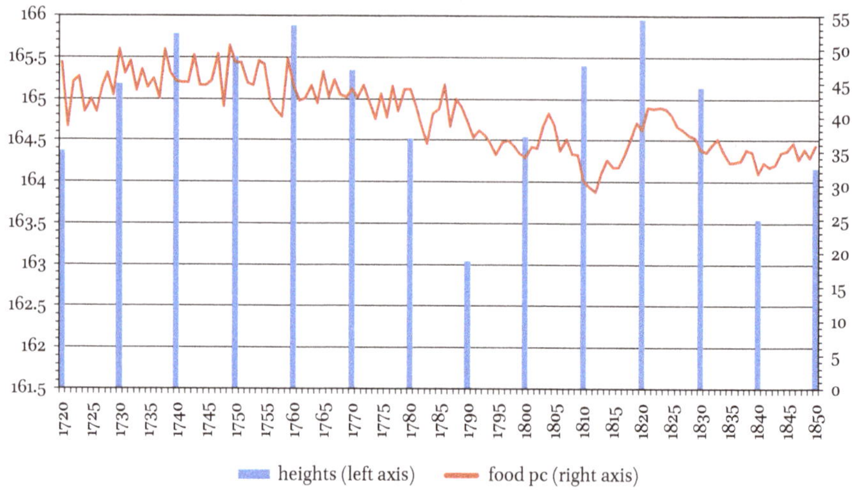

FIGURE 6.5 *Average heights and food supply per capita (1720–1850)*
SOURCE: FOR AVERAGE HEIGHTS, STOLZ ET AL., 2013; FOR FOOD SUPPLY, SEE FIGURE 6. NOTE: STATURE MEASUREMENT TAKEN 20 YEARS AFTER PER CAPITA FOOD CONSUMPTION OBSERVED.

consistently provided a level of consumption which was close to accepted subsistence requirements and thus made it possible for the population to withstand Malthusian pressures and keep on expanding. We may thus conclude that our findings are not an underestimate of gross agricultural output.

Per capita food intake offers us a second opportunity to corroborate our initial output findings, this time using evidence from the field of anthropometric history.[52] A recent study regarding the evolution of the height of Portuguese males between 1720 and 1980 has shown that the standard of living and the level of human capital of this population were, in the long run, the principal determinants of stature (Stolz *et al.* 2013). The first of these was measured by these authors using the real wage level, but in the present instance we employ average real food consumption instead. This choice is, in fact, an even better solution since we use a variable which has a more direct impact on corporal growth than wages, only part of which are spent on the nutrients which promote physical development. On the other hand, it has the disadvantage for us of covering only the years between 1720 and 1850.

Even so, setting these two variables side by side, as in Figure 6.5, provides encouraging support for our estimates of the agricultural trend during the last

52 The link between food consumption and heights has been explored in the same way for Italy by Federico and Malanima (2004).

century and a half of the period under observation. In the forty years from 1720 to 1760, steadily rising food supplies per person were matched in Portugal by the upward evolution in the height of the military recruits born in those decades. During the 1760s to 1790s, this co-movement was reversed, after which the two variables together entered another upward swing until the 1820s. The last period for which we have data, from 1820 to 1850, witnessed another decline in both food consumption and stature.

The preceding approaches to the consistency problem have considered only the long-term behavior of our estimated variable. Looking at its short-term fluctuations can also help us assess the reliability of this indicator. The method, in this case, is to compare qualitative records of agricultural disturbance – pestilence, bad weather, earthquakes, war and civil commotions – with episodes of severe harvest failure such as were identified in Figure 6.2 above. The source for these accounts is an annual "compilation of calamities" located in the PWR-Portugal site. It is based on a variety of contemporary descriptions and chronicles, as well as historical studies, covering the period from 1309 to 1909.[53] The expectation is that "bad agricultural years" should coincide with or at least be close to years in which "calamities" occurred. The former are identified here as those in which gross agricultural production fell by more than 10 percent relative to trend. As we saw earlier, the total number of these incidents amounts to 22, which were unevenly spread out over three and a half centuries.

The conclusion is that pronounced momentary downturns in our output estimate were indeed associated with years in which Portugal was visited by misfortune, as observed by contemporaries. They were usually linked to "food shortages", "famine" or "disastrous crops". On the other hand, we must not ignore the fact that not all the years in which such disasters were reported were attended by severely insufficient harvests. There are several reasons why this match might not always have occurred. One of them concerns years when, owing to lack of data, interpolations had to be made in order to fill gross agricultural product gaps. This could lead to a "smoothing" effect in the time-series and could make what was in effect a bad year "disappear" from the statistical curve. Another is that adverse circumstances in a particular year might harm a certain crop but not necessarily all of them. Since our main indicator is constructed from several weighted averages, this could lead to an "ironing out" of signs of a sudden drop in production. It is

53 See PWR-Portugal, under "Chronology_of Calamities (earthquakes, famines, food crises, hunger, pandemics, pestilences, plagues, starvations, wars, ...)".

interesting to note that almost two thirds of these bad years were related to sharp rises in both wheat and wine prices. If one of these crops alone had been stricken e.g. by weather or quakes, it would have taken a much more severe shock in the other one to show up in the aggregate output statistics. Finally, we have to consider that statistical smoothing of the output curve may also have taken place as a result of efforts by the crown or municipal authorities to combat shortages, in particular when they involved measures to restore decimated food supplies either by importing grain or by stimulation of its overland trade.[54]

Conclusion

This chapter provides for the first time a much-needed consistent metric for Portugal's real gross agricultural output during the Early Modern period. It uses a standard indirect method based on a consumption function for agricultural products and a rich recently mined database for its construction.

Our year-by-year estimation allows us to clarify a number of important issues. Overall, Portugal's pre-industrial agriculture did not stagnate during the period considered, contrary to what has often been thought. Indeed, it attained a reasonable rate of expansion, similar to that of many other European countries. In fact, it succeeded in sustaining a fast-growing population using a fixed supply of land thanks to a process which involved significant shifts in the structure of production, in the intensity of land use and in increased labor effort. This did not prevent a permanent fall in the level of food consumption compared to that enjoyed by its population at the onset of our period of study. This agrarian regime transition occurred in the course of a succession of stages in which growth alternated with stagnation or even reversion, and which we are now able to define with greater precision than before.

The principal aim of the present study has been to draw a profile of the most important sector of the economy of Portugal during the Early Modern period using quantitative tools. It does not attempt to reach a deeper and more focused kind of analysis than this, something which is done in the earlier chapters of this book. The objective here is simply to lay the groundwork for and facilitate future research in the field of Portuguese agrarian history.

54 There is a good deal of anecdotal evidence in this respect but it has yet to be used systematically in order to allow an encompassing analysis of this problem.

Statistical Appendix

	Gross Agricultural Output, constant prices (1500 = 100)	Food per capita, constant prices (1500 = 100)
1500	100	100
1501	92	92
1502	92	90
1503	92	90
1504	74	72
1505	89	86
1506	92	87
1507	92	87
1508	89	83
1509	89	83
1510	90	83
1511	88	81
1512	94	85
1513	69	63
1514	69	62
1515	91	81
1516	82	73
1517	74	65
1518	78	68
1519	82	71
1520	88	75
1521	30	26
1522	26	22
1523	66	55
1524	56	47
1525	58	48
1526	58	47
1527	57	47
1528	60	49
1529	61	49
1530	64	52
1531	60	48
1532	61	48
1533	78	62

(cont.)

	Gross Agricultural Output, constant prices (1500 = 100)	Food per capita, constant prices (1500 = 100)
1534	80	64
1535	73	58
1536	69	55
1537	60	47
1538	66	52
1539	69	54
1540	70	54
1541	69	53
1542	70	53
1543	69	52
1544	65	49
1545	63	47
1546	71	53
1547	75	55
1548	90	66
1549	91	66
1550	83	60
1551	79	57
1552	71	51
1553	90	65
1554	80	58
1555	67	48
1556	59	42
1557	65	47
1558	73	52
1559	73	52
1560	82	56
1561	88	60
1562	88	60
1563	84	57
1564	95	64
1565	85	57
1566	90	61
1567	93	62
1568	87	59

	Gross Agricultural Output, constant prices (1500 = 100)	Food per capita, constant prices (1500 = 100)
1569	83	57
1570	86	59
1571	87	59
1572	90	61
1573	89	60
1574	96	64
1575	89	58
1576	80	52
1577	78	51
1578	94	62
1579	84	54
1580	72	46
1581	83	53
1582	98	61
1583	98	60
1584	88	54
1585	92	55
1586	95	56
1587	95	55
1588	90	51
1589	73	40
1590	87	48
1591	79	43
1592	80	43
1593	99	53
1594	96	51
1595	102	54
1596	86	45
1597	78	41
1598	78	41
1599	75	40
1600	102	55
1601	101	54
1602	120	65
1603	82	44
1604	83	45

(cont.)

	Gross Agricultural Output, constant prices (1500 = 100)	Food per capita, constant prices (1500 = 100)
1605	88	47
1606	90	48
1607	92	49
1608	85	46
1609	85	45
1610	87	48
1611	93	51
1612	85	47
1613	80	43
1614	82	44
1615	87	46
1616	93	49
1617	80	42
1618	83	43
1619	79	41
1620	94	48
1621	97	49
1622	96	48
1623	85	42
1624	84	41
1625	89	43
1626	89	43
1627	89	43
1628	86	41
1629	94	43
1630	92	42
1631	84	38
1632	81	37
1633	87	41
1634	97	45
1635	98	46
1636	78	37
1637	83	39
1638	86	41
1639	91	43

	Gross Agricultural Output, constant prices (1500 = 100)	Food per capita, constant prices (1500 = 100)
1640	89	42
1641	99	46
1642	85	39
1643	103	48
1644	95	43
1645	92	42
1646	93	43
1647	90	41
1648	89	40
1649	89	40
1650	92	42
1651	86	39
1652	87	39
1653	89	40
1654	92	41
1655	94	42
1656	81	36
1657	98	44
1658	94	42
1659	86	38
1660	86	39
1661	98	44
1662	77	34
1663	87	39
1664	95	43
1665	94	43
1666	82	37
1667	77	35
1668	88	40
1669	94	43
1670	89	40
1671	89	40
1672	88	40
1673	90	40
1674	96	43
1675	89	40

(cont.)

	Gross Agricultural Output, constant prices (1500 = 100)	Food per capita, constant prices (1500 = 100)
1676	90	40
1677	95	42
1678	97	42
1679	83	36
1680	92	40
1681	94	40
1682	95	41
1683	95	40
1684	90	38
1685	85	36
1686	90	38
1687	89	38
1688	96	40
1689	108	45
1690	98	41
1691	96	39
1692	99	40
1693	95	38
1694	98	39
1695	93	36
1696	90	35
1697	94	37
1698	95	37
1699	98	38
1700	95	37
1701	94	37
1702	89	35
1703	88	34
1704	96	37
1705	109	42
1706	97	37
1707	102	39
1708	103	39
1709	113	43
1710	91	35

	Gross Agricultural Output, constant prices (1500 = 100)	Food per capita, constant prices (1500 = 100)
1711	98	37
1712	103	39
1713	111	43
1714	120	46
1715	109	42
1716	106	41
1717	108	41
1718	110	42
1719	100	38
1720	128	48
1721	103	39
1722	122	45
1723	124	46
1724	111	41
1725	116	43
1726	111	41
1727	121	44
1728	128	47
1729	119	43
1730	139	50
1731	130	46
1732	135	48
1733	123	44
1734	132	47
1735	125	44
1736	128	46
1737	120	43
1738	141	50
1739	131	47
1740	128	45
1741	127	45
1742	127	45
1743	138	49
1744	125	45
1745	126	45
1746	129	45

(cont.)

	Gross Agricultural Output, constant prices (1500 = 100)	Food per capita, constant prices (1500 = 100)
1747	140	49
1748	117	42
1749	143	51
1750	136	48
1751	135	48
1752	127	45
1753	126	45
1754	136	48
1755	135	48
1756	119	42
1757	115	41
1758	112	40
1759	137	49
1760	126	45
1761	120	42
1762	122	43
1763	128	45
1764	121	42
1765	136	47
1766	126	43
1767	133	46
1768	127	43
1769	127	43
1770	130	44
1771	126	43
1772	131	45
1773	125	42
1774	118	40
1775	129	44
1776	119	40
1777	133	45
1778	123	41
1779	133	44
1780	133	44
1781	125	42

	Gross Agricultural Output, constant prices (1500 = 100)	Food per capita, constant prices (1500 = 100)
1782	116	39
1783	109	36
1784	124	41
1785	127	41
1786	138	45
1787	119	39
1788	132	43
1789	129	41
1790	123	39
1791	117	37
1792	120	38
1793	119	37
1794	116	36
1795	111	34
1796	117	36
1797	119	37
1798	117	36
1799	113	35
1800	111	34
1801	116	36
1802	116	35
1803	127	39
1804	134	41
1805	128	39
1806	116	35
1807	122	37
1808	115	35
1809	115	35
1810	103	31
1811	100	30
1812	98	29
1813	108	32
1814	114	34
1815	111	33
1816	112	33
1817	118	34

(cont.)

	Gross Agricultural Output, constant prices (1500 = 100)	Food per capita, constant prices (1500 = 100)
1818	126	37
1819	135	39
1820	132	38
1821	144	41
1822	143	41
1823	144	41
1824	144	41
1825	141	40
1826	136	39
1827	134	38
1828	132	37
1829	131	37
1830	125	35
1831	125	35
1832	129	36
1833	132	37
1834	125	35
1835	120	33
1836	121	33
1837	122	34
1838	128	35
1839	128	35
1840	116	32
1841	122	33
1842	121	33
1843	122	33
1844	129	35
1845	130	35
1846	135	36
1847	126	34
1848	132	35
1849	128	34
1850	135	36

SOURCES AND NOTES: SEE TEXT

References

Allen, Robert C. (2000). "Economic structure and agricultural productivity in Europe, 1300–1800". *European Review of Economic History*, 4, pp. 1–25.

Allen, Robert C. (2001). "The great divergence in European wages and prices from the Middle Ages to the First World War". *Explorations in Economic History*, 38, pp. 411–447.

Allen, Robert C. (2009). *The British Industrial Revolution in Global Perspective*. Cambridge: Cambridge University Press.

Allen, Robert, Murphy, Tommy and Schneider, Eric (2012). "The colonial origins of the divergence in the Americas: a labour market approach". *Journal of Economic History*, 72, pp. 863–894.

Álvarez Nogal, Carlos and Prados de la Escosura, Leandro (2007). "The decline of Spain (1500–1850): conjectural estimates". *European Review of Economic History*, 11, pp. 319–366.

Álvarez Nogal, Carlos and Prados de la Escosura, Leandro (2013). "The rise and fall of Spain (1270–1850)". *Economic History Review*, 66, pp. 1–37.

Amorim, Inês (1997). *O Mosteiro de Grijó. Senhorio e Propriedade: 1560–1720 (Formação, Estrutura e Exploração do seu Domínio)*. Braga: Barbosa e Xavier.

Andrés Ucendo, J.L. and Lanza García, R. (2014). "Prices and real wages in seventeenth-century Madrid". *Economic History Review*, 67, pp. 607–626.

Angeles, Luis (2008). "GDP per capita or real wages? Making sense of conflicting views on pre-industrial Europe". *Explorations in Economic History*, 45, pp. 147–163.

Balbi, Adrien([1822] 2004). *Essai statistique sur le royaume du Portugal et d'Algarve*. Lisbon: Imprensa Nacional-Casa da Moeda.

Broadberry, S.N., Campbell, B.M.S., Klein, A., Overton, M., and Leeuwen, B. van (2015). *British Economic Growth, 1270–1870*. Cambridge: Cambridge University Press.

Cabral, Manuel Villaverde (1976). *O Desenvolvimento do Capitalismo em Portugal no Século XIX*. Porto: A Regra do Jogo.

Campos, Maria do Rosário Castiço (1989). *Foz de Arouce no Século XVIII: Economia Agrária e Reconversão Agrícola*. Coimbra: Universidade de Coimbra (PhD dissertation).

Capela, José Viriato (1987). *Entre-Douro-e-Minho, 1750–1830. Finanças, Administração e Bloqueamentos Estruturais no Portugal Moderno*. Braga: Universidade do Minho (PhD dissertation).

Clark, Gregory (2002). "Land rental values and the agrarian economy: England and Wales, 1500–1914". *European Review of Economic History*, 6, pp. 281–308.

Clark, Gregory (2007). *A Farewell to Alms: a Brief Economic History of the World*. Princeton: Princeton University Press.

Costa, Leonor and Reis, Jaime (2015). *The Chronic Food Deficit of Early Modern Portugal: Curse or Myth?* Working Papers of GHES, 2016/58.

Costa, Leonor Freire, Lains, Pedro and Miranda, Susana Münch (2016). *An Economic History of Portugal 1143–2010*. Cambridge: Cambridge University Press.

Costa, Leonor, Palma, Nuno and Reis, Jaime (2015). "The great escape? The contribution of the empire to Portugal's economic growth, 1500–1800". *European Review of Economic History*, 19, pp. 1–17.

Dodds, Ben (2007). *Peasants and Production in the Medieval North-east: The Evidence from Tithes 1250–1536*. Newcastle: Boydell Press.

Drelichman, Mauricio and Gonzalez-Agudo, David (2014). "Housing and the cost of living in early modern Toledo". *Explorations in Economic History*, 52, pp. 27–47.

Edvinsson, Rodney (2011). "New estimates of Swedish GDP by activity 1665–2010". *Stockholm Papers in Economic History*, 12.

Federico, Giovanni and Malanima, Paolo (2004). "Progress, decline, growth: product and productivity in Italian agriculture, 1000–2000". *Economic History Review*, 55, pp. 437–464.

Ferreira, Jaime (1995). *Origem do Proteccionismo da Cerealicultura em Portugal (de 1814 a 1820 – Hesitações e Afirmação)*. Coimbra: Boletim de Ciências Económicas.

Fonseca, Helder Adegar and Reis, Jaime (2011). "The limits of agricultural growth in a fragile eco-system: total factor productivity in Alentejo, 1750–1850". In Matts Olsson and Patrick Svensson (Eds). *Agricultural Production and Productivity in Europe*. Turnhout: Brepols, pp. 167–194.

Franzini, Marino Miguel (1843). *Considerações Ácerca da Renda Total da Nação Portugueza e sua Distribuição por Classes com algumas Reflexões sobre o Imposto da Décima*. Lisbon: Imprensa Nacional-Casa da Moeda.

Gil, Maria Olímpia da Rocha (1965). *Arroteias no Vale do Mondego durante o Século XVI*. Lisbon: Instituto de Alta Cultura.

Godinho, Vitorino Magalhães (1955). *Prix et Monnaies au Portugal 1750–1850*. Paris: Armand Colin.

Godinho, Vitorino Magalhães (1963–1965). *Os Descobrimentos e a Economia Mundial*. Lisbon: Arcádia.

Godinho, Vitorino Magalhães (1968). *Ensaios*. Lisbon: Sá da Costa.

Henriques, António (2015). "Plenty of land, land of plenty: the agrarian output of Portugal (1311–20)". *European Review of Economic History*, 19, pp. 149–170.

Hoffman, Philip T. (1996). *Growth in a Traditional Society. The French Countryside, 1450–1815*. Princeton: Princeton University Press.

Justino, David (1988). *A Formação do Espaço Económico Nacional. Portugal, 1810–1913*. Lisbon: Vega.

Kelly, Morgan and O'Grada, Cormac (2013). "*Numerare est errare*: agricultural output and food supply in England before and during the industrial revolution". *Journal of Economic History*, 73, pp. 1132–1163.

Krantz, Olle and Schön, Lennart (2012). "The Swedish economy in the early modern period: constructing historical national accounts". *European Review of Economic History*, 16, pp. 529–549.

Lains, Pedro (2003). "New wine in old bottles: output and productivity trends in Portuguese agriculture, 1850–1950". *European Review of Economic History*, 7, pp. 43–72.

Lains, Pedro (2009). *A Herança do Antigo Regime, 1750–1807* (unpublished manuscript).

Macedo, Jorge Borges de (1951). *A Situação Económica no Tempo de Pombal*. Porto: Portugália.

Magalhães, Joaquim Romero de (1988). *O Algarve Económico, 1600–1773*. Lisbon: Estampa.

Magalhães, Joaquim Romero de (1993). "As estruturas da produção agrícola e pastoril". In José Mattoso (Ed.), *História de Portugal*. Lisbon: Círculo de Leitores, pp. 263–267.

Magalhães, Joaquim Romero de (2008). *O Algarve Económico: 1600–1773*. Lisbon: Estampa.

Maia, Paula (1991). *O Mosteiro de Bustelo e a Produção Agrícola. O Antigo Regime – 1710–1821*. Penafiel: Câmara Municipal de Penafiel.

Malanima, Paolo (2009). *Pre-modern European Economy. One Thousand Years (10th–19th Centuries)*. Leiden-Boston: Brill.

Malanima, Paolo (2011). "The long decline of a leading economy: GDP in central and northern Italy, 1300–1913". *European Review of Economic History*, 15, pp. 169–219.

Malanima, Paolo (2013). "When did England overtake Italy? Medieval and early modern divergence in prices and wages". *European Review of Economic History*, 17, pp. 45–70.

Marques, A.H. de Oliveira (1973). *História de Portugal desde os Tempos mais Antigos até ao Governo do Sr. Marcelo Caetano*. Lisbon: Edições Ágora.

Martins, Conceição Andrade (1998). *Vinha, Vinho e Política Vinícola em Portugal. Do Pombalismo à Regeneração*. Évora: Universidade de Évora (PhD dissertation).

Medeiros, Carlos Alberto de (1993). "Environnement, agriculture et élevage au Portugal à l'époque des découvertes maritimes". In Robert Durand (Ed.). *L'Homme, l'animal domestique et l'environnement du Moyen Age au XVIIIè Siècle*. Nantes: Université de Nantes, pp. 307–331.

Monteiro, Nuno Gonçalo (1998). *A Corte, as Províncias e as Conquistas: Centros de Poder e Trajectórias Sociais no Portugal Restaurado (1668–1750)*. Lisbon: Colibri.

Monteiro, Nuno Gonçalo (2005). "A ocupação da terra". In Pedro Lains and Álvaro F. da Silva, (Eds.) *História Económica de Portugal*. Lisbon: Instituto de Ciências Sociais, vol.1, pp. 67–91.

Morrison, Christian and Snyder, Wayne (2000). "The income inequality of France in historical perspective". *European Review of Economic History*, 4, pp. 59–83.

Mota, Salvador Magalhães (2006). *Cistercienses, Camponeses e Economia Rural no Minho na Época do Antigo Regime: o Mosteiro de Santa Maria de Bouro e o seu Domínio*. Lisbon: Imprensa Nacional-Casa da Moeda.

Neto, Margarida Sobral (1997). *Terra e Conflito: Região de Coimbra, 1700–1834*. Viseu: Palimage.

Neto, Margarida Sobral (2012). "A crise da agricultura portuguesa no século XVII". In Álvaro Garrido, Leonor Freire Costa e Luís Miguel Duarte (Eds.) *Economia, Instituições e Império (Estudos em Homenagem a Joaquim Romero de Magalhães)*. Coimbra: Almedina, pp. 263–277.

Oliveira, Aurélio de (1979). *A Abadia de Tibães 1630–1813: Propriedade, Exploração e Produção Agrícolas no Vale do Cávado durante o antigo regime*. Porto: Universidade do Porto (PhD dissertation).

Oliveira, Aurélio de (1982). "Contabilidades monásticas e produção agrícola durante o Antigo Regime: os dízimos do mosteiro de S. Tirso, 1621–1821". In *Actas do Colóquio de História Local e Regional*. S. Tirso: s.n., pp. 131–172.

Oliveira, Aurélio de (2007). "Vinhos em Portugal (1300–1820). Um sector de sucesso na agricultura do Antigo Regime". *Douro. Estudos e Documentos*, 22, pp. 255–271.

Oliveira, João Nunes de (1990). *A Produção Agrícola de Viseu entre 1550 e 1700*. Viseu: Câmara Municipal de Viseu.

Oliveira, João Nunes de (2002). *A Beira Alta de 1700 a 1840. Gentes e Subsistências*. Viseu: Palimage.

Palma, Nuno and Reis, Jaime (2016). "From convergence to divergence: Portuguese Demography and Economic Growth". Groningen Growth and Development Centre Research Memorandum, 161.

Pedreira, Jorge (1994). *Estrutura Industrial e Mercado Colonial : Portugal e Brasil (1780–1830)*. Linda-a-Velha: Difel.

Pfister, Ulrich (2009). *Economic Growth in Germany, 1500–1850*. Unpublished manuscript presented at the "Quantifying long-run economic development" Conference. University of Warwick, Venice.

Pfister, Ulrich, Riedel, Jana and Uebele, Martin (2012). *Real Wages and the Origins of Modern Economic Growth in Germany, 16th to 19th Centuries*. EHES Working Papers in Economic History no 17.

Reis, Jaime (2000). "How poor was the European periphery before 1850? The Mediterranean vs. Scandinavia". In Sevket Pamuk and Jeffrey G. Williamson (Eds.). *The Mediterranean Response to Globalization before 1950*. London: Routledge, pp. 17–44.

Reis, Jaime (2005). "O trabalho". In Pedro Lains and Álvaro F. da Silva (Eds.). *História Económica de Portugal*. Lisbon: Imprensa de Ciências Sociais, vol. 2, pp. 119–151.

Ribeiro, Orlando (1986). *Portugal, o Mediterrâneo e o Atlântico*. Lisbon: Sá da Costa.

Ribeiro, Ana Isabel (2012). *Nobrezas e Governança. Identidades e Perfis Sociais (Coimbra, 1777–1820)*. Coimbra: Faculdade de Letras.

Rodrigues, Teresa Ferreira (Ed.) (2008). *História da População Portuguesa*. Porto: Edições Afrontamento – Cepese.

Salvado, João Paulo (2010). *Nobreza, Monarquia e Império: A Casa Senhorial dos Almotacés-mores do Reino (Séculos XVI–XVIII)*. Lisbon: Universidade Nova de Lisboa (PhD dissertation).

Santos, Rui (1998). "A sociedade rural". In Diogo Ramada Curto (Ed.) *O Tempo de Vasco da Gama*. Lisbon: CNCDP, pp. 135–155.

Santos, Rui (2003). *Sociogénese do Latifundismo Moderno. Mercados, Crises e Mudança Social na Região de Évora. Séculos XVII a XIX*. Lisbon: Banco de Portugal.

Serrão, José Vicente (1993). "O quadro económico. Configurações estruturais e tendências de evolução". In J. Mattoso (Ed.). *História de Portugal*. Lisbon: Círculo de Leitores, vol. 3, pp. 71–117.

Serrão, José Vicente (2005). "A agricultura". In Pedro Lains and Álvaro F. da Silva, (Eds.). *História Económica de Portugal*. Lisbon: Imprensa de Ciências Sociais, vol. 1, pp.145–175.

Serrão, José Vicente (2009). "Land management responses to market changes. Portugal, seventeenth-nineteenth centuries". In Vicente Pinilla (Ed.). *Markets and Agricultural Change in Europe from the Thirteenth to the Twentieth Centuries*. Turnhout: Brepols, pp. 47–73.

Silbert, Albert (1966). *Le Portugal meditérranéen à la fin de l'Ancien Régime. Contribution à l'histoire agraire comparée*. Paris: S.E.V.P.E.N.

Silva, Célia Taborda da (1994). *O Mosteiro de Ganfei. Propriedade, Produção e Rendas no Antigo Regime (1629–1683 e 1716–1822)*. Lisbon: Fragmentos.

Soares, Fernando de Brito (2005). "A agricultura". In Pedro Lains and Álvaro F. da Silva (Eds.). *História Económica de Portugal*. Lisbon: Imprensa de Ciências Sociais, vol. 3, pp. 158–183.

Sousa, Fernando de and Alves, Jorge (1997). *Alto Minho. População e Economia nos Finais de Setecentos*. Lisbon: Presença.

Stolz, Yvonne, Baten, Joerg and Reis, Jaime (2013). "Portuguese living standards, 1720–1980, in European comparison: heights, income, and human capital". *Economic History Review*, 66, pp. 545–578.

Vries, Jan de (2008). *The Industrious Revolution: Consumer Behavior and the Household Economy, 1650 to Present*. Cambridge: Cambridge University Press.

Wrigley, E.A. (1985). "Urban growth and agricultural change: England and the Continent in the early modern period". *Journal of Interdisciplinary History*, 15, pp. 683–728.

Wrigley, E.A. and Schofield, R.S. ([1981] 1989) *The Population History of England 1541–1871*. Cambridge: Cambridge University Press.

Zanden, Jan Luiten van and Leeuwen, Bas van (2012). "Persistent but not consistent: the growth of national income in Holland 1347–1807". *Explorations in Economic History*, 49, pp. 119–130.

PART 3

Growth, Structural Change and Economic Policy

CHAPTER 7

Growth, Institutional Change and Innovation, 1820–1930

Amélia Branco and Ester Gomes da Silva

Introduction

As happened with most European countries, in Portugal the period broadly located between the Liberal Revolution (1820) and the end of the First Republic (1926) was marked by an overall positive evolution of the agrarian sector. Output and labor productivity increased considerably, along with the cultivated area, whereas employment remained high. The evolution was not linear, though. Progress was faster between 1870 and 1902, whereas the first decades of the 20th century were marked by a significant growth slowdown. Moreover, despite the broad positive evolution, Portuguese agriculture remained backward by European standards, making limited use of modern tools and techniques and presenting rather low productivity levels (Federico 2005).

This was a period of major political and economic unrest. Liberal ideals were only incorporated in the 1822 Constitution, after two civil wars and significant economic destruction. The independence of Brazil in the same year made the political and economic environment even more unstable. In the 1850s, the emergence of a new government and the subsequent implementation of a number of constitutional reforms and of an ambitious program of national development gave rise to the period known as *Regeneração*. However, despite the reformist impetus and the observed acceleration of economic growth, by the end of the century a set of events threatened the stability of the kingdom, leading ultimately to a military coup and to the end of the monarchic regime in 1910. The effects of the First World War, combined with the financial crisis of 1923 and deep political instability, created the conditions for another military coup in 1926. This led to the *Estado Novo* dictatorship, whose consolidation in the early 1930s was facilitated by the effects of the Great Depression and the spread of authoritarian ideals across Europe.

The political and economic turmoil substantially accentuated the country's peripheral position: in 1860, the Portuguese GNP represented 86 percent of the average GNP of developed countries, whereas in 1913 the corresponding figure was only 45 percent (Reis 1993). A few structural changes did nevertheless

 | DOI 10.1163/9789004311527_009

take place. The share of agriculture in total GDP declined, whereas industry increased its importance, reaching about one third of total production in 1930, a figure close to that of agriculture (Lains 2009). Although the pace of change was relatively slow, some progress was made in the agrarian sector, most notably in the cultivated area, which increased substantially. Some improvements in cultivation methods and techniques were also introduced, such as the use of chemical fertilizers, draft animals and mechanical threshing.

In this chapter an analysis is made of these changes, taking into account the influence of both institutional and market factors. Although public and private spheres of action were to some extent intertwined, an effort is made to disentangle their effects on the advances registered in agriculture. After having described the main stylized facts of agriculture growth, we provide an analysis of changes in the institutional background, covering the changes from old rules of property rights and land tradability and the policies undertaken by liberal governments in the agriculture domain. An investigation of the actions developed by agriculture associations is also made, in an attempt to assess their role in the process of technology adoption and in the modernization of agriculture.

Land, Output and Productivity

Quantitative data on Portuguese agriculture for the 19th century are rather scarce, particularly for the first half of the century.[1] Output data for the sector as a whole are available only from 1848 (Lains 1995; Lains and Sousa 1998; Batista *et al.* 1997), although some evidence has been provided for earlier periods in time with regard to specific agricultural products (Valério 1983; Mateus and Mateus 1985; Justino 1988). Reliable data on land use, on the other hand, are available only from the 1860s (Mateus and Mateus 1985; Lains 1995).

Evidencing the backwardness of the sector (and of the country as a whole), agriculture remained the most important sector of the economy until very late, accounting for about one third of total production and more than 60 percent of total employment as late as in 1930 (Table 7.1). Compared to other European countries, these figures are extremely high (Zanden 1991). The importance of agriculture in the economy started to decline only after 1900, and even then the rate of decay was rather slow.

1 Official data covering the production of the main crops, land and livestock date back to the 1840s, but their publication on an annual basis has been provided only since 1915.

TABLE 7.1 *Agriculture shares in GDP and employment (percent)*

	GDP	Employment[1]
1850	47.8	67.5
1870	40.8	65.9
1900	40.1	66.4
1930	31.5	60.9

NOTES: 1 Data on employment refer to 1841 and 1862, rather than 1850 and 1870, respectively.
SOURCE: LAINS 2009: 335

TABLE 7.2 *Growth of output, labor force and labor productivity in agriculture, 1850–1930 (annual average growth rates, percent)*

	Output	Labor force	Labor productivity
1850–1970	−0.50	0.70	−1.20
1870–1900	1.62	0.72	0.90
1900–1930	0.82	0.59	0.23
1850–1930	0.79	0.67	0.12

SOURCE: LAINS 2009: 337 AND AUTHORS' COMPUTATIONS

Still, between 1850 and 1930, Portuguese agriculture witnessed considerable progress, growing at an average annual rate of 0.8 percent (Table 7.2). The growth dynamics varied over time, however, with three phases of growth usually being considered (Lains 2003b, 2009): from 1850 to 1970, agricultural output decreased, reflecting the decline in cereal production, due to adverse weather conditions, and the spread of a severe viniculture disease (*oidium*) (Cabral 1979; Pereira 1979; Justino 1988); the subsequent period (1870–1900) was marked by recovery and sustained growth at a relatively fast rate (1.62 percent a year, on average); from the turn of the century a considerable slowdown occurred in output growth, which lasted until 1930.

The periodization of agricultural output growth shows close resemblance with the pace of expansion of the cultivated land. Accompanying strong demographic growth (from the end of the 18th century up to 1930, the population

TABLE 7.3 *Land use (percent)*

	1867	1902	1929
1. Under Crop	21	34.6	36.5
2. Fallow and pasture	23.5	21.4	17.4
3. Agriculture (1 + 2)	44.5	56	53.9
4. Forest	13.8	21.8	25.9
5. Productive (3 + 4)	58.3	77.8	79.9
6. Unused. but fit	37.2	16.9	14.7
7. Unfit	3.2	4.1	4.3
8. Inland watercourses	1.2	1.2	1.2
Total (5 + 6 + 7 + 8)	100	100	100

SOURCE: LAINS 2009: 339

more than doubled, reaching a total of 6.8 million in the early 1930s), the land devoted to agriculture and forestry increased substantially. Based on regional data and qualitative historical evidence, some authors have located the enlargement of the cultivated area already in the 1830s–1840s, with the expansion of the cultivation of maize and rye across the northern regions of the country (Pereira 1983; Mateus and Mateus 1985). Based on stronger quantitative evidence, Lains (2009) shows that the land devoted to productive purposes (agriculture and forest) increased at a very fast rate between 1867 and 1902, thereafter showing relative stagnation (Table 7.3). The period of fastest increase in output thus coincides with the increase in cultivated land.

With regard to agricultural land, the observed increase reflects the rise in crop cultivation (a slight decline occurred in the proportion of land used in fallow and pasture). The impressive rise in the forest area, on the other hand, is mostly related to the production of pine and cork oak, which became increasingly important export products (Lains 2003; Câmara 2005). Cork production, in particular, increased considerably, presenting the highest growth of all agricultural products between 1870 and 1930 (Table 7.6). Such an expansion reflects the high economic return associated with its exploitation, due to the increasing relevance of cork stoppers for sealing bottles and to the development of the cork agglomerate industry. The development of pine, on the other hand, reflects essentially the expanding external demand for pinewood for mine and railway construction.

TABLE 7.4 *Growth accounting, 1865–1927 (annual average growth rates, in percent)*

	Output growth	Contribution of inputs			TFP growth	Labor productivity	Land productivity
		Land	Labor	Capital			
1865–1902	1.41	0.23	0.29	0.16	0.73	0.69	0.75
(in percent)	100.0	16.4	20.4	11.2	52.0		
1902–1927	0.77	-0.05	0.24	0.22	0.37	0.18	0.91
(in percent)	100.0	-6.4	30.6	27.9	47.8		

SOURCE: LAINS 2009: 340

The nature of the changes underlying agriculture output growth between 1850 and 1930 can be better understood with recourse to the standard neoclassical decomposition of growth (Table 7.4). The growth accounting exercise shows that inputs and TFP growth accounted for roughly equal parts of output growth during the period under study. In the period of fastest growth, between 1865 and 1902, agricultural output benefited from the increase in all inputs, as more land was put into agricultural use, along with capital and labor (Table 7.2). Nevertheless, the contribution of capital was rather low – capital intensity has actually declined – which factor is related to the slow introduction of mechanization in Portuguese agriculture (Reis 1993).[2] The deceleration of growth in the subsequent period, on the other hand, is partly related to the land input, whose contribution to growth became negative after 1902, and to the deceleration of multifactor productivity growth. Moreover, the decrease in labor productivity growth reflects the decline in both TFP growth and in the amount of land per worker, which were partially compensated for by an increase in capital deepening. Land productivity growth, on the other hand, has increased, revealing that technical change was more land-saving than of a labor-saving nature, which makes sense, taking into account that the amount of uncultivated but fit land had declined substantially.

Taken as a whole, these findings reveal a positive evolution of Portuguese agriculture, although one that was not sufficient to promote significant catch-up with more developed economies. Portugal's agricultural output growth rate between 1870 and 1913 was only slightly above the western and southern

2 Despite the increase in land in use, there was also a slight decline in land per worker, as the labor force increased faster.

TABLE 7.5 *Structure of agriculture production (percent)*

	1861–70	1900–9	1935–9
Cereals and potatoes	38.9	35.4	40.6
Wheat	10.8	11	17.5
Other cereals	23.6	19.9	14
Potatoes	4.5	4.5	9.1
Wine	21.8	23.3	13.5
Olive oil	5.8	8.1	7.9
Fruit and vegetables	7.8	6.5	6.5
Animal products	24.8	23.1	28
Cork	0.7	3.8	3.2
	100	100	100

SOURCE: LAINS 2009: 343

European average growth rates for the same period (Federico 2005) and productivity growth rates during the first three decades of the 20th century were considerably below those of countries such as Spain, Denmark or France (Yamada and Ruttan 1989; Bringas 2000). Taking into account that labor and land productivity levels were rather low, especially when compared with the most industrialized countries (Lains 1995; Yamada and Ruttan 1989), the potential to stimulate growth by incorporating capital (through mechanization) and knowledge from abroad was very high. Why this potential has not fully materialized is an issue to which we will return later in this chapter.

Moreover, the deceleration of growth after 1900 indicates that, faced with greater difficulties in achieving growth by increasing land use (by 1902, the percentage of "unused but fit land" was only 17), farmers were unable to stimulate significant improvements in resource efficiency, by making more substantial investments in the mechanization of production or by promoting rapid structural change, one of the most relevant sources of TPF growth during this period (Federico 2005). In fact, in a period lasting almost a century, changes in the composition of agricultural production were rather modest (Table 7.5). Cereals and wine remained the most relevant products, accounting for more than a half of total output until 1910. Animal products (meat, dairy and wool) kept their importance relatively unchanged (low by European standards), whereas the share of fruit and vegetables remained relatively small.

Given the predominance of grains and wine, the evolution of agricultural output as a whole reproduces to some extent the dynamics observed in these products. More precisely, as indicated earlier, the agriculture crisis of 1850–1970 partially reflects the sharp decline in wine production following the spread of the *oidium* disease. Moreover, the increase in output between 1870 and 1900 is mainly related to the growth of wine, olive oil and cork production, in that order. The recovery of wine production from *oidium* and *phylloxera* diseases since the 1880s, followed by high growth rates in subsequent years, even led to wine overproduction at the turn of the century, which prompted a number of policy measures designed to regulate wine production.[3] By the same token, the relative stagnation observed during the first three decades of the 20th century is partly explained by the decrease in the production of wine and some grains, most notably maize and rye. In fact, although there was an increase in the production of wheat and some other crops (especially potatoes), growth in agricultural output slowed down.

Looking at the finer detail, some changes can also be found with regard to the structure of grains' production. In the early 1800s, maize was the main grain cultivated in Portugal, especially in the north (Martins 2005).[4] During the 19th century, the production of wheat and other grains increased considerably, partly because of the expansion of the cultivated area, partly due to the increase in yield per hectare, although technological improvements were relatively slow.[5] Nevertheless, maize remained the most important grain during the second half of the 19th century, being the most popular cereal among small farmers on the coastal plains to the north of Lisbon, and expanding to other districts, namely Évora and Beja, in the south (Pereira 1983). The higher yields compared to wheat, along with the possibility of cultivation in either wet or dry lands and its high versatility, it being used for both human and animal food, explain the spread of cultivation across the country.

3 One of the implemented measures was the reestablishment of the official recognition of the Douro Wine Region, a terroir first defined in 1756 and then abandoned in the 1850s. During the period of overproduction, the State also defined other new wine terroirs: two for fine wines (Setúbal and Madeira) and five for table wines (Bucelas, Colares, Carcavelos, Dão, Vinho Verde). The definition of wine regions, a relatively pioneering policy in Europe, was supported by the general view that there was a need to produce wines in lesser quantity, but of greater quality (Freire 2010).

4 See Chapter 5 in this volume.

5 There is some evidence of the country's relative self-sufficiency in grain during the first half of the 19th century, until the emergence of the "cereal crisis" (Martins 2005).

TABLE 7.6 *Agricultural output growth by product (percent)*

	Wheat	Maize	Rye	Rice	Potato	Wine	Olive oil	Meat	Cork	Total
1870–1900	1.10	0.48	1.89	−1.87	1.41	3.49	2.60	0.85	2.30	1.62
1900–1930	1.83	−1.11	−1.06	3.11	3.25	−0.61	−0.01	0.57	2.73	0.55
1870–1930	1.46	−0.32	0.40	0.59	2.33	1.42	1.29	0.71	2.51	1.08

SOURCE: LAINS 2003B: 62 AND AUTHORS' COMPUTATIONS

From the turn of the 20th century there was, however, a negative trend in the price of maize which led to a decline in its production (Table 7.6). Moreover, the improved yields of maize relative to wheat, due to the introduction of more productive seed varieties and the widespread use of irrigation, led to a marked reduction in the area under maize cultivation. In contrast, the area occupied by wheat grew steadily between 1870 and 1930, reflecting lower productivity levels and the impact of the protectionist laws implemented in 1888 and 1899 (Reis 1979). The ratio between the land under maize cultivation and that under wheat fell sharply from the turn of the century, and by 1920 the production levels of these two crops had converged.

Along with maize and wheat, the main grains under cultivation, rye, barley, oats and rice, were also typical crops. Before the expansion of maize, the Portuguese diet was mostly based on wheat in the central and southern regions, and on rye, the preferred grain for making bread, in the north. Barley and oats, grown in rotation with other crops, were used mostly for animal feed. Rice, on the other hand, was destined for human consumption. Between 1870 and 1930, there was a significant decline in the production of this cereal, following the decline in the cultivated area determined by the new legislation on land and water uses (Justino 1988; Vaquinhas 1991; Martins 2005)

In contrast, the production of potatoes increased considerably, often replacing bread as a source of calories. After a long confinement to the north-west region, its cultivation reached the coastal areas, following food subsistence crises (Justino 1988). By 1930, it was already widespread across the country, forming part, along with other vegetables and olive oil, of typical Portuguese meals. The consumption of meat and other animal proteins, on the other hand, was less frequent. Nevertheless, there was an improvement in livestock until the beginning of the 20th century, especially in the Alentejo and Estremadura regions

TABLE 7.7 *Share in Portuguese exports of agricultural products (percent)*

	1842–1849	1850–1859	1860–1869	1870–1879	1880–1889	1890–1899	1900–1909	1905–1914	1914–XX
Raw cork	2,1	2,5	3,6	4,4	8,9	9,3	8,8	9,4	
Olive oil	2,8	5,1	4,3	2,5	0,7	1,2	1,8	1,8	
Port wine	37,7	37,7	37,1	35,3	26,6	24,4	17,7	17,8	
Madeira wine	7,1	3,7	1,3	2,1	2,6	2,8	2,3	1,9	
Ordinary wine	6,6	10,5	6,5	7,9	22,8	16,4	13,5	14,7	
Total	56,3	59,5	52,8	52,2	61,6	54,1	44,1	45,6	

SOURCE: LAINS 2003A: 75 AND AUTHORS' COMPUTATIONS.

(Fonseca 1996), as well as an increase in its share of agricultural production: from 13 percent in 1853/62 to 23.1 percent in 1900/9 (Lains 1995: 53). Some of this boost was in mules, which were widely used for farm work and transportation, and sheep, whose increase in 1906 and 1925 was associated with the expansion of grain in the Alentejo region, where they grazed after the harvest and in fallows (Reis 1979). It would be long before steam engines and tractors replaced mules. The unevenness of some of the land and the small size of properties, particularly in the north, help to explain this delay (Radich 1996; Radich and Monteiro 2000).

Changes in the composition of agricultural exports were also modest. Being the most important sector of the economy until at least the 1930s, Portuguese exports were dominated by a few agricultural products, and most particularly, by wine, which still accounted for about 44 percent of total exports at the turn of the century (Table 7.7). Between the mid-19th century and 1930 the main changes in the structure of exports involved a sustained decline in the wine share after 1890 (mainly Port) and, contrastingly, an increase in cork. With the exception of these products, in which a significant part of the production was directed to the external market,[6] the export intensity of the agriculture sector remained small, with most of the production being absorbed by domestic consumption. In fact, the strong orientation towards the domestic market posed serious demand constraints, since real wages were low and displayed constant

6 Between 1880 and 1913 wine exports represented about one third of total production, whereas for cork the corresponding figure was about 80 percent (Fonseca 1996: 142).

or even decreasing trends since 1862 (Valério 2001), which refrained farmers from increasing the scale of production.

During the phase of fastest growth, agricultural exports expanded rapidly, but at the turn of the century there was a decline in external demand (Lains 2003b), particularly from the English market. Such a decline was a result of increasing competition from new producing countries (e.g., the US and Argentina) that, given the development of transport and refrigeration systems, placed their products in Europe at lower prices. It reflected also the country's inability to promote significant structural change, shifting to products with higher demand elasticities (Lains 1995). This decline could not be compensated for by a redirection in trade towards the Portuguese colonies, given the low incomes and the reduced number of trade settlers of these markets. Brazil, the major Portuguese colony, was an exception, but its independence in 1822 implied the removal of old trade privileges, particularly affecting wine and olive oil exports. Between 1864 and 1884, wine exports to Brazil still experienced an increase, but since then they have gradually declined.

The deceleration of agricultural exports' growth was reflected in the reduction of the share of agricultural products in total Portuguese exports. Conversely, the share of agricultural products in total imports increased between 1842 and 1865, reflecting the effects of the agricultural crisis and the subsequent rise in cereal imports, particularly wheat (Reis 1979; Lains 2003b). Even after the implementation of protectionist measures in 1888 and 1899 and the substantial rise in wheat production, cereal imports remained high. Demographic growth and the increasing preference for wheat bread explains the persistent shortage of domestic production relative to consumption levels (Reis 1979).

Institutional Change

After a long period of civil wars and political unrest, the second half of the 19th century witnessed a progressive enlargement of State intervention, involving the modernization of institutions, the provision of public infrastructure, most notably roads and railways, and a general orientation towards a more efficient functioning of markets. With regard to agriculture, a broad consensus emerged on the necessity to promote its development and modernization. The heightened interest in the problem led to the creation of a number of public services to assist the government in the formulation of agriculture policy, which were centralized in the Ministry of Public Works, Trade and Industry in

1852. The political importance of agricultural affairs was reinforced in 1886 with the creation of the General Department for Agriculture within the Ministry of Public Works, Trade and Industry and, later, during the First Republic, with the establishment of an autonomous ministry devoted to agriculture in 1918.

The growing concern about agriculture materialized in a number of policies and legislative interventions aimed at solving several deficiencies of the sector, namely, the restrictions on property and land tradability inherited from the *Ancien Régime,* the lack of an appropriate institutional setting for the provision and diffusion of agricultural knowledge, and the chronic shortage of financial capital. With regard to the last, acknowledging the insufficiency of Common Barns (*Celeiros Comuns*),[7] several measures were implemented to increase the amount of credit available to farmers, allowing institutions such as *Misericórdias* (charitable institutions), *Irmandades* (brotherhoods) and *Confrarias* to allocate funds to the creation of agricultural credit banks.[8] However, only three banks were created in provincial towns (Viseu, Viana do Castelo and Faro) and from this group, one only – the Banco Agrícola e Industrial Visiense – showed some signs of prosperity in 1912 (Santos 2012). The general conditions of credit access remained rather difficult: there was a significant lack of specialized institutions awarding loans; the existing banking institutions were concentrated in the most urbanized areas of the country; the available capital was in short supply and interest rates were high; most of the farmers who were tenants or partners could not access credit, since they could not use land as a security for loans; loans against credit letters were restricted to large landowners, who could provide better guarantees. The creation of the legal basis of the modern agricultural credit system would only be accomplished after the fall of monarchy, through the creation of the Agricultural Credit Board and the Special Fund for Agricultural Credit and the implementation of several laws on mutual agricultural credit. Following the creation of these laws, 85 *Caixas de Crédito* (local banks) were established between 1911 and 1929, mostly located in the agro-towns surrounding Lisbon, Alentejo and Algarve, in which a stronger concentration of large landowners existed (Santos 2012).

At the same time, a number of initiatives were carried out by the State to stimulate education and knowledge diffusion in agriculture. The Agricultural Institute for higher education and three intermediate schools were created

7 *Celeiros Comuns*, created in the 16th century, were entrusted with the function of providing seeds on credit to farmers in years of scarce production.

8 Decrees 4-10-1834, 24-3-1843, 8-2-1845 and 22-6-1867.

in the 1850s, providing the central and regional services with experts in the areas of agriculture, forestry and livestock. Much of the scientific knowledge produced on rural technology, chemistry, animal production and agronomic policies came from these schools, being developed by reputed specialists (e.g., Ferreira Lapa and Silvestre Bernardo de Lima) and spread to the "civil society" through training courses and public conferences (Madureira 2002). Later, during the First Republic, the foundations of the modern higher education system were established, through the creation of *Instituto Superior de Agronomia*, awarding degrees in agronomy and forest engineering, and the Higher School of Veterinary Medicine.

Liberal governments also attempted to lay the foundations of a modern tax structure, abolishing several tributes and taxes. Nevertheless, the pace of change was slow, and the strong prevalence of indirect taxation and, particularly, of customs duties as sources of fiscal revenue has even increased (Esteves 2004). With regard to agricultural products, customs affected mostly cereal production, especially wheat, in line with several other European countries during this period.[9] The governments of *Regeneração* have gradually introduced some liberalization measures, abolishing the restrictive legislation related to the production of Port wines in 1852 and 1865 and promoting a relative liberalization of grain trade between the 1850s and the 1880s, following the cereal crisis. However, towards the end of the 19th century, the combination of a number of forces – the deep international crisis, the escalating external debt, and the strong pressures exerted by absentee-landlord farmers from the south – compelled governments to move back to protectionism, specially targeted to wheat, and lasting from 1889 to 1914 (Reis 1979; Lains 2006). The protectionist measures undertaken followed the general European trend, but in the case of Portugal, they were mainly implemented through the establishment of minimum domestic prices, introduced first in 1889 and reinforced ten years later by the "Hunger Law". These laws explain the increase in land devoted to wheat cultivation and the substantial rise in production levels, which was particularly intense after 1889 (Reis 1979).

Political intervention in the domain of agriculture was, however, especially relevant with regard to the abolition of old rules on land property and tradability. Between the Liberal Revolution and the emergence of the Republican regime in 1910, a number of policy measures were implemented to eliminate market restrictions inherited from the *Ancien Régime* (Silbert 1966; Silveira

9 The rapid expansion of wheat exports from overseas (United States and Argentina), particularly after 1870 led many countries to use tariff protection against wheat trade. See Federico (2006) and Junguito (2006) for an analysis of the Italian and Spanish cases, respectively.

1980; Fonseca 1996; Amaral 2012). Despite the political volatility of the epoch, the most prominent features of the old institutional arrangements related to agricultural property and land use were slowly being abandoned.

In this regard, the most relevant policy actions – the conversion of crown lands to national assets, and the reform of the municipal rights charts – were already established by 1821/22. However, the military rebellion of 1823 and the subsequent abolition of the Constitution of 1822 by the restored absolutist regime impeded their application. In a context of deep political instability, which culminated in a civil war between liberals and absolutists, the application of the most significant pieces of legislation was delayed to 1834. Known as the *Law of Mouzinho da Silveira*, this legislation included, among other measures, the removal of several civil and religious obligations related to land property and taxation, most notably: *Morgadios* (land to be inherited that could not be sold or divided), *Foral* laws (local censuses and rations), *dízimos* (local ecclesiastical tithes) and *sisas* (taxes on property transfers). The nationalization of the properties of religious orders (disentitlement), following their abolition in 1834 and their subsequent sale to the market, amplified the changes in property that were under way (Silveira 1980; 1993; Amaral 2012).

Responding to the political pressures of the elites, who defended the prevalence of private property rights, a number of laws were also directed to the property and use of commons (*baldios*) (Gralheiro 1990; Neto 1990, 2012 and 2013). Like crown and religious orders' lands, commons were long held responsible for many underdevelopment problems of the Portuguese agricultural sector. The liberal regime began to alter the rules governing the commons in 1822: under the *Lei dos Forais* (1822), different types of common lands were unified, falling under the remit of local municipalities and gaining the ability to be privately owned. In 1850, a new classification of common land was applied, separating parish from municipal assets, but the disentitlement of these lands took place only after 1866 with the implementation of the new Civil Code, which consolidated the legislative changes started in 1822. During the First Republic, some efforts were also made to regulate the use of commons with the publication of a number of bills after 1911 (Gralheiro 1990). Globally, the legislative trend was oriented towards the division and privatization of land, which led often to protracted tensions and conflicts within local communities (Freire 2004).

The aforementioned changes in property rights led to a substantial rise in the number of landowners, which, in the mid-19th century, was rather small: the ratio of landowners to inhabitants was 1:11 in Portugal, whereas in Belgium and France it was 1:6 and 1:4.7, respectively (Silva 1868: 259). Between 1895 and 1916, the number of landowners increased from about 1.5 to 2.2 million, with

TABLE 7.8 *Regional distribution of landowners and average area per landowner, 1868*

Regions	Landowners (%)	Average area per landowner (ha)
North	52	13
Centre	33	17
Lisbon and Tejo Valley	8	43
Alentejo	4	163
Algarve	3	37

SOURCE: SILVA 1868: 260

a considerable rise being observed in large landowners, as shown by official statistics. Nevertheless, the distribution of land tenure within the country did not change significantly, and the regional asymmetries already apparent in the mid-19th century remained: in the South, especially in interior Alentejo districts, large properties prevailed, whereas in the rest of the territory, mainly in coastal areas, the process of fragmentation of land was largely dominant (Table 7.8).[10] The sale of nationalized assets, in particular, seems to have had only a minor influence on the structure of property, despite its importance in State revenue (Silveira 1993). The increase in the number of landowners led to a reduction in the national acreage average from 21 hectares to 4.7 hectares per owner near the turn of the century. This relatively small figure can help explain the low levels of labor productivity of Portuguese agriculture.

Between 1910 and 1930, the acreage of agricultural land per owner increased, reversing the trends of property division prevailing since the mid-19th century (Basto 1934). In the northern regions, such evolution can be explained by the high levels of emigration during the previous decades, especially to Brazil, and to the rise of local families' incomes due to remittances from abroad. In the

10 An example of land fragmentation in coastal areas took place in the moor sands linking the banks of the Tejo and the Sado rivers. Between the late 19th century and the 1930s, thousands of migrants, originating mainly from the populated villages in the north and central regions, became tenants of land plots, measuring between 5 and 7 hectares. After clearing the land, the farmers started the production of wine, grain, and other products to be sold at the market (Freire 2011; Prata 2013). However, agricultural incomes were seldom sufficiently rewarding, and many farmers were compelled to find complementary sources of income (Baptista 1993). Although the process contributed to the "democratization" of access to land, it also led to the impoverishment of many small and medium-sized landholders, who became more dependent on other sources of income.

southern regions, less populated and attracting migrants from other regions, the increase in acreage is related to the growing prosperity of small landowners, who were able to purchase some of the land that became liberated from the *Ancien Régime*'s institutional limitations.

All in all, and despite the advances made, the economic impact of institutional change on agriculture was somewhat limited. The transition from the old rules of the *Ancien Régime* to the new liberal ones was slow, taking several decades to fully materialize (Amaral 2012). Although changes in property rights and land tradability allowed the extension of cultivated land, which, as seen before, constituted an important source of agricultural output growth, the country's agrarian structure remained basically unaltered. Moreover, initiatives aimed at facilitating access to credit had a circumscribed impact, with no proper agricultural credit existing until the onset of the First Republic. Educational advances and the spread of novel ideas on cultivation methods and techniques, on the other hand, had limited effects in a largely illiterate population (Reis 1984).

Innovation

Accompanying the changes introduced by governments in the institutional background, a number of improvements were also introduced by several scientific, economic and cultural associations aimed at the modernization of agriculture. Reflecting the spirit of the epoch, many of these initiatives were the result of the intertwined action of public and private spheres (Martins 2005). In this regard, an important initiative was the creation of regional associations (*associações distritais*) by liberal governments during the first half of the 19th century to facilitate the access to agriculture experts and accelerate agrarian modernization. In 1836, a law was introduced requiring each district to have an agricultural association aimed at assisting the government in the implementation of modernization measures in the agricultural sector. This measure had, however, little impact, due to the political instability of the epoch and to the cycle of relative agricultural prosperity that was in place, which was more favorable to actions aimed at satisfying private ends and interests rather than collective goals.

Between 1844 and 1877 new agricultural societies were promoted with some impact at the regional level. Along with the spread of knowledge and agricultural improvements, these societies also provided the government with reports and proposals on topics of regional and national interest, regarding trade policy, rural credit, agricultural education and public infrastructure. They also sponsored several agricultural shows and conferences, and disseminated useful information in the local media. Some of these societies were very dynamic, being particularly helpful in periods of agricultural crisis (e.g., the Agricultural Society of Porto, during the powdery mildew plague). This regional network helped the formation of the first specialized administrative elite and, by 1876, each district already had an agronomist appointed by the government coming from the Lisbon high school.

Also, following the trends witnessed in other countries, from the second half of the 19th century many agricultural associations and societies were created in Portugal without direct State intervention. The role of agricultural associations was more politically proactive than that of societies, especially the *Real Associação Central de Agricultura Portuguesa* (RACAP), created in 1860. Bringing together landowners (especially large ones), farmers and agricultural experts, the main goal of this association, as expressed in its statutes, was to "challenge established routines", although the key factor underlying its creation was probably the defence of agriculture protectionism (Reis 1979). In order to promote the modernization of agriculture, the RACAP intended to introduce and disseminate new techniques and cultivation processes, modern machinery and better agricultural tools, selected seeds and breeds. It promoted conferences, exhibitions, competitions and the publication of agronomic studies, relying on the skills and expertise of its members, many of whom were teachers and former students of the Agricultural Institute, as well as officials in the ministries dealing with agrarian issues. It also relied on the political influence of its associates in government and in parliament to defend its goals.[11] The RACAP influenced the creation of national networks of different types of organizations, including unions, social cooperatives and credit cooperatives, with several degrees of success and territorial presence. Cooperatives had a lower adhesion and were negatively viewed by farmers' unions (Freire and Pereira forthcoming). Farmers' unions had more State support: for instance, in 1892, the wine crisis led the government to create a law that laid the basis for the creation of wine cooperatives, responding to the demands expressed by several unions. The government also granted the right to set up credit banks, connected to associations and cooperatives. However, as indicated earlier, many of these organizations, though formally created, ultimately did not start working or were short-lived.

---

11 An example of an important political action was the pressure exerted upon the government to pursue protectionism measures against wheat imports in the 1880s (Reis 1979).

Between the second half of the 19th century and the beginning of the *Estado Novo*, associations thus became one of the axes of the triangle in defence of farmers' interests, which was composed also by public services, including regional networks, and the High Institute of Agronomy. This network operated at the national level, but was particularly active in the centre and south of the country, promoting technological innovations as well as the discussion of agrarian policies. During this period, the technical and political elites circulated across different organizations, strengthening the presence of agriculture in government decision-making.[12]

At the same time, the foundation since 1852 of the leading tenants' and landowners' associations, alongside public services, regional networks, and the establishment of agricultural schools and research institutions, approached the country with the *agrarian revolution* underway in other European countries, and promoted the spread of agricultural innovations. Although the lack of credit remained an obstacle to agrarian improvement, several innovations were introduced, reflecting both the capacity of Portuguese experts to spread scientific and technical improvements and the entrepreneurship of many landowners. The gradual erosion of the old rules from the *Ancien Régime* led some landowners to relocate financial capital to the enlargement and modernization of farms (Fonseca and Reis 1987; Martins 1992). It also led to the acquisition of small properties by the urban middle class and by former tenant farmers (*foreiros*), who became landowners and, in some cases, even large landowners. Changes in the legislation relating to land thus brought about the emergence of a new class of agricultural entrepreneurs, who were keen on promoting land improvements and adopting technological innovations. The introduction and dissemination of these innovations was uneven, favoring some regions to the detriment of others.

One of the best documented cases of technology adoption concerns the introduction of the steam harvester in Portugal, in 1856. According to Reis (1982), when compared to other countries, Portugal was not particularly late. England was the pioneer in the use of the steam engine harvester, first fixed during the 1830s, then mobile during the 1840s. By 1880 only a small proportion of the grain produced in England was not threshed mechanically. France imported this technology from England, and the same happened in the US. In Portugal

12 From the 1930s onwards, the consolidation of the dictatorship changed these institutional interrelations and the RACAP lost its ability to influence technical and political initiatives (see Chapter 8 in this book).

the spread of the steam harvester was slow, and its use was rather limited until the late 19th century. Between 1900 and 1930, the spread of this technology increased considerably, including to small and medium-sized landowners, partly as a result of the Law on Agricultural Mobilization (1917). This measure aimed at the promotion of easier access to seeds, machinery and steam engines, along with the promotion of the cultivation of common and uncultivated lands. In the same year (1917), the first tractors with internal combustion engines were introduced into the country. The following years witnessed the progressive mechanization of agriculture, aiming at the improvement of the competitiveness of important productive activities, such as wine and livestock (Marques and Rollo 1991).

Along with mechanization, there was also a progressive spread of the use of fertilizers in agriculture. Following a number of dissemination initiatives promoted by several public services and institutes, fertilizers were well accepted by Portuguese farmers, particularly during the last quarter of the 19th century. Wine growers, who made widespread use of fertilizers, benefited from special State measures during the phase of economic recovery following the *phylloxera*. The same happened with the production of grain, since the protectionist legislation of the end of the 19th century was crucial for the diffusion of superphosphates in land of poor quality. Between 1903 and 1914, the consumption of fertilizers was of 100,000 tons per year (Lains 2003b), although Portugal was listed last for the consumption of fertilizer per acre (about 0.611 tons per acre in 1912/13).

Another important technique was related to irrigation. Being a Mediterranean country, irrigation was crucial to improving Portuguese agriculture's productivity. Innovations in irrigation devices were however very few until the last quarter of the 19th century, reflecting mainly the substitution of wood by iron, and the use of a rather small number of steam centrifugal pumps (Radich 1996). According to the geographer Orlando Ribeiro (1967: 73–78), irrigation was spread across about one tenth of the country and a fourth of the agricultural area.

In the region between the Douro and Minho, as well as other areas of the Douro valley, irrigation was based on the communal use of wells and contour channels (Raposo 1994). At the time of the proclamation of the Republic, some dams and canals already existed. The creation of the Agriculture Development Board and the Agriculture Development Fund enabled the introduction of subsidies for agricultural irrigation works from 1924 (Marques and Rollo 1991). Even though during the first decades of the 20th century the small amount of irrigated land in the total cultivated area was a major concern for policymakers, only in the 1930s did the State start investing in dams and irrigation infrastructures.

From the second half of the 19th century onwards, the focus of public services experts was to expand agricultural and forest lands and change farming methods. To achieve these goals they promoted the clearing of wasteland (*maninhos*) and the forestation of dunes and mountains (Branco 2011); the control of forestry practices; support for livestock reproduction and the control of animal diseases, and the provision of products (chemicals or otherwise) for the reactive and preventive treatment of plants. Public services' activities included the collection and organization of statistics and other information on agriculture, in order to evaluate existing problems and to take the necessary steps to overcome them.

To disseminate new agrarian methods and knowledge, the services promoted agricultural conferences, exhibitions and shows, the dissemination of training to farmers about new machinery or labor techniques; rewarded producers for improvements in production; supported the establishment of agricultural societies or associations, as well as the creation of agrarian high- and medium-level education. Though experts considered these initiatives the more appropriate to modernize Portuguese agriculture, the persistent lack of State financial resources often limited their implementation. There is in fact some qualitative evidence of a renewed capacity in the placement of agricultural products in different markets, and of the use of more efficient solutions and new technologies (Nunes 2001). The pace of adoption and diffusion of technology was, however, extremely low, increasing the gap relatively to the most developed countries. Until the end of the 19th century, agricultural output grew mostly as a result of the expansion of land under cultivation, without a great use of mechanization.

The late development of metalworking, directly related to the production of agricultural machinery, conditioned the development and modernization of Portuguese agriculture. On the other hand, the high price of machinery (most of which was imported) and the high cost of its maintenance negatively affected the expansion of mechanization. Furthermore, the agrarian structure and the forms of land tenure were not favorable to intensive mechanization, except in some southern regions.[13] Even there, in some cases it was simply not profitable to introduce modern technologies and machines.

13 Nevertheless, the fragmentation of property in the north could not constitute such a significant bottleneck, if it were combined with a strong development of cooperatives, as happened in other European countries, such as Denmark, France and Germany (Reis 1984).

In this regard, one of the best documented cases refers to the use of the steam harvester (Reis 1982). Being introduced in the country in the 1850s, as happened in other countries (e.g., France, Russia and the US), its use remained insignificant until the end of the century, whereas in several developed countries most of the grain produced was already threshed mechanically. In Portugal the steam harvester was only applied in a significant way between 1900 and 1930, when increases in the scale of production following wheat protectionist laws made it more profitable than the more traditional techniques. The new technology was adopted first by large farms, but later it was extended to small and medium-sized ones, as the opportunities for sharing the acquisition and use of the new machines were gradually being created (Reis 1982).

Novel fertilization methods were also slowly adopted by farmers. Fertilizers were well accepted by wine growers, who benefited from special State measures during the phase of economic recovery following the *phylloxera*. The use of fertilizers was, however, conditioned by the incipient development of the domestic chemical industry and the high international prices charged, which were incompatible with low endowed capital farmers. The use of chemical fertilizers in the production of grain only became significant in the last decade of the 19th century, after wheat protectionist laws have been adopted. The diffusion of superphosphates in land of poor quality was the result of the more profitable environment created by these laws, combined with a substantial decline in fertilizers prices after 1880 (van Zanden 1991). Between 1903 and 1914 the consumption of chemical fertilizers increased markedly, but it still presented rather low levels by European standards (Lains 1995).

In sum, increases in the productivity of the land and labor were limited by the late diffusion of irrigation and fertilizers, and the incipient use of machines and modern tools. Although some innovations were introduced, the weak and late development of some industries related to agriculture, such as metalworking, directly related to the production of agricultural machinery, and the chemical industry, producing fertilizers and disinfectants, conditioned the development and modernization of agriculture.

The evolution of the Portuguese economy during this period stands as an example of slow structural change, both between economic sectors and within the agricultural sector. The development of industrial activities was insufficient to generate higher labor mobility and a stronger demand for agricultural products. The relative inertia observed in the country, when compared to other countries, reflects the rather low levels of investment in both physical

The creation of cooperatives would enable farmers to benefit from economies of scale and from a more rapid diffusion of knowledge.

and human capital, and the slow pace at which the institutional changes introduced by liberal governments were gradually adopted. The absence of a true agrarian reform, and the subsequent unequal distribution of land, unassisted by the creation of a significant network of cooperatives in the fragmented part of the country, did not allow for the concentration of the resources required to finance the adoption of innovative techniques; whereas in the south, large landowners had few incentives to modernize and mechanize, due to the access to a large pool of cheap labor. Production was essentially directed to the domestic market, with exports playing a minor role in agricultural development. In contrast with other countries, exports were not an important source of improvement and growth for Portuguese agriculture.

Conclusion

Despite the institutional advances made and the progress broadly achieved, the gap between Portuguese agriculture productivity and the already high levels found in other European countries has widened between 1820 and 1930. The low development level of the agriculture sector was visible in the enormous importance of the sector in employment and production: in 1930 the agricultural sector still accounted for almost 60 percent of the total active population and about one third of Portuguese GDP.

Although an institutional structure was being created which would provide a favorable environment for the functioning of markets and the diffusion of scientific knowledge, the reforms implemented took many decades to produce effects, and the pace of modernization and technology adoption was low. The incipiency of the manufacturing sector delayed the absorption of labor and postponed the modernization of the agricultural sector.

Why was the country incapable of reproducing the changes implemented in other economies, some of which were as poor as the Portuguese? Important constraints to the adoption of innovation were related to the country's environmental conditions (soil and climate) and general illiteracy: in 1910, the illiteracy rate was 70 percent, one of the highest rates among European countries (Reis 1993). The generalization of the ability to read scientific and technical works occurred late in time (Matos 2000), hampering the effects of the aforementioned developments in schooling and in the spread of technical advances. Access to scientific and technical newspapers was difficult, due to high prices and the difficult (transport) access to the countryside, although the number and the diversity of these publications had increased since the second half of the 19th century (Freire 2013). In the European context, the strong presence of a workforce with a very high percentage of illiteracy represented a serious obstacle to further expansion of the agricultural sector.

Moreover, structural changes associated with modern economic growth were slow. Forest areas expanded, boosted by international demand, but exports of vegetables, fruit, and meat decreased, due to the increasing competition from other countries. The reallocation of resources to more productive crops (e.g., livestock production and horticulture), as happened with other European countries, was rather slow. Protectionist measures favored the cultivation of cereals to meet the demands of a growing population, preventing a specialization based on comparative advantages and leading to an even more unbalanced ownership structure between the south and the north.

During the 20th century, technical progress was the main driver of agricultural output growth, but in the Portuguese case, until the 1930s, technical progress was limited to some improvements in the use of fertilizers and agricultural tools, the introduction of new plant varieties and infrastructure advances, with the construction and improvement of roads and development of the railway network at the end of the nineteenth century. Private investments were constrained by the lack of agricultural credit, which inhibited the adoption of some technologies and land improvements, such as irrigation.

The institutional framework of a modern agricultural sector was created, but its practical implementation faced several obstacles, namely the isolation of certain regions, the small size of properties, labor-intensive agriculture and the lack of credit. In addition to political instability, the post-World War I period was accompanied by an economic and financial crisis in the country. The pace of the modernization process initiated in the second half of the 19th century declined, and as a result the agricultural sector stagnated. At the onset of the 1930s, Portugal had an agricultural sector that was still economically backward and unable to support industrial development, an essential factor for modern economic growth.

References

Amaral, Luciano (2012). "Institutions, property and economic growth: back to the passage from the Ancien Régime to liberalism in Portugal". *Análise Social*, 202, XLVII (1), pp. 28–55.

Baptista, Fernando Oliveira (1993). *A Política Agrária do Estado Novo*. Porto: Afrontamento.

Basto, Eduardo Alberto Lima (1934). *Inquérito Económico-Agrícola*. Lisbon: Universidade Técnica de Lisboa.

Batista, Dina et al. (1997). *New Estimates for Portugal's GDP: 1910–1958*. Lisbon: Banco de Portugal.

Branco, Amélia (2011). "El plan de repoblación forestal (1938–1968): una medida del Estado Novo por evaluar". In Dulce Freire and Daniel Lanero (Eds.). *Agriculturas e Innovacíon Tecnológica en la Península Ibérica (1946–1975)*. Madrid: Ministerio de Medio Ambiente y Medio Rural y Marino, pp. 135–166.

Bringas, Miguel Ángel (2000) *La productividad de los factores en la agricultura española (1752–1935)* Madrid: Banco de España.

Cabral, Manuel Villaverde (1979). *Portugal na Alvorada do Século XX. Forças Sociais, Poder Político e Crescimento Económico de 1890 a 1914*. Lisbon: Regras do Jogo.

Câmara, Benedita (2005). "Relações económicas com o exterior". In Pedro Lains and Álvaro F. Silva (Eds.), *História Económica de Portugal 1700–2000*. Lisbon: Imprensa de Ciências Sociais, vol. 2, pp. 337–356.

Esteves, Rui Pedro (2004). "Finanças públicas". *In* Pedro Lains and Álvaro F. da Silva (Eds.), *História Económica de Portugal, 1700–2000, O Século XIX*, Lisbon: Imprensa de Ciências Sociais, vol. 1, pp. 305–335.

Federico, Giovanni (2005). *Feeding the World. An Economic History of Agriculture, 1800–2000*. Princeton: Princeton University Press.

Federico, Giovanni (2006). "Protection and Italian economic development: much ado about nothing". In Jean-Pierre Dormois and Pedro Lains (Eds.), *Classical Trade Protectionism 1815–1914*. London & New York: Routledge, pp. 193–218.

Fonseca, Hélder Adegar (1996). O *Alentejo no Século XIX. Economia e Atitudes Económicas*. Lisbon: Imprensa Nacional-Casa da Moeda.

Fonseca, Hélder Adegar and Reis, Jaime (1987). "José Maria Eugénio de Almeida, um capitalista da Regeneração". *Análise Social*, 99, XXIII (4), pp. 865–904.

Freire, Dulce (2004). "Os baldios da discórdia: as comunidades locais e o Estado". *Congresso Mundo Rural: Transformação e Resistência na Península Ibérica no Século XX*. Lisbon: Edições Colibri, pp. 191–224.

Freire, Dulce (2010). *Produzir e beber. A Questão do Vinho no Estado Novo*. Lisbon: Âncora Editora.

Freire, Dulce (2011). "Ensayos de ingeniería social: reforma agraria y modernización de la agricultura en las últimas décadas del Estado Novo (1954–1974)". In Daniel Lanero and Dulce Freire (Eds.). *Agriculturas e Innovación Tecnológica en la Península Ibérica (1946–1975)*. Madrid: Ministerio de Medio Ambiente y Medio Rural y Marino, pp. 213–241.

Freire, Dulce (2013). "Modernising ambitions: agronomists in action between dictatorship and democracy (Portugal, 1957–1986)". In Ana Delicado (Ed.). *Associations and other Groups in Science: an Historical and Contemporary Perspective*. Newcastle: Cambridge Scholars Publishing, pp. 92–107.

Freire, Dulce and Pereira, Joana Dias (forthcoming). "Consumer co-operatives in Portugal: debates and experiences from the 19th to the 20th century". In Mary Hilson, Silke Neunsinger and G. Patmore (Eds.). *A Global History of Consumer Co-operation since 1850: Movements and Business*. Leiden: Brill.

Gralheiro, Jaime (1990). *Comentário à(s) Lei(s) dos Baldios*. Coimbra: Livraria Almedina.

Junguito, Antonio Tena (2006). "Spanish protectionism during the *Restauración*, 1875–1930". In Jean-Pierre Dormois and Pedro Lains (Eds.), *Classical Trade Protectionism 1815–1914*. London & New York: Routledge, pp. 265–297.

Justino, David (1988). *A Formação do Espaço Económico Nacional. Portugal 1810–1913*. Lisbon: Vega.

Lains, Pedro and Sousa, Paulo Silveira e (1998). "Estatísticas e produção agrícola em Portugal, 1848–1914". *Análise Social*, 149, XXXIII (5), pp. 935–968.

Lains, Pedro and Sousa, Paulo Silveira e (1998). "Estatística e produção agrícola em Portugal, 1846–1915", *Análise Social*, 149, XXXIII (4), pp. 935–968.

Lains, Pedro (1995). *A Economia Portuguesa no Século XIX. Crescimento Económico e Comércio Externo 1815–1913*. Lisbon: Imprensa Nacional-Casa da Moeda.

Lains, Pedro (2003a). *Os Progressos do Atraso. Uma Nova História Económica de Portugal, 1842–1992*. Lisbon: Imprensa de Ciências Sociais.

Lains, Pedro (2003b). "New wine in old bottles. Output and Productivity trends in Portuguese agriculture, 1850–1950". *European Review of Economic History*, 7 (1), pp. 43–72.

Lains, Pedro (2009). "Agriculture and economic development in Portugal, 1870–1973". In Pedro Lains and Vicente Pinilla (Eds.). *Agricultural Development in Europe since 1870*. London: Routledge, pp. 333–352.

Madureira, Nuno Luís (Ed.) (2002). *História do Trabalho e das Ocupações*. Oeiras: Celta Editora, vol. 3.

Marques, A.H. de Oliveira and Rollo, Fernanda (1991). "Agricultura, pecuária e pescas". In Joel Serrão and A.H. de Oliveira Marques (Eds.). *Nova História de Portugal*. Lisbon: Editorial Presença, XI, pp. 65–114.

Martins, Conceição Andrade (1992). "Opções económicas e influência política de uma família burguesa oitocentista: o caso S. Romão, José Maria dos Santos". *Análise Social*, 116–117, XXVII (2), pp. 367–404.

Martins, Conceição Andrade (2005). "A agricultura". In Pedro Lains and Álvaro F. da Silva (Eds). *História Económica de Portugal, 1700–2000*. Lisbon: Imprensa de Ciências Sociais, vol. 2, pp. 219–258.

Mateus, Margarida and Mateus, Abel (1985). "Technological change, trade regimes and the response of agriculture in Portugal during the 19th century". *Working Paper* no. 52, Lisbon: Universidade Nova de Lisboa.

Matos, Ana C. (2000). "Os agentes e os meios de divulgação científica e tecnológica em Portugal no século XIX". *Scripta Nova. Revista Electrónica de Geografia Y Ciencias Sociales* 69 (29).

Neto, Margarida Sobral (1990). "As estruturas agrárias: a força da tradição". *Revista de História*, 10, pp. 129–135.

Neto, Margarida Sobral (2012). "Conflits entre entités seigneuriales et municipalités à propos des communaux". *Revue du Nord. Hors Collection Art et Archéologie*, 18, pp. 1–4.

Neto, Margarida Sobral (2013). "Propriedade, usos comunitários e sustentabilidade das comunidades camponesas (Olhares historiográficos)". *Atas do I Encontro Internacional de História Ambiental Lusófona*. Coimbra: Universidade de Coimbra, Centro de Estudos Sociais.

Nunes, Maria de Fátima (2001). *Imprensa Periódica Científica (1772–1852). Leituras de "Sciencia agricola" em Portugal*. Lisbon: Edições Estar.

Pereira, Miriam Halpern (1979). *Política e Economia (Portugal nos séculos XIX e XX)*. Lisbon: Livros Horizonte.

Pereira, Miriam Halpern (1983). *Livre-Câmbio e Desenvolvimento Económico. Portugal na Segunda Metade do Século XIX*. Lisbon: Sá da Costa Editora.

Prata, Cristina (2013). *Palmela. Chão Que Dá Uvas. A Terra e o Trabalho das Gentes (1945 a 1958)*. Lisbon: Colibri.

Radich, Maria Carlos (1996). *Agronomia no Portugal Oitocentista. Uma discreta desordem*. Lisbon: Celta Editora.

Radich, Maria Carlos and Alves, A. Monteiro (2000). *Dois Séculos da Floresta em Portugal: Séculos XIX e XX*. Lisbon: Edições CELPA.

Raposo, José Rasquilho (1994). *História da Rega em Portugal*. Lisbon: Instituto da Água, Ministério do Ambiente e Recursos Naturais.

Reis, Jaime (1979). "A 'Lei da fome': as origens do proteccionismo cerealífero, 1889–1914". *Análise Social*, 60, XV (4), pp. 745–793.

Reis, Jaime (1982). "Latifúndio e progresso técnico: a difusão da debulha mecânica no Alentejo, 1860–1913". *Análise Social*, 71, XVIII (2), pp. 371–433.

Reis, Jaime (1993). *O Atraso Económico Português em Perspectiva Histórica, 1850–1930*. Lisbon: Imprensa Nacional-Casa da Moeda.

Ribeiro, Orlando (1967). *Portugal, o Mediterrâneo e o Atlântico*. Lisbon: Livraria Sá da Costa.

Santos, Dina (2012). *Do Mutualismo ao Mercado – A Banca Mutualista em Portugal, Estudo de Caso: o Crédito Agrícola Mútuo*. Lisbon: Faculdade de Ciências Sociais e Humanas da Universidade Nova de Lisboa (Master's Dissertation).

Silbert, Albert (1966). *Le probléme agraire portugais au temps des premiéres cortès libérales*. Paris: Fundação Calouste Gulbenkian.

Silva, Luís Augusto Rebelo da (1868). *Memoria sobre a População e a Agricultura de Portugal desde a Fundação da Monarchia até 1865*. Lisbon: Imprensa Nacional-Casa da Moeda.

Silveira, Luís Espinha da (1980). "A venda de bens nacionais (1834–43): uma primeira abordagem". *Análise Social*, 61–62, vol. XVI (1), pp. 87–110.

Silveira, Luís Espinha da (1993). "La desamortización en Portugal". *Ayer*, 9, pp. 29–60.

Valério, Nuno (Coord.) (2001). *Estatísticas Históricas Portuguesas*. Lisbon: Instituto Nacional de Estatística.

Valério, Nuno (1983). "O produto nacional de Portugal entre 1913 e 1947: uma primeira aproximação". *Revista de História Económica e Social*, 11, pp. 89–102.

Vaquinhas, Irene Maria (1991). "Um espaço em transformação: a extensão da cultura do arroz nos campos do Mondego, 1856–88". *Análise Social*, 112–113, XXVI (3), pp. 689–703.

Yamada, S. and Ruttan, V.W. (1989). "Comparaciones internacionales de la productividad agraria". In *La modernización de la agricultura española, 1956–1986*. Madrd: Mapa, pp. 73–159.

Zanden, J.L. van (1991). "The first Green Revolution: The growth of production and productivity in European agriculture, 1870–1914". *The Economic History Review*, New Series, 44 (2), pp. 215–239.

CHAPTER 8

Agricultural Policy, Growth and Demise, 1930–2000

Luciano Amaral and Dulce Freire

Introduction

During the period from 1930 to the end of the 20th century, the course of Portuguese agriculture was affected by many political and economic changes. Two aspects deserve special attention. One was the structural change in the Portuguese economy, through which agriculture lost its role as the main contributor to GDP and principal source of employment. Another was Portugal's accession to the European Communities (EC) in 1986, which led to the insertion of its agrarian sector within the institutional mechanisms of the Common Agricultural Policy (CAP). These changes took place in a few decades, between 1960 and 1990, closing a long-term pathway that, as shown by the previous chapters of this book, dates back to the formation of Portugal as a national entity around the year 1000. So until the second half of the 20th century, agriculture was at the core of economic, social and political decisions.

Portugal throughout this period changed political regime. In 1974, a revolution overthrew the old *Estado Novo* dictatorship and paved the way for democracy. Unsurprisingly, this political change had impacts on the institutional framework and the orientations of public policies for agriculture, but studies have shown that there was some continuity between the two regimes. In fact, some of the most important changes in agriculture are not directly connected with the nature of the political regime, as shown by the State's market intervention strategies. Although the country witnessed some market deregulatory changes after the 1960s, until the accession to the EC the general political objectives were maximizing domestic production in order to supply the population with national agricultural goods. While in practice achieving these objectives has become increasingly difficult, the idea continues to influence political discourse.

Since 1986, the implementation of the CAP has implied successive readjustments of Portuguese agrarian production and marketing logics. On the one hand, the incentives have been directed to regions and products that offer more comparative advantages in the context of the European common market, forcing the painful processes of agricultural conversion or the abandonment of much cultivated land. On the other hand, the mechanisms of the CAP slowed down the process of structural change in Portuguese agriculture

 | DOI 10.1163/9789004311527_010

as compared with the other Mediterranean countries. The result was a divergence of Portuguese agriculture in relation to Western Europe in terms of labor productivity. Thus the mechanisms supporting the CAP, associated with the peculiar features of Portuguese agriculture, introduced a system of incentives leading farmers to use more labor than land and capital, in comparison with countries at a similar level of economic development.

In a period of just over 30 years, the Portuguese economy has passed from presenting pre-modern features, in which agriculture took up about 50 percent of the working population and 30 percent of GDP, to display those of a modern industrial one, subject to the rules of the European Communities. The processes of rapid economic, social and institutional change occurring in recent decades motivated intense political debates and mobilized different interest groups. As agriculture was at the heart of these discussions and dynamics, the structural changes affecting the sector were subject to different, and sometimes contradictory, interpretations. Several authors writing during the 1960s and 1970s, who followed closely the structural changes in the economy, considered that many of the negative impacts affecting agriculture could be avoided if more efficient public policies and institutions framed the sector. These authors, many of them agronomists, believed that Portuguese agriculture was not taking advantage of the *green revolution* innovations and, because of this, was failing to respond to the new demands for food products. In the 1970s and 1980s the perception that Portuguese agriculture was backward, unable to react to modernization stimuli and, therefore, that it had undermined the overall performance of Portuguese economic growth, became generalized (Freire 2013). These interpretations influenced the historiography of the 1980s and 1990s which, putting the State's action at the core of the analysis, pointed to the lack of institutional changes, in particular the political incapacity to promote agrarian reform, as a key factor to explain the loss of agricultural relevance.[1] Still during the 1990s, several studies helped to re-evaluate the changes in agriculture and rural society, while seeking to broaden the comparison of the Portuguese case in the European and the global framework. These studies, applying different theoretical perspectives and empirical data from multiple sources, provide several interpretations of the changing paths of Portuguese agriculture during the 20th century.[2] While some of the latter are more focused on the examination of the macroeconomic performance, others offer interdisciplinary

1 See Baptista (1993) and Rosas (1994).

2 For the 20th century the official statistics allow a detailed analysis of agrarian performance. The central statistical offices were created in the 1930s, but regular statistics for agriculture have been produced since the mid-19th century.

analyses interrelating social, institutional or agro-ecological factors to explain changes in specific value chains and regions.[3] Crossing quantitative and qualitative data and expanding comparisons with other countries, the authors presented new interpretations about constraints and possibilities in Portuguese agricultural growth.

The chapter will proceed as follows. The next section discusses the evolution of the main economic indicators, comparing Portugal with other Mediterranean and European Union countries. Then we analyze the impact of the public policies followed during the last five decades when agriculture was dependent on national government decisions. The third part is focused on the national reception and the effects of the common agricultural policy, from the late 1970s onwards, when negotiations for the accession of Portugal to the European Communities began. The last section examines how, since the 1930s, the interrelations between the State, corporatist and civil society organizations framed the execution of public policies and the changes in agrarian activities.

The Growth of Output and Productivity

Between 1930 and 2000, the Portuguese economy entered into the process of structural change normally associated with modern economic growth. The contribution of agriculture to GDP fell from a level of around 30 percent in 1930 to a residual figure of roughly 3 percent in 2011. Most of the decline occurred from the 1950s onwards, accompanying the also typical two-step path of industrialization, first, and, later, deindustrialization, followed by the increasing importance of services (Figure 8.1). In terms of employment, the evolution was similar, but with a twist in recent years. Agricultural employment showed a clear declining trend until the early 1990s (from around 56 percent in the 1930s to around 13 percent), but from then on practically stabilized, as it still represented more or less 10 percent of employment in 2011 (Figure 8.2). Industrialization, deindustrialization and the growth of services are also clear in this dimension.

As shown by Figures 8.3 and 8.4, much of this high level of agricultural employment in recent years is a specifically Portuguese feature. In these figures, Portuguese agriculture is compared to that of the large Mediterranean countries (Greece, Italy, and Spain), taking the European Union-15 as a benchmark,

3 Among others see Amaral, (2002); Soares (2005); Lains (2009); Baptista (1993); Rosas (1994). See also Brito, Baptista, Pereira (1996); Simões (2006); Sobral (1999); Freire, (2007); Lanero and Freire, (2011).

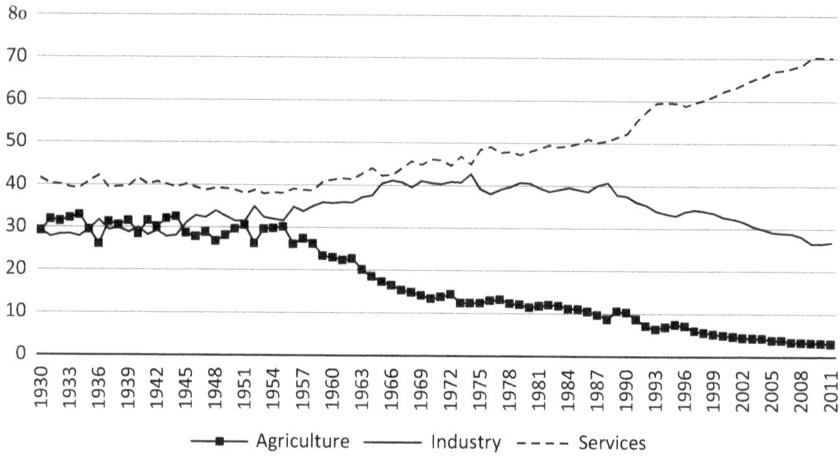

FIGURE 8.1 *Portugal, structure of* GDP, *1930–2011 (percent)*
SOURCE: 1930–1952: BATISTA ET AL., (1997); 1953–1995: PINHEIRO (1997); 1996–2011: AMECO

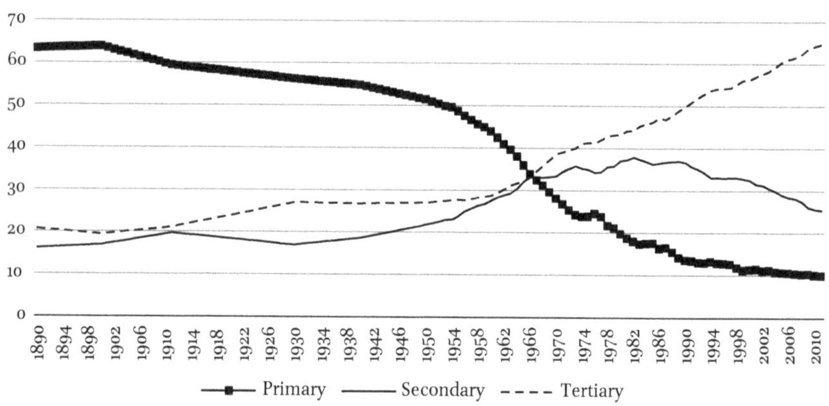

FIGURE 8.2 *Portugal, structure of employment, 1930–2011 (percent)*
SOURCE: 1930–1952: NUNES (2001), CORRECTED TO SPLICE WITH PINHEIRO (1997); 1953–1995: PINHEIRO (1997); 1996–2011: EXTRAPOLATED FROM PREVIOUS SERIES WITH GROWTH RATES FROM AMECO

from 1970 to 2011. In terms of contribution to GDP (Figure 8.3), the Portuguese situation converged quite closely with that of the EU-15, despite some slowdown in the early 21st century. In terms of employment, however, the picture is very different (Figure 8.4). Portuguese convergence was clear only until the early 1990s. From then on, it slowed down considerably, and in 2011 it was equal to Greece's. The proportion of agricultural employment in both countries is still very high today in the context of the developed European economies.

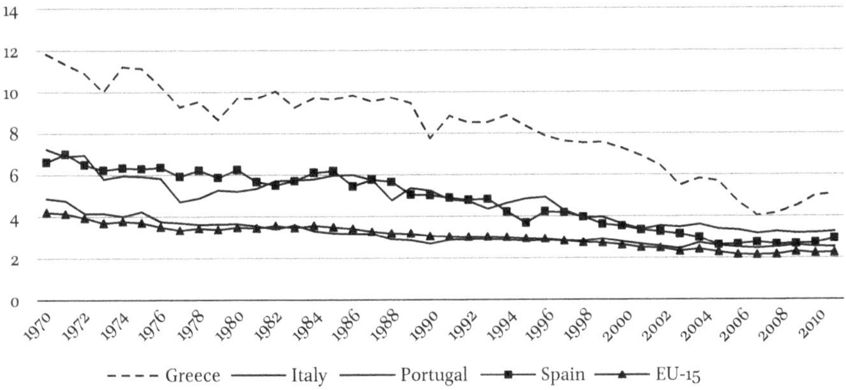

FIGURE 8.3 *Weight of agriculture in* GDP, *1970–2011 (percent)*
SOURCE: 2000 VALUE AND 1970–2000 GROWTH RATES: OECD; GROWTH RATES 2000–2011: EUROSTAT, EXCEPT PORTUGAL: GROWTH RATES 1970–1995: PINHEIRO (1997), AND 1995–2011: EUROSTAT

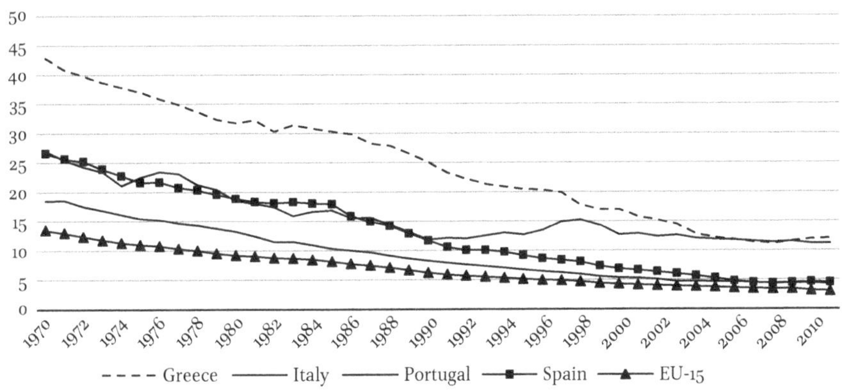

FIGURE 8.4 *Weight of agriculture in employment, 1970–2011 (percent)*
SOURCE: SEE FIGURE 8.3

The dissociation between the contribution to GDP and the level of employment points to a structural problem of low labor productivity in Portuguese agriculture. Table 8.1 shows the growth rates of labor productivity since 1930 for each decade until 2011. Up to the 1980s, the growth of labor productivity was interesting but lower than in most comparable countries. Figure 8.5 compares the level of labor productivity in Portuguese agriculture with the same measure in Italy, Spain and Greece, all of them as a percentage of labor productivity in Western Europe between 1950 and 2005.[4] The figures show that, unlike in Spain and Italy, labor productivity in Portuguese agriculture diverged sharply during

4 See Martín-Retortillo and Pinilla (2012).

TABLE 8.1 *Annual growth rates labor productivity, Portugal, 1930–2011 (percent)*

1931–1940	**1.81557**
1941–1950	4.04
1951–1960	4.29
1961–1970	4.41
1971–1980	4.64
1981–1990	6.88
1991–2000	-0.50
2001–2011	0.16
1930–2011	3.18

Source: GDP: 1953–1995 (Pinheiro 1997, extrapolated backwards with rates from Batista *et al.*, 1997 and forward with rates from OECD); employment: 1930–1952 (Nunes 2001, corrected to splice with Pinheiro 1997); 1953–1995 (Pinheiro 1997, corrected with Amaral 2009); 1996–2011 (OECD).

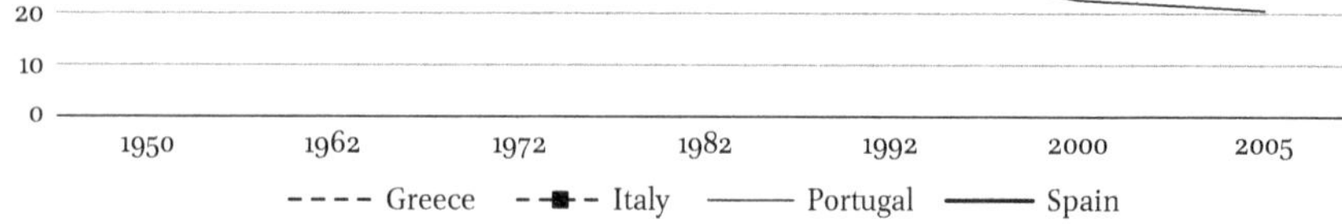

FIGURE 8.5 *Productivity of labor in agriculture, agricultural value added per person employed as a percentage of Western Europe (international USD 1999–2000), 1950–2005*
SOURCE: MARTÍN-RETORTILLO AND PINILLA (2012), EXCEPT PORTUGAL, AUTHOR'S CALCULATION

this period. Already low in 1950, at a level of roughly 45 percent of Western Europe, it fell to a mere 20 percent in 2005. Spanish agriculture, which had a similar level in 1950, converged until it reached more than 50 percent of the European level in 2005.

Labor productivity, which is a ratio between output and labor, can be seen as the product of two other ratios, output and land (land productivity) and the land-labor ratio

$$Y/L = Y/A \times A/L$$

where Y is output, L is labor, and A is land.

Thus if labor productivity was low in Portuguese agriculture throughout this period, the reason must be found in either low land productivity or a low land-labor ratio. Now, land productivity in Portugal has been comparatively high in historical terms. Figure 8.6 shows, in the first place, that it was much higher in 1950 than labor productivity, at a level of around 60 percent of Western Europe. Then it fell fast, until reaching roughly 30 percent in 1982. After that year, however, it recovered much of the previously lost ground, reaching 70 percent of Western Europe's level in 2005. In that year, land productivity in Portugal was higher than in Spain and quite close to Greece's.

If land productivity was high in relative terms, then the explanation for low labor productivity must lie in the land-labor ratio. Figure 8.7 shows just that. The figures cover only the period since 1970, but we can see that the ratio was very low in 1970, although still comparable to Italy's and Greece's at the time. Nevertheless, from then on it diverged both from the European standard and from the other Mediterranean countries. Since the early 1990s it has even declined, so much so that in 2009 it had returned to about the same value as in 1970, whereas for all other European countries it had practically trebled. Land and labor productivity could have grown comparatively, despite the declining land-labor ratio, if there had been an increase in the capital intensity of the sector. However, Figure 8.8 shows that the capital-labor ratio has grown slightly

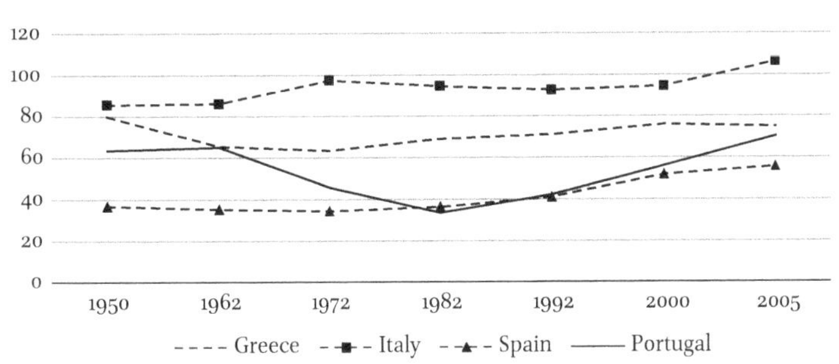

FIGURE 8.6 *Productivity of land, agricultural value added per hectare as a percentage of Western Europe (international USD 1999–2000), 1950–2005*
Note: land = arable land + meadows and pastures + permanent crops
SOURCE: MARTÍN-RETORTILLO AND PINILLA (2012)

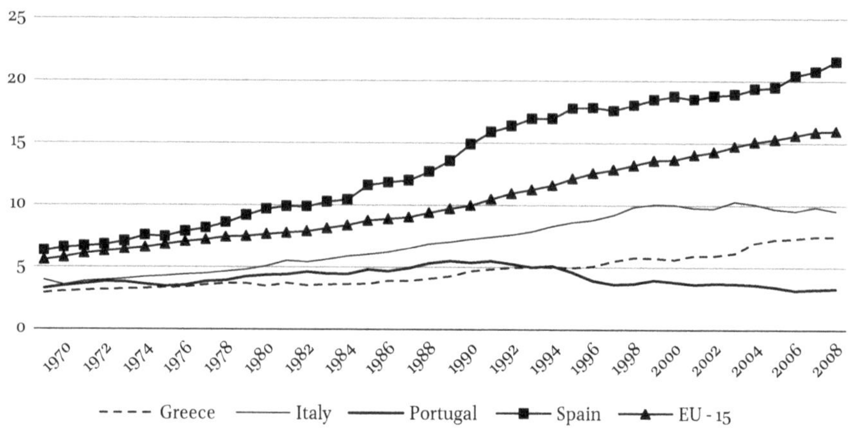

FIGURE 8.7 *Land – labor ratio, 1970–2009 (hectares per employed person)*
Note: land = arable land + meadows and pastures + permanent crops
SOURCE: LAND: FAOSTAT; LABOR: 2000 VALUE AND GROWTH RATES: OECD, EXCEPT PORTUGAL GROWTH RATES: 1970–1995: PINHEIRO (1997) AND 1996–2011: EUROSTAT

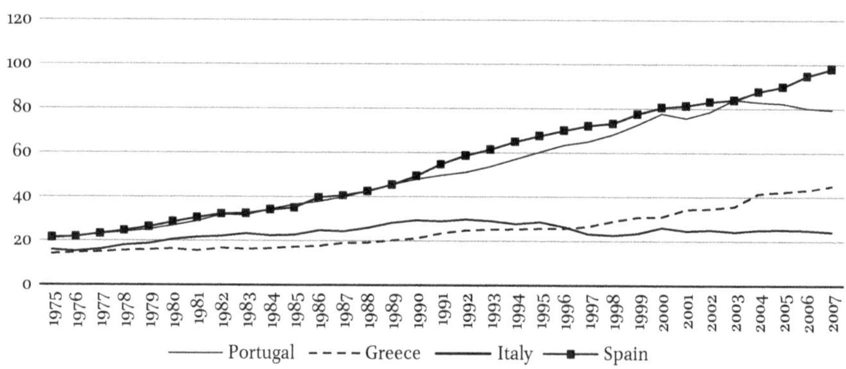

FIGURE 8.8 *Capital – labor ratio, stock of capital per unit of labor (,000 2005 USD), 1975–2007*
SOURCE: CAPITAL: FAOSTAT; LABOR: 2000 VALUE AND GROWTH RATES: OECD, EXCEPT PORTUGAL GROWTH RATES: 1970–1995: PINHEIRO (1997) AND 1996–2011: EUROSTAT

since the mid-1970s, and has even declined from the early 1990s onwards, in stark contrast to Italy and Spain.

State Protection

Since during the 20th century the State deepened the intervention mechanisms in the economy, and particularly in agriculture, it is possible to establish some connections between the quantitative evolution and several political decisions taken by different governments. The period from the 1930s to the 1980s represents the last 50 years in which Portuguese agriculture was subject to national public policies. In many cases, political decisions regarding agriculture could be seen as random, as attempts to meet the wishes of different pressure groups, and lacking clear planning, showing that often governments failed to follow an effective development strategy for agriculture, even when there were previously defined plans.[5]

While contradictory measures were taken and with different objectives, some concerns were common across several governments and even across the different political regimes that ruled the country during those years. Together, those measures and decisions were crucial to understanding the sector's behavior during and after the period covered by this chapter. The analysis of several official decisions reveals three main orientations. One was to maximize the use of available natural resources, especially land, to increase the production of food and raw materials. Another was developing policies to ensure abundant labor supply to execute the various tasks related to agricultural production. Finally, to regulate the domestic market mainly to favor national agricultural products and, in some cases, the colonial ones, which implied controlling importations and domestic consumption levels.

Actually, the growing State intervention in agriculture has been justified by its responsibility in *feeding the nation*. If the concern of *rationally exploring* all useful natural resources for producing foodstuffs or raw materials already existed in previous centuries, the political importance of the subject grew during the 19th century. Many Portuguese authors argued that the main objective of national agriculture should be to produce goods to satisfy the needs of the country's population.[6] These arguments made seem irrelevant the proposals of those arguing that the State should encourage the exploration of comparative advantages, by promoting productive specialization and exporting products with competitive capacities in the international market. During the first half of the 20th century, many European countries developed different kinds of protectionist policies, especially the ones ruled by authoritarian regimes, such as the Portuguese *Estado Novo*. Although, after the Second World War there

5 In fact, between 1953 and 1974, the State defined several development plans, presenting public action programs and financial investments for infrastructures and agricultural modernization. Although agronomists took part as technical advisers in the drafting of these five-year development plans, the government did not help implement all the measures planned for the different regions of the country. See Baptista (1993), Rosas (1994) and Lanero (2014).

6 See, for example, Campos (1922); Jesus and Campos (1923) and Basto (1936).

was a trend towards market liberalization, concerns over domestic agricultural production continued to mark the policies of different countries. In this sense, Portuguese public policies followed similar guidelines to those of neighboring countries (Tracy 1989; Federico 2008). In this sense, Portuguese public policies followed similar guidelines to those of neighboring countries.

Even considering that in Portugal, between the 1930s and the 1980s, a food self-sufficiency logic prevailed, several stages can be identified. One, marked by a stronger protectionism, can be established between the early 1930s, in the aftermath of the Great Depression, and the 1960s. At this time, the national and international dynamics combined to lift several protectionist measures, attending to market liberalization and stimulating some reconfiguration of regional agrarian productions. In the mid-1970s, at the time of the international economic crisis and the Portuguese revolutionary governments that provided the transition from dictatorship to democracy, there was a return of a discourse of food self-sufficiency supported by some public policies to maximize agricultural production. But in the late 1970s this sense had faded again, and diluted further with the trade agreements that prepared the accession of Portugal to the European Communities. Taking into account these phases, it is important to refer what were the main policies for agriculture, rural employment and markets.

In late 1920s was still believed that Portugal had vast areas of underused fertile lands which, properly explored, could help to boost agricultural productivity. To promote the exploitation of these resources, the governments took several decisions. At the same time, to evaluate what those resources were and where they were located, in the 1930s the *Estado Novo* promoted a survey on Portuguese soils, assessing the most suitable ones for intensive and extensive farming or for forestation purposes. The survey showed that there were just over a million hectares of land not properly explored.

One of the dictatorship's first political measures was the so-called Wheat Campaign. Beginning in 1929 and continuing during the 1930s, it allowed the cultivation of many dry lands, especially in the Alentejo region, and an increase in wheat production. During the early years, the favorable weather conditions and several political stimuli (namely the introduction of new seeds and fertilizers, appealing subsidies and prices) helped obtain good yields (Saraiva 2010). For example, the record production of 1934 was of more than 800,000 tons. In the following years, wheat crops were more uncertain but the State continued to control the entire system of grain production, transformation and marketing.

If wheat helped to increase the amount of cultivated areas in the Southern regions, in the North the *Estado Novo* developed a strict forestation programme targeting local commons (Branco 2011). In the second half of the 1940s, the forestation of around 400,000 hectares of mountainous areas began. The *baldios*, which in many cases were already part of the common heritage of local inhabitants during the *Reconquista* times, were captured by the State and forested with maritime pine. This species allowed developing some industries (chemicals and furniture), but the consolidation of the forest value chain was sustained mainly in eucalyptus trees to produce paper. Throughout the 20th century, the forest recorded higher growth rates than agriculture and livestock. Between 1915 and 1958 the annual average growth rate of forest GDP was 2.2 percent, and from 1976 to 1980 this figure was 9.85 percent (Brito 2005: 159–160). These policy measures have contributed to building the industry of forest products, a major contributor to the GAP during the next decades.

Even though strategies to expand dryland farming were being promoted, it was recognized that irrigation could increase the total factors' productivity. However, given the geographical and ecological characteristics of the country, naturally irrigated lands were scarce, and the State would have to make large financial investments to build irrigation systems to allow expansion of the areas allocated to intensive agriculture. In the second half of the 1930s, the State began the construction of several dams to expand irrigation in different country regions. Some of these dams, especially around the major rivers of the Centre and the South (Mondego, Tejo and Sado rivers), started working during the Second World War. These public works allowed irrigate 56,000 hectares of land spread over various regions of the country (Baptista 1993: 73–76). In the late 1950s, the State presented the *Plano de Rega do Alentejo* (irrigation plan for the Alentejo area), which included the construction of several dams and many miles of canals, to irrigate about 170,000 hectares. The public works started in the following years and led to the construction of several dams and the expansion of areas with new irrigation systems cultivated with vegetables (especially tomatoes for the agro-industry and exportation). However, the dams were constructed slowly (in 2011 the same plan was still being implemented and was unfinished) and in many cases water has not been used by the property owners to develop more intensive production. When all the infrastructures are completed, the country will have 265,000 hectares of land irrigated by "modern systems". To these should be added about 40,000 hectares of land irrigated by the so called "traditional" systems, which still exist in various regions.

The *Estado Novo* promoted some land reform measures, namely through internal colonization projects, but these measures have had little impact on land ownership and farming activities. The State created some settlements, especially in the North and the Centre of the country, in poor soils and without easy access to water for irrigation. The large latifundium of the Alentejo region remained untouched. Opposing to what the agronomists wanted, in Portugal irrigation was not associated to property division and colonization. Irrigation connected to land reform, as it was being used by Franco's regime in Spain, was not possible during the *Estado Novo* (Freire and Lanero 2013).

Contrary to what was happening in other countries of Western Europe after the Second World War, the Portuguese governments did not promote extensive programs of technical assistance and rural extension. In fact, a rural extension service, considered essential by a number of influential agronomists to promote the modernization of Portuguese agriculture since 1916, was missing until 1977. In the 1950s, American experts who visited Portugal in the context of the Marshall Plan also recognized the urgent need for a rural extension service (Rollo 2007). This kind of public services were not created during the dictatorship, however, the recognition of the productive outcomes associated to the *green revolution*, and the increasing integration of Portuguese technicians in international debates contributed to redraw the proposals regarding agricultural modernization. As the possibilities for expansion of extensive agriculture were exhausted, agronomists and economists advocated the intensification of the exploitation of natural resources. For an efficient use of available inputs, they defended that State action should at first lead to two major changes. One was to change the conditions of access to land through an agrarian reform. The second was to expand areas of irrigated crops, by building a large irrigation system in areas dominated by dry lands.

In fact, during those years several governmental decisions contributed to the expansion of the cultivated land, reaching a maximum in the 1950s of 4.7 million hectares under cultivation and almost 3 million hectares covered by forests. From the 1960s onwards, the extent of uncultivated land began to rise again. By the end of the 20th century the cultivated area was about 3 million hectares, a similar amount to that recorded in 1902 (Lains 2009: 339).

Apart from the expansion of cultivated land, the *Estado Novo* promoted policies aimed at ensuring a vast supply of labor for agrarian activities, thus keeping wages low. Until the 1960s, as the population continued to grow (rose from about 7 million in 1930 to 9 million recorded in 1960 and 1970) and few employment alternatives existed outside agriculture, many workers had great difficulty in finding work. Since emigration was prohibited, this historical way to mitigate the effects of rural unemployment was not available or became a very risky solution. The number of unemployed workers tended to be higher in the regions of Alentejo and Ribatejo, dominated by larger properties, but the historiography has shown that, in fact, this was a general problem affecting all the country, even in the North and Centre, dominated by small and medium-sized properties. It is estimated that between the 1930s and the 1970s the unemployment rate was relatively low, ranging between 2 and 4 percent, but official statistics are very poor.[7] After 1974, unemployment rates increased, ranging from 6 to 10 percent (Amaral 2005: 85). However, as mentioned in the previous section of this chapter, the relative weight of the agricultural labor force was considerably reduced from the 1960s.

During the 1930s, when the effects of Great Depression demanded extraordinary measures in Western countries, Portuguese government created, in 1932, the *Comissariado do Desemprego* (Unemployment Office) to manage unemployment. This office, set up with a transitory purpose, worked until the end of the dictatorship, indicating that its functions continued to be relevant during all those decades. The *Comissariado do Desemprego* received regular information on rural unemployed and programmed the workers' occupation in public works, following strict rules about salaries and the rotation of the workers, who worked only in the periods in which there were fewer agricultural activities. Thus, although these workers were counted in official statistics as agricultural labor force, in fact they worked regularly in the secondary or tertiary economic sectors. The *Comissariado do Desemprego* developed activities in conjunction with the regional and local corporatist bodies network and the largest farmers of each parish. Under this coordination it was possible to establish wages, working conditions and the number of workers needed for each farm.

Despite the regime's repression, in several villages and agro-towns the workers developed clandestine organizations or social movements challenging the working conditions imposed by the State and farmers. During the 1940s and the 1950s, because of the need to prevent social tensions and the difficulties in securing jobs in public works for all the available unemployed workers, the State forced the landowners to receive even the surplus workers, which faced strong opposition from farmers. On the one hand, they preferred replacing labor by adopting technological innovations. On the other hand, they wanted to continue to choose only the best workers. The tensions between the State and farmers were more pronounced in regions, such as Ribatejo and Alentejo, with better geographical and economic conditions for using tractors and harvesters. For example, in 1950, the largest farmers in the Ribatejo region wrote directly to the head of government (the dictator Oliveira Salazar) explaining the reasons why they preferred machines instead of workers (Freire 2007: 511). Farmers' concerns increased as the economic dynamism, the growing illegal emigration, the proliferation of industries, and the colonial wars led to rural

7 Although some official service had collected data about rural unemployment, official statistics do not offer this kind of information until the 1960s.

exodus and a rise of agrarian labor wages. The rapid depopulation of rural areas promoted the abandonment of many cultivated lands and the decline of remote rural regions.

As predicted by neoclassical theory, the reduction labor force accelerated the mechanization and motorization of many agricultural activities. By and large, the substitution of labor by capital was financed by State subsidies and private investment, particularly by the remittances sent by migrants to the families who remained in the villages of origin.[8] As analyzed below, the market conditions for agricultural inputs and outputs guaranteed by the State also favored the replacement process that lasted until the late 20th century. In the mid-1970s, the combination of the international economic crisis and the demographic dynamics associated with the end of the *Estado Novo* in April 1974 (decolonization, end of the colonial wars, return of emigrants) increased the supply of labor. Unemployment had become a threat that revolutionary governments sought to mitigate, facilitating employment in agricultural and rural activities. Among the policy measures taken by the revolutionary governments was for instance the so-called "emergency agricultural credit", which could be used to finance salaries or farms' inputs, and that remained available up to 1980.[9]

About trade, the first decades of Oliveira Salazar's regime were marked by the redefinition of protectionist policies applied to key agricultural products. In general, the principle followed by dictatorship had two basic objectives. One was to ensure that farmers were getting good prices for a number of agricultural products that the government considered essential. Among these products were grain (especially wheat), olive oil, milk, meat, but also wine. Products such as fruit and vegetables did not benefit from these governmental guarantees and could be traded in the "free market". Another objective was to prevent fluctuations in supply and demand that can have negative effects on consumers' purchasing power and wages. The intervention mechanisms in the market were different depending on the agricultural product. For example, in practice, since the early 1930s wheat was the crop that had deserved more protection, but the State influenced the market prices of almost all products (from fuels to fish, etc.). The network of corporatist bodies

8 In the first decades, emigration was mainly masculine. Many women stayed behind to care for their small children, and family farms that dominated the Northern regions. It is estimated that between 1900 and 1950 migrated 1 million. But between 1950 and 1970, 2 million people left the country, of whom about 40 percent during the 1960s (Veiga 2005: 59). The rural exodus also expanded the cities of the coast, see maps A11 and A12 in the Appendix.

9 See Mansinho (1980) according to whom there were several attempts to end this form of credit, but they were ineffective.

and economic coordination organizations provided bureaucratic procedures, required by different types of market and price intervention. In 1947, the State joined several existing financial funds in the powerful *Fundo de Abastecimentos* (Supply Fund), ensuring that price regulation was centralized and more efficient. This fund, which collected various fees charged by the corporatist organizations in the different stages of the product value chain sold in Portugal, was operational until the mid-1980s, and was abandoned only when Portugal joined the European Communities.

However, between 1930 and 1980 several changes affected official price policies. One of the major changes occurred in the early 1960s, when the State carried out some liberalization of prices and lifted other restrictions in domestic and foreign markets. These changes resulted, on the one hand, from Portugal's accession to the European Free Trade Association and, on the other hand, reflected the economic and social changes that were underway in the country. The State ensured compensating prices for farmers and, at the same time, made them more flexible for consumers, as the gap between the two prices could be compensated by the supply fund. In many cases, changes in price policies accelerated the abandonment of some productions, such as wheat and olive oil, which were important to sustain economic activities in some inland rural regions. With the 1974 Revolution the intervention policy in agricultural prices was abandoned, and, with rising wages, prices at the consumer level were formed freely from the producer price guarantee (Lobão 1977: 23). However, from 1976 new mechanisms to control prices were implemented, returning to a system of established maximum prices of many food products. Until the mid-1980s, the State continued to intervene directly and indirectly in price formation and, as analysed in the next section, this tendency continued after Portuguese accession to EU.

In general, agricultural prices were favorable to producers, but due to regional specializations, the positive impacts of State policies spread unevenly through the territory. Additionally, between 1960 and 1980, the evolution of the terms of trade benefited agriculture. The terms of trade were measured by the ratio between the prices received by farmers for agricultural products and the prices paid for inputs (such as fertilizers and machinery). In fact, public policies stimulated the production of some agricultural products and, from the 1960s onwards, promoted technological changes. In a first moment, the most benefited subsector was farming, followed by the livestock. In a second phase, there was a suppression of prices of inputs, especially fertilizers, coupled with subsidies for the purchase of machinery, which became necessary due to the acceleration of the rural exodus. The combination of these measures contributed to the increase in the number of tractors, harrows or seeders available on farms (Estácio Ed. 1985).

In those years the Portuguese agriculture continued to show low use of industrial inputs, such as chemical fertilizers and improved seeds. For example, in 1968 these represented approximately 50 percent of wheat and barley seeds used, but this value drops to 15 percent in the case of maize, rye and oats (Soares 2005: 174–175). From 1968 to 1999 there was more dynamism in replacing labor by machinery, especially cultivators (multiplied by 40) and tractors (multiplied by 10).

By the 1970s the combination of subsidies allocation to purchase of agricultural inputs, which would compensate the wages increase resulting from rural exodus, with market policies focused on prices was still visible. In this way, prices and subsidies policies "leading to an evolution of the trading terms of Portuguese agriculture, generally favorable to agricultural producers, contributed positively to the increase in the numbers of tractors, fences, planters and cattle in the farms" (Estácio Ed. 1985) that are recorded in the agricultural censuses of the 1950s and 1960s. However, as pointed out by agronomists, economists and other authors, even if the mechanization of agriculture was being promoted, the several production factors were not used efficiently in order to ensure the desirable growth of land productivity or to increase the production of goods demanded by urban consumers.

Despite all the changes witnessed in agriculture to demonstrate the experts' fears, in 1961 the State Secretary of Agriculture said that agriculture was in the middle of a war, which "must be won by the primary sector or it will create a situation impeding the progress of the country and the harmonious economic and social development of the Portuguese population" (cited by Freire 2007). During these years, the perception that agriculture was not experiencing a modernization process became recurrent. These perceptions helped to deepen the idea of strong discrepancies between the *immobility*, imposed by the dictatorship, and the constant renewal of modernizing opportunities offered by the *green revolution*, which were being exploited by other countries. The data collected by agronomists and economists were interpreted as a demonstration of the inability of the *Estado Novo* to change agriculture and to promote the country's development. In their view agriculture was not contributing to economic growth. Rather than spending exchange on imports of essential goods (e.g. fodder, meat, milk and vegetables), the State should, they argued, stimulate national production. These importations were the origins of the inflationary process, which widened from the second half of the 1960s. (Girão 1980).

In the last years of the *Estado Novo* it seemed clear that agriculture had lost its opportunities to modernize and to contribute to economic growth. In the fourth Development Plan (IV Plano de Fomento), launched in 1974 and interrupted by the 1974 Revolution, "the urgent need to provide agriculture with recovery conditions" was put forth (Presidência do Conselho 1974: 333). These statements seemed to indicate that the dictatorship was unable to "feed the Nation", leaving one of its main tasks unfulfilled. Until the early 1960s, the prevailing economic autarky meant that the Portuguese population had to live with what national agriculture could produce. The population growth, the limited natural resources and import restrictions led to several constraints on the quantity and quality of available food. It was estimated that only from the mid-1960s, the Portuguese started to have access to the 2600 per capita daily calories that already constituted the dietary pattern of Mediterranean European countries before the Second World War. Importantly, at this time the consumption pattern in Northern European countries was of about 3000 per capita daily calories (Freire 2011a). The enlargement of food supply was achieved through the increase in importations rather than by improvements in Portuguese agricultural productivity.

The change of political regime in 1974 occurred at a time when the *green revolution* was still at its peak in Western Europe and expectations were heightened. In the months that followed the military coup, many of the legislative measures enacted aimed at introducing changes that the dictatorship had been postponing. Much of the published literature on the period tends to accentuate the differences between the political measures regarding agriculture. An overall analysis of this legislation however reveals different influences, which resulted either from modernizing proposals suggested in previous years, or from the national and international economic conjunctures. In this way, in the months that followed the military coup, many of the legislative measures aimed to introduce changes that had been put off by the dictatorship.

The statistical evidence confirms the perceptions of contemporary witnesses. As stressed by Soares (2005), during the first half of the 20th century (from 1915 to 1958), excluding the wars years, the Portuguese agriculture performed well, with an average of 1.9 percent a year. From the 1960s onwards, the growth trend of the sector changed. The declining share of the agricultural output in GDP was sharp, from 31 to 20 percent, from 1950 to 1960, and further down to 12 percent in 1970, and 9 percent in 1980. Between 1959 and 1995, the average growth rate was 0.3 percent for the year. The agrarian sector had a growth due primarily to forestry and livestock, but not to crop such as grain, fodder, fruit and vegetables. Although the process of forestation had negative social and ecological impacts, the forestry industry proved to be the most successful economic subsector. And, increased livestock (mainly cattle, poultry, eggs and milk) sought to match the rise of domestic demand, but became more import fed without constituting a stimulus to the domestic production of fodder.

An overview of the entire agrarian sector shows that crop production had difficulties in responding to the needs of the expanding livestock, but also to correspond to the changes in the demand raised by the acceleration of urbanization, which opened the market for many fresh products (especially fruits and vegetables). Thus, agriculture did not follow "the growth of the rest of the Portuguese economy during the so-called 'golden age' of the 1960s" (Soares 2005).

The Effects of the Common Agricultural Policy

From about 1975 to the end of the 1990s Portuguese agriculture experienced a particularly complicated phase due to the superimposition of three processes of an often contradictory nature. One was the transition from the agricultural policy of the *Estado Novo* to that of the 1974–1976 revolutionary period. The second process was the transition from the revolutionary period to the new democratic regime and its progressive conformity with CAP. The final process, occurring after Portugal became an EU member, was the transition of CAP itself, which was then shifting from a phase of high guaranteed prices into a phase of low ones. The result of the combination of these processes was a sequence of rapid shifts in the system of incentives offered to Portuguese agricultural agents. To understand the pathways followed by Portuguese agriculture during the last decades, three aspects that defined the background in which it was inserting need to be taken into consideration. The first aspect is the changing objectives of the Common Agricultural Policy. CAP was part of the foundational document of the EC, the Treaty of Rome, and was based on two pillars: a set of measures aimed at intervening in agricultural prices and markets; and another set of measures pointing at the structural problems of farms and rural areas. From the beginning, however, European decisions favored the first pillar, making of prices the main CAP intervention instrument. Structural reform has received much less attention, despite many discussions on the issue.[10]

Initially, CAP had three main objectives. The most important one was to increase agricultural production by productivity improvements, with the ultimate purpose of ensuring European self-sufficiency. Among its goals was a) stabilization of agricultural markets and b) guaranteeing to the population

10 The first major "structural reform" plan was the 1968 *Mansholt Plan*, whose main purpose was to promote the merging of scattered farms and increase their average size in order to increase their productivity and make them less burdensome to the EC budget (Tracy 1986).

working in agriculture a reasonable standard of living. These should be achieved through a system of high support prices combined with protection against foreign producers (which implied a mechanism of "community preference", i.e. the buying of most agricultural goods within the EEC) and export subsidies.[11] This was very much within the tradition of agricultural policy in European countries, including Portugal, since at least 1930.

Thus, in the mid-1970s and early 1980s, the main problem became one of surpluses rather than scarcity, implying a growing financial burden (OECD 2011). Consequently, European politicians started then pondering measures to restrict production and gave more attention to the problems that were affecting the development of rural areas. In 1992, the *McSharry Reform* imposed lower subsidies for agricultural prices.[12] Additionally, it included a set-aside scheme, according to which payments were made to farmers leaving a percentage of their arable land uncultivated. More or less at the same time, the negotiations of the Uruguay Round on agriculture also determined a decrease in the level of protection of European agriculture (OECD 2011). The MacSharry Reform was followed by a set of other "reforms" during the early 21th century. Another important component of the European policy, most strongly adopted after the 2003 reform, was "decoupling", or the dissociation between direct payments and production (European Commission 2010a; OECD 2011). The main thrust of all these reforms was the lowering of guaranteed prices, the increase in direct payments and the promotion of structural change. This meant that CAP policy guidelines were totally inverted in relation to previous periods and that prices were strongly reduced.

The second aspect to bear in mind is that the transition specific to Portuguese agriculture took place precisely during this period. Portugal demanded accession to the European Communities in 1977 and entered in 1985. The general negotiations between Portugal and the Communities started in 1978, but the negotiations pertaining specifically to agriculture only started in 1980 (Varela 1991 and 2007). As was shown before, the type of policy existing in Portugal was similar to that prevailing in Europe: protection at the border, guaranteed prices, and some (but little) structural reform. In this respect, there was, thus, no major conceptual adaptation to be made. Nevertheless, there were two main problems: the differences in guaranteed prices and the lower structural development of Portuguese agriculture. The Portuguese approach

11 See Tracy (1986); Fennel (1997); Federico (2008).

12 Namely, the *Agenda 2000* reform; the *2003 Luxembourg* reform; the 2004 reform of the cotton, hop, olive oil, and tobacco regimes; the 2007 fruit and vegetable reform; the 2008 wine reform; and the 2009 Health Check (European Commission 2010a; OECD 2011).

to negotiations was to make few demands on price levels but, at the same time, ask for strong support on structural reform (Varela 1991). At the end of the negotiations a double transition process on prices was agreed between the parties. One was a classical transition, lasting seven years (1986–1993) and applying to products where no major difficulties existed in terms of price harmonization between Portuguese agricultural policy and CAP (Varela 1986). This type of transition affected goods for which Portuguese prices were lower than the CAP's or where the price gap was not too pronounced. The proportion of products with lower prices in Portugal was just 15 percent of Portuguese agricultural output (Cunha 2010). Each year Portuguese and European common prices were forced to converge by a certain amount until reaching complete harmonization. The second type of transition affected the remaining products and lasted for 10 years, between 1986 and 1996. In this transition, no price harmonization would take place in the first five years, and Portuguese agriculture would benefit from funds for the reorganization of marketing schemes. In the second half of the period, prices would converge along lines similar to the classical transition (Varela 1986).

At the same time, other types of measures to promote "structural reform" were negotiated to be implemented in 10 years. This structural reform program sought investment and resource reorganization.[13] Together with this, a "pre-accession package" was signed, according to which Portugal benefited from two EEC payments, in 1980 and 1983, devoted to the modernization of various aspects of Portuguese agriculture. A third and last point should be mentioned. Throughout this period, two other important events also affected the pace of harmonization of Portuguese prices. The first was the creation of the European Single Market, in 1993. The Portuguese government decided that agriculture should be no exception to price harmonization, meaning that the second five-year phase of the 1986–1996 transition process was reduced to three years, even if compensated by some amounts in the form of direct payments (Avillez *et al.* 2004; Cunha 2010). The other event was the decision of the government, in order to combat inflation, to revalue the Portuguese currency (the *escudo*) by 30 percent, between 1989 and 1993.[14] This led to a strong decline in the support prices for Portuguese agriculture in *escudo* terms. In this way, between 1987 and 1993, real prices for Portuguese agricultural goods declined by 28 percent (Avillez *et al.* 2004).

---

13 Was the so-called *Programa Específico de Desenvolvimento da Agricultura Portuguesa* (PEDAP).

14 As the anti-inflationary result of this policy took some time to take effect, the inflation differential between Portugal and the other members of the community was kept at a wide level (between 5 and 10 percent) for a number of years, see Abreu (2001).

European integration was associated with deep changes in society, the economy and politics. In this way, the challenges faced by Portuguese farmers and other agents intervening in agricultural value chains were part of these global processes. The CAP nevertheless imposed some specific features that redefined market conditions and the country's regional production. During the negotiation period, most products were subsidized with administrative prices, with a few exceptions. For example, from 1974 to 1978, the prices paid to Portuguese producers of wheat and animal feeds remained below world prices, and imports were subsidized. This was mostly due to the concern of the governments of the day with keeping the price of basic consumer goods low. From then on, prices started rising and were no longer below the world level. These increases came alongside the lifting of an input support policy in 1983, which for decades had been crucial to ensure farmers' incomes. In fact, the CAP did not have an input price policy and its output prices were generally lower than the Portuguese ones until 1985.[15]

Despite the specific access conditions negotiated by Portugal, between 1987 and 1999 the level of protection to Portuguese agriculture was substantially reduced (Avillez *et al.* 2004). This was only partially compensated for by an increase in direct support to production and in "decoupled" support. The result was that overall support mechanisms for Portuguese agriculture declined 49.5 percent in that period. In 1987, support for prices corresponded to 93 percent of all the support for Portuguese agriculture through CAP, while direct support represented only 2 percent and "decoupled" support 5 percent. In 1999, support through prices had declined by 69 percent and represented only 58 percent of CAP support to Portuguese farmers, while direct support to production had increased by 528 percent and represented 23 percent, and "decoupled" support had increased by 109 percent and represented 19 percent. This shows how far the incentives present in the support system of CAP changed drastically in the period: the overall decline was large, and the composition of support was altered. In 1999, support through prices and direct support to production were much lower in Portugal than in the rest of the member states on average, measured as a ratio to agricultural land, support to Portuguese farmers through price interventions was 69 percent that of the EU-15; support through payments directly to production was 70 percent of the EU-15; support totally "decoupled" from production was 230 percent of the EU-15. In Baptista's (1994) expression, Portuguese agriculture became a "subsidized" sector rather than a "productive" one, meaning that support to Portuguese agriculture was not essentially directed to improve better productive conditions of the sector, but rather to keep farmers' incomes without promoting structural transformation.

15 For all of this see Pearson *et al.* (1987) and Varela (1986).

This was something radically different from the historical roles played by farmers and agriculture, which sought solutions (better seeds, better technology, more land, etc.) to increase production and productivity, even though often those objectives were not fully achieved.

The bias towards supporting the income of farmers rather than production was, thus, particularly strong in Portugal. According to Avillez (1998), this was linked with a series of factors. Firstly, half of agricultural income depended on transfers from CAP. Secondly, despite the changes brought by the successive reforms, price support remained the central policy element of CAP. Finally, price support was very much concentrated on a small number of crops, most of which had a small share in the product mix of Portuguese agriculture. In fact, in 1996, 75 percent of income transfers to European Union farmers were absorbed by four types of products: grain, oil seeds, milk and beef. Even if these products represented only 35 percent of the products existing in the agricultural sector of the European Union, they received the main percentage of CAP's budget. In contrast, the typically *Mediterranean* products (fresh fruit, horticultural goods, olive oil, tobacco, and wine), despite representing 26 percent of the production, received only 15 percent of the support. In practice, this distribution of CAP support to the different agricultural products affected mainly Southern European countries, where grain, oil seeds, milk and meat had a historically lower weight in the agricultural production structure.

In 1996, Portuguese agriculture was analysed in detail by the European Commission, which concluded that the major effects of these policy shifts had been manifold. According to the first Cohesion Report (European Commission, 1996), CAP had an "anti-cohesion" effect in Portugal for three main reasons. The first was that, since the policy was financed by consumers through higher prices, Portugal was a net contributor to farmers in other countries via imports of goods that would otherwise have been cheaper if bought in the world market. The second factor was that as intervention prices declined massively in terms of the national currency, Portuguese firms were very seriously constrained in their operation. Finally, as the number of Portuguese crops which benefited from high price support was much lower than the European average, due to the national product-mix, the overall support to the sector by CAP was much lower than in most other countries. As noted by the European Commission in 2003, of the four cohesion countries (Portugal, Spain, Greece, and Ireland), Portugal was the only net contributor (European Commission, 2003). The same report also noted that "decoupling" improved the Portuguese situation, but with a twist: support was mostly directed to farmer's income rather than to production.

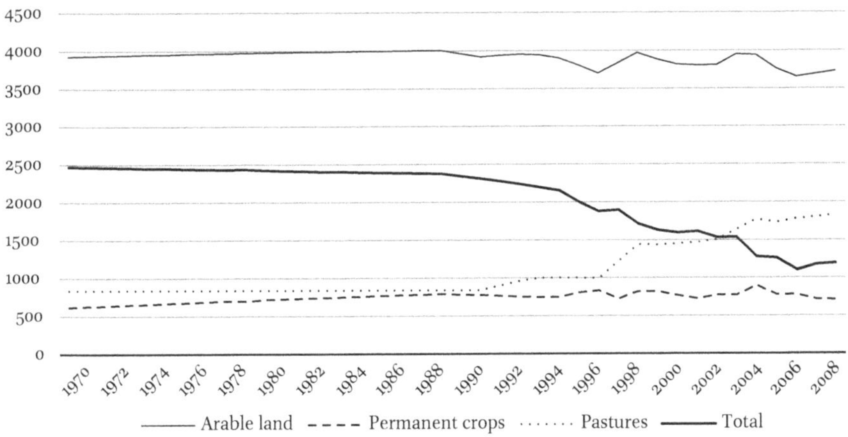

FIGURE 8.9 *Land use (,000 hectares), 1970–2009*
SOURCE: FAOSTAT

Support mechanisms had a strong impact on land use and the product mix. The amount of land dedicated to agriculture did not change much in the period, but the use made of it did. As shown by Figure 8.9, in 1970 arable lands occupied about 2,500 million hectares and pasture less than 1,000 million. However, in 2009 the situation was almost the reverse, with pasture occupying close to 2,000 million hectares and arable land little less than 1,000 million. The area dedicated to permanent crops remained practically unaltered. Agriculture became much more extensive than it used to be, with a large percentage of the soil no longer dedicated to temporary or permanent crops. Under these circumstances, the product-mix that characterized Portuguese agriculture changed significantly.

Aside from these changes, we argue that CAP mechanisms also introduced incentives leading Portuguese farmers to keep a much more labor-intensive factor mix than in most developed countries. Portuguese agriculture received large amounts of funding for "structural reforms", through either PEDAP or the European Structural Funds, affecting the size of farms, cattle intensity, and the level and type of investment that was made. The amount of land in use declined only slightly (3.4 percent between 1989 and 1999), but the number of farms declined significantly (30.5 percent in the same period). This means that, following a tendency for land concentration already witnessed in previous decades, the average size of agrarian parcels rose from 6.7 hectares to 9.3 hectares. Despite the visible evolution, this was still low at the EU level: in 1999, the average size of farms was practically the double, 18.4 hectares. At the same time, the number of workers per farm changed little, from 1.6 in 1989 to 1.3 in 1999,

TABLE 8.2 *Farm structure, Portugal, 1952–1999 (number of farms per class size)*

	<1 ha	1–5 ha	5–20 ha	20–50 ha	50–100 ha	>100 ha
1952–1954	50%	38%	10%	1%	0.5%	0.5%
1979	44%	42%	11%	2%	1%	1%
1989	30%	53%	13%	2%	1%	1%
1999	27%	52%	16%	2%	1%	2%

SOURCE: 1952–1954 (FREITAS *ET AL.*, 1976); 1979 (CAVACO 1986); 1989 AND 1999 (AVILLEZ *ET AL.*, 2004).

which compares with 0.95 in 1999 as an average for the EU. Despite the preference revealed by European politicians for larger farms, CAP had practically no effect on the secular pattern dual structure of Portuguese farms. Large farms prevailed in the South, while small ones dominated in the North. Table 8.2 shows that the years of CAP saw little change in what concerns the smaller farm units (between less than 1 hectare and 5 hectares). If the average size of farms had increased it was mostly due to medium-sized farms (5–20 hectares) and very large ones (>100 hectares). The European common policy appears to have helped to perpetuate the smallest and most traditional type of farms.

It thus seems that ultimately there was considerable structural change in Portuguese agriculture, although not in the sense traditionally envisioned by Portuguese reformers, and not in a way favorable to faster productivity convergence with the rest of Europe. Paradoxically, the main agents of structural change were not the programs explicitly devoted to structural change, but rather the guaranteed prices system and its reforms since 1992. As a consequence of the essential permanence of the farm structure, the capital-intensity of the sector increased but much less than in the rest of the European Union, and Portugal's agrarian labor productivity diverged.

State, the Corporatist System and other Organizations

During the 70 years analysed in this chapter, profound changes occurred in the institutional structure that frames the various stages of the food value chains. Thus, two phases can be established. The first one runs from about 1935 to 1975, when agriculture was framed by State agencies and the corporatist system's services. The second phase, from 1975 onwards, witnessed a process of reducing State services and greater incentives to the formation of civil society organizations. It is also necessary to take into account the years of transition between these institutional models that marked the decades of 1930 and 1940, and again the decades of 1970 and 1980. During these transition phases, uncertainty and conflicting decisions often had negative impacts on the behavior of production and trade activities.

In the 1930s, while stemmed dictatorship consolidation, many civil society organizations (cooperatives, associations, trade unions), which had multiplied since the second half of the 19th century, were dissolved. Some of the skills and activities previously performed by these organizations were transferred to the corporatist system that the *Estado Novo* regime started building in the mid-1930s. Thus, since then agriculture was framed by a network of public services, which were part of the State's structure and by corporatist bodies, forming a *parastatal* system made up of thousands of entities.

In fact, following the trend launched during the second half of the 19th century (see Chapter 7 in this book), the network of public services continued to grow in the following decades. From late 1931 onwards, the reorganization of the Ministry of Agriculture adapted it to the goals of the new regime's policies. This Ministry was divided into three general directorates comprising agriculture, livestock, forestry and agricultural hydraulics. Each one of these general directions had the responsibility, in their respective fields of action, of fighting crops diseases and pests; guiding the use of new seeds and animal breeding; providing technical assistance and monitoring the use of fertilizers and pesticides. In 1936, a new directorat joined the Ministry: the *Junta de Colonização Interna*. Inspired by fascist Italy, this new service was the public body in charge of implementing land reform and population redistribution (Freire 2011b). Along with several central departments, each general direction was provided with a regional or local independent network. In 1940, the Ministry of Agriculture was merged with the Ministry of Commerce and Industry and brought together in to the super-Ministry of Economy. Although in the following decades several requests were made to separate the ministries again, this did not happen during the dictatorship. A reorganization of the ministries took place only after the 1974 Revolution, allowing the creation of a new Ministry of Agriculture in 1975.

The execution of market regulation measures affecting the more important agricultural sectors became the responsibility of the new corporatist system created by the *Estado Novo* since the 1930s. As in other countries – for example, England, France, Denmark (Ingersent and Rayner 1999) – in Portugal severel organizations were created to intervene in the market, responding to the impacts of Great Depression. Some of these bodies started working soon thereafter, but after 1936 these first ones were reorganized and the State merged others. In this way, the main agrarian products (grain, olive oil, wine, cork and other forest products, livestock, fruit and vegetables) came under the supervision of the economic coordination bodies. All had to articulate with the broader corporatist system

The *Estado Novo* withdrew from the ministries most of their intervention functions in trade and agro-industrial activities, assigning them to different corporatist entities. The complex corporatist networks, based on local farmers' guilds (*Grémios da Lavoura*), had as their top the economic coordination organisms established from 1936 onwards. As mentioned in the law that created them, these institutions emerged to "coordinate and regulate the economic and social life of the activities directly related to import and export products".[16] The corporatist and the economic coordination organisms began to be under the supervision of the Trade and Industry Ministry, passing after to the Ministry of Economy. As part of the corporatist system, these corporatist institutions had a different legal status from the State's public services that continued to oversee the different areas of agricultural production. As shown by Manuel de Lucena (Lucena 1976), corporatist institutions were ambiguous, as they incorporated private interests, but never ceased to be under the State's protection. Through the corporatist and economic coordination organisms, the State entered the import, export and domestic trade circuits. The State intervened in prices, ensuring profits to farmers, supported agricultural inputs, lowering farm workers' wages, and contributing to increase production (Rosas 1986). The State conceded monopolies to some corporatist bodies (namely along the grain value chain), but more often it gave to the corporatist organisms privileges over a specific value chain (wine, olive oil, fruits, livestock).

Some corporatist institutions had, from the beginning, strong intervention power in the market, such as the *Federação Nacional dos Produtores de Trigo* (the national federation of wheat producers), an organization in which all wheat producers had to be affiliated. Other organisms (*Junta Nacional do Vinho* and *Junta Nacional do Azeite*, for example, collecting together wine and olive oil producers, respectively) only had the ability to influence prices, and they usually reserved these interventions for situations where the market was not able to ensure the farmers' incomes. Finally, other corporatist bodies (such as the *Junta Nacional das Frutas*, putting together fruit producers) avoided intervening in domestic prices, rather investing in storage infrastructures, thus ensuring export conditions, and in incentives to increase agro-industries. Although some changes were introduced, the responsibilities of the several corporatist organizations were not substantially altered during the long years of the dictatorship (1933–1974). Thus several corporatist services with uneven functions and different intervention capacities coexisted, and were subjected to the markets fluctuations and lobbies actions. In fact, even if the corporatist

16 As explained in the law (Decree n° 26757, 08.07.1936).

system has been exhausted after the April 1974 revolution, most of the economic coordination organisms continued to operate until the mid-1980s (Freire and Ferreira 2013). Thus, during the early years of democracy some market intervention mechanisms that had been implanted at the beginning of the dictatorship continued to operate.

While the State gave corporatist system bodies the power to regulate trade of the agrarian products (by deciding prices, giving subsidies, etc.), it continued to entrust public services with the tasks of promoting innovation among farmers and creating the conditions to improve production. For example, the most important measures aimed at maximizing the exploration of natural resources and the intervention in the land structure were within the realm of public services. Many agronomists and other experts saw this institutional separation, between trade and production, as a very conflictive situation, giving rise to contradictory decisions that delayed the development of Portuguese agriculture. Despite the criticisms, the *Estado Novo* did not change this institutional situation. Both in the organizational chart of the State and in the capacity to implement political decisions, the corporatist institutions, which were responsible for agro-industries and trade activities, were separated from the public services related to general directorates, which had responsibilities in the production areas. This separation resulted in constant difficulties in the relationship and coordination between the different entities, which contributed to a deficient implementation of public policies aimed at the modernization and facilitated the development of some proposals that were contradictory or had inconsistent goals.

After the 1974 Revolution, this corporatist system was slowly dismantled and it became legally possible to form civil society organizations. This was however a slow process, for two main reasons. On the one hand, the democratic governments did not immediately dissolve all the organizations connected to the corporatist system. Among others, the powerful organizations of economic coordination, dependent on the State, which were responsible for managing the processes of transformation and marketing of major agricultural products, were still in operation. These bodies were not closed down until 1986, when Portugal joined the European Communities. On the other hand, when the revolutionary enthusiasm was gone, the emergence of civil society organizations was slow. Equally rare were farmers' associations, cooperatives and other organizations meant to mediate the relationship between economic agents and the State and the market. These organizations have since become more numerous, largely because of the need to meet the challenges posed by the EU.

With Portugal's accession to the EU, although the design of policies for agriculture ceased to be national, the implementation of the measures continued to be the responsibility of national organizations. The implementation of the CAP required a profound change in the State's relationship with farmers and firms. Alongside the extinction of economic coordination organizations, there was a reduction in State services, and these functions were entrusted to civil society organizations. In this sense, with strong encouragement from the EU, there was a proliferation of associations and cooperatives.

Conclusion

During the second half of the 20th century, Portuguese agriculture followed a path of structural change associated with industrialization and modern economic growth, with changes in the composition and the use of land and output, the abandonment of marginal land and a decline in the workforce. However, there were some relevant differences with regard to other Mediterranean or North European countries. Changes in the sector led to political disputes among interest groups and a series of policy measures that impacted on the composition of agricultural output and on its distribution among different regions of the country. The main sectors of Portuguese agriculture were focused on meeting the needs of the internal market or the requirements of foreign markets. In the first case, the sectors involved were grain, olive oil and a part of wine production. In the second, we may include cork and high quality wines. Other sectors, such as fruits and vegetables, olive oil and livestock production, had little weight in the overall economy, even if they could sustain export channels for some niche markets. Faced with the food shortages and low incomes of the Portuguese population, the incentives to the expansion of domestic production were directed to sectors with more economic and social relevance.

Along with the protection of the internal market, from the late 1920s the State promoted a number of measures to maximize the exploration of natural resources, aiming to increase food and raw materials production. Nevertheless, agricultural labor productivity was low and about half of the active working population depended on the primary sector. After Portugal joined the European Communities, labor's abandonment of agriculture slowed down and the sector became more labor-intensive than in its Mediterranean counterparts. The financial incentives of the Common Agricultural Policy helped to reduce the economic expression of agriculture, favoring the promotion of rural development activities in large parts of the territory. Those financial incentives led to slower structural change, in terms of employment, when compared with similar countries. This has had important consequences. To this day, the sector is more labor-intensive than that of its European counterparts (including the Mediterranean ones). The result in comparative terms was a divergence of Portuguese agriculture in relation to the EU-15 in terms of labor productivity.

References

Abreu, Marta (2001). "Da adesão à Comunidade Económica Europeia à Participação na União Europeia e Monetária: a Experiência Portuguesa de Desinflação no período 1984–1998", *Boletim Económico do Banco de Portugal*, Dezembro 2001, pp. 17–38.

Amaral, Luciano (1992). "O plano inclinado do socialismo: sobre o intervencionismo económico do Estado Novo". In J.L. Cardoso and A. Almodôvar (Eds.), Actas do Encontro Ibérico sobre História do Pensamento Económico. Lisbon: CISEP, pp. 373–394.

Amaral, Luciano (2002). *How a Country Catches-up. Explaning the Economic Growth in Portugal in the Post-war Period.* Florence: European Institute of Florence (PhD Dissertation).

Amaral, Luciano (2005). "O trabalho". In Pedro Lains and Álvaro F. da Silva (Eds.). *História Económica de Portugal 1700–2000. O Século XX*. Lisbon: Imprensa de Ciências Sociais, vol. III, pp. 65–90.

Amaral, Luciano (2009). "New series of Portuguese population and employment, 1950–2007: implications for GDP per capita and labor productivity". *Análise Social*, 193, XLIV (4), 767–791.

Avillez, Francisco (1998). *Os Efeitos Redistributivos da Política Agrícola Comum.* Lisbon: Instituto Superior de Agronomia.

Avillez, Francisco *et al.*, (2004). *Rendimento e Competitividade Agrícolas em Portugal. Evolução Recente. Situação Actual e Perspectivas Futuras.* Lisbon: Almedina.

Baptista, Fernando Oliveira (1993). A Política Agrária do Estado Novo. Porto: Edições Afrontamento.

Baptista, Fernando Oliveira (1994). "A agricultura e a questão da terra – do Estado Novo à Comunidade Europeia". *Análise Social*, 128, XXIX (4), 907–921.

Basto, A. Lima (1936). *Inquérito Económico-Agrícola. Alguns aspectos da agricultura em Portugal.* Lisbon: Universidade Técnica de Lisboa, vol. IV.

Brito, J. Pais de, Fernando O. Baptista, Benjamim Pereira (Eds.) (1996). *O Voo do Arado.* Lisbon: Museu Nacional de Etnologia/Ministério da Cultura.

Batista, Dina, Martins, Carlos, Pinheiro, Maximiano and Reis, Jaime (1997). *New Estimates for Portugal's GDP, 1910–1958.* Lisbon: Banco de Portugal.

Branco, Amélia (2011). "El Plan de Repoblación Forestal (1938–1968): una medida del Estado Novo por evaluar". In Daniel Lanero and Dulce Freire (Eds.). *Agriculturas e Innovación Tecnológica en la Península Ibérica (1946-1975)*. Madrid: Ministerio de Medio Ambiente y Medio Rural y Marino, pp. 135–165.

Campos, Ezequiel (1922). *Lázaro! ... Subsídios para a Política Portuguesa.* V.N. Famalicão: Tipografia Minerva, vol. I

Cavaco, Carminda (1986). *Estruturas Agrárias e Mecanização Agrícola em Portugal.* Lisbon: Centro de Estudos dos Povos e Culturas de Expressão Portuguesa – Universidade Católica Portuguesa.

Cunha, Arlindo (2010). "Impactos da adesão na agricultura e nas pescas". *Revista Europa Novas Fronteiras*, 26–27, pp. 159–167.

Estácio, Fernando (Ed.) (1985). *Importância das Políticas Macroeconómicas no Comportamento do Sector Agrícola. Contribuição para a sua Análise em Portugal no Período de 1960–1980*. Oeiras: Instituto Gulbenkian de Ciência/Centro de Estudos de Economia Agrária.

European Commission (1996). *Cohesion Report*. Brussels: European Commission.

European Commission (2003). *Agricultural Situation in Portugal*. Brussels: European Commission.

European Commission (2010). *Situation and Prospects for EU Agriculture and Rural Areas*. Brussels: European Commission.

EUROSTAT, http://epp.eurostat.ec.europa.eu/portal/page/portal/eurostat/home.

FAOSTAT, http://faostat.fao.org/beta/en#home.

Federico, Giovanni (2008). *Feeding the World: an Economic History of Agriculture, 1800–2000*. Princeton: Princeton University Press.

Fennel, Rosemary (1997). *The Common Agricultural Policy: Continuity and Change*, Oxford: Oxford University Press.

Freire, Dulce (2007). *Portugal e a Terra. Itinerários de Modernização da Agricultura na segunda metade do século XX*. Lisbon: Universidade Nova de Lisboa (PhD Dissertation).

Freire, Dulce (2011a). "Produzir mais e melhor: Estado, agricultura y consumo alimentario en Portugal (1926–1974)". *Ayer. Revista de Historia Contemporánea*, 83 (1), pp. 101–126.

Freire, Dulce (2011b). "Ensayos de ingeniería social: reforma agraria y modernización de la agricultura en las últimas décadas del Estado Novo (1954–1974)". In Daniel Lanero and Dulce Freire (Eds.). *Agriculturas e Innovación Tecnológica en la Península Ibérica (1946–1975)*. Madrid: Ministerio de Medio Ambiente y Medio Rural y Marino, pp. 213–241.

Freire, Dulce (2013). "Modernising ambitions: agronomists in action between dictatorship and democracy (Portugal, 1957–1986)". In Ana Delicado (Ed.). *Associations and Other Groups in Science: An Historical and Contemporary Perspective. Newcastle: Cambridge Scholars Publishing*, pp. 92–107.

Freire, Dulce and Ferreira, Nuno E. (2013). "From dictatorship to democracy: the Portuguese corporatist system in transition". Paper presented at the *Workshop on Portuguese History and Political Science. Frontier Research at ICS*. 11 April 2013, at ICS-University of Lisbon (unpublished).

Freire, Dulce and Lanero, Daniel (2013). "The Iberian dictatorships and agricultural modernisation after the Second World War." *Integration through Subordination. The Politics of Agricultural Modernisation in Industrial Europe*. Turnhout: Brepols Publishers, pp. 183–203.

Freitas, Eduardo de, Almeida, João Ferreira de and Cabral, Manuel Villaverde (1976). *Modalidades de Penetração do Capitalismo na Agricultura – Estruturas Agrárias no Portugal Continental (1950–1970)*. Lisbon: Editorial Presença.

Girão, José A. (1980). *Natureza do Problema Agrícola em Portugal (1950–73): uma Perspectiva*. Oeiras: Instituto Gulbenkian de Ciência.

INE, *Área, Produção e Produtividade das Principais Culturas*, https://www.ine.pt/xportal/xmain.

Ingersent, Ken A., Rayner, A.J. (1999). *Agricultural Policy in Western Europe and the United States*. Cheltrnham: Edward Elgar.

Jesus, Quirino and Campos, Ezequiel (1923). *A Crise Portuguesa, Subsídios para a Política de Reorganização Nacional*. Porto: Gráfica Porto.

Lains, Pedro (2009). "Agriculture and economic development in Portugal, 1870–1973". In Pedro Lains and V. Pinilla (Eds.). *Agriculture and Economic Development in Europe since 1870*. London: Routledge, pp. 333–352.

Lanero, Daniel (2014). "The Portuguese *Estado Novo*: programs and obstacles to the modernization of agriculture, 1933–1950". In L. Fernandez Prieto, J. Pan-Montojo and Miguel Cabo (Eds.). *Agriculture in the Age of Fascism Authoritarian Technocracy and Rural Modernization 1922–1945*. Turnhout: Brepols Publishers, pp. 85–111.

Lanero, Daniel and Freire, Dulce (Eds.) (2011). *Agriculturas e Innovación Tecnológica en la Península Ibérica (1946–1975)*. Madrid: Ministerio de Medio Ambiente y Medio Rural y Marino.

Lobão, António Cortez (1977). *Algumas Considerações acerca da Política de Preços em Portugal e na CEE. Separata da Revista de Ciências Agrárias*, 2.

Lucena, Manuel (1976). *A Evolução do Sistema Corporativo Português*. vol. I. *O Salazarismo*, vol. II. *O Marcelismo*. Lisbon: Perspectivas & Realidades.

Martín-Retortillo, Miguel and Pinilla, Vicente (2012). "Why did agricultural labor productivity not converge in Europe from 1950 to 2005?" *European Historical Economics Society Working Paper*, no. 25.

Nunes, Ana Bela (2001). "Economic activity of the population". In Nuno Valério (Ed.), Portuguese *Historical Statistics*. Lisbon: Instituto Nacional de Estatística, pp. 149–196.

OECD, *OECD Historical Statistics* in the OECD Statistical Compendium online, http://libraries.fe.unl.pt/index.php/e-resources/statistical/item/oecd-statistical-compendiu.

OECD (2011). *Evaluation of Agricultural Policy Reforms in the European Union*. Paris: OECD.

Pearson, Scott R. (1987). *Portuguese Agriculture in Transition*. Ithaca and London: Cornell University Press.

Pinheiro, Maximiano (1997). *Séries Longas para a Economia Portuguesa, Pós-II Guerra Mundial*, vol. I – *Séries Estatísticas*. Lisbon: Banco de Portugal.

Presidência do Conselho (1974). *IV Plano de Fomento 1974–1979*, vol. I, *Metrópole*. Lisbon: Imprensa Nacional Casa da Moeda.

Rollo, Fernanda (2007). *Portugal e a Reconstrução Económica do Pós-guerra. O Plano Marshall e a Economia Portuguesa dos anos 50*. Lisbon: Ministério dos Negócios Estrangeiros.

Rosas, Fernando (1986). *O Estado Novo nos Anos Trinta (1928–1938)*. Lisbon: Editorial Estampa.

Rosas, Fernando (1994). *O Estado Novo*, Vol. VII. In José Mattoso (Ed.) *História de Portugal*. Lisbon: Editorial Estampa.

Saraiva, Tiago (2010). "Fascist landscapes: geneticists, wheat and the landscapes of fascism in Italy and Portugal". *Historical Studies in the Natural Sciences*, 40 (4), pp. 457–498.

Simões, Orlando (2006). *A Vinha e o Vinho no século XX*. Oeiras: Celta Editora.

Soares, Fernando Brito (2005). "A agricultura". In Pedro Lains and Álvaro F. da Silva, (Eds.), *História Económica de Portugal, 1700–2000. O Século XX*. Lisbon: Imprensa de Ciências Sociais, vol. III, pp. 157–183.

Sobral, José M. (1999). *Trajectos: o Presente o Passado na Vida de uma Freguesia da Beira*. Lisbon: Imprensa de Ciências Sociais.

Tracy, Michaël (1986). *L'État et l'agriculture en Europe occidentale, crises et réponses au cours d'un siècle*. Paris: Economica.

Tracy, M. (1989). *Government and Agriculture in Western Europe 1880–1988*. 3rd ed. New York: New York University Press.

Varela, J.A. Santos (1986). *A Política Agrícola Comum e a Agricultura Portuguesa: A Política de Preços e Mercados*. Lisbon: Publicações D. Quixote.

Varela, J.A. Santos (1991). *As Negociações com a CEE e a Agricultura Portuguesa*. Lisbon: Publicações D. Quixote.

Varela, J.A. Santos (2007). *A Agricultura Portuguesa na PAC, Balanço de Duas Décadas de Integração, 1986–2006*. Lisbon: Almedina.

Veiga, Teresa R. (2005) "A transição demográfica". In Pedro Lains and Álvaro F. da Silva (Eds.). *História Económica de Portugal 1700–2000. O Século XX*. Lisbon: Imprensa de Ciências Sociais, vol. III, pp. 37–63.

PART 4

Lessons from the Second Millennium

CHAPTER 9

Agriculture and Economic Development on the European Frontier, 1000–2000

Pedro Lains

Introduction

Studies in economic history need to be fed with information on the evolution of aggregate output, labor, capital, land, natural resources, as well as on the institutional context, but the data is not always readily available, if at all, and increasingly less so as we go back in time. Yet, when the right questions are posed and an appropriate research focus is defined, such as the understanding of growth and cycles of convergence and divergence across regions or nations, economic historians have often proved to be quite able to build indicators of qualitative or quantitative substance. Moreover, the levels and rates of growth of income per capita or labor productivity can be inferred from the evolution of wages and prices for which there are some historical data.[1] Thanks to recent work, we are now able to provide a clearer picture of the evolution of the Portuguese agricultural sector across the second millennium, using qualitative as well as quantitative data. With that long run perspective we may go further in answering major questions about economic predicaments and advances in Portugal, and add this case to the broader discussion about the fundamentals of long-run development in Europe and maybe in other parts of the world. The aim of the present chapter is to present the new research on Portuguese agriculture and set it in the context of the wider debate on divergence and convergence within the European economy.

In order to study the evolution of the agricultural sector, we take into account the evolution of the institutional setting, including the changes that occurred in property rights and forms of landownership; the changing composition of agricultural output and demand for foodstuffs and agricultural inputs, looking at the introduction of new and more productive crops, some originated in the empire; the transformation of the patterns of domestic trade; the evolution of methods of production and the technology employed; and the evolution of foreign trade. To complement these studies we also have new indicators for the evolution of output and labor productivity, for the period

1 See, first of all, Allen (2000) and Hoffman (2000).

 | DOI 10.1163/9789004311527_011

from 1500 to 1800, which can be linked to earlier estimates for the 19th and 20th centuries. Part of the information we are using here has been available for some time in a rather disperse and somehow unrelated forms, and it has been assembled here in a way that allows a better description of the evolution of Portugal's agrarian sector.

What can we learn from the enquiry into the agricultural sector of a peripheral and for most of the period here considered backward country of the European continent? How does such a study help explaining major themes in European economic history and economic growth at large? Were conditions for growth different in Portugal? Was it the case that opportunities for faster change were there but were not grasped because of institutional failure? What was the role of the central state, the government and public policies? Did the foundation and expansion of the overseas empire hinder or favour economic growth, or was it irrelevant for the evolution of agriculture at large? Did the Portuguese economy expand according to its growth potential or was there some degree of failure? What are the lessons for European economic history that we may derive from the study of long-run economic performance in this peripheral country? We hope that this study provides enough scope to discuss such a wide range of questions. Eventually, the answers are also in large numbers and varied and explanatory factors that may apply to one period of time or a region may not apply to other periods or regions. But we are better endowed now to understand economic growth in topics such as the importance of domestic and external factors in shaping economic transformation, the relevance of institutional consolidation, the contribution of natural endowments, demographic transitions and the evolution of human capital, the role of technology, markets, foreign trade, investment in infrastructures, and the role of the state in promoting or hindering growth.

By studying a period that goes well back in time, before the first industrial revolution and the industrialization of Europe, we can take into account other sources of growth besides technological change, in manufacturing and transport, or long-distance trade. In fact, we need to deal with a more complex reality where growth and its sources may be found in less spectacular ways, where changes in demand may be more relevant, where "proto-industrialization" or an "industrious revolution" may occur, where weather conditions may matter more, and where growth and stagnation are contemporary phenomena. These themes also show up in the modern period, although they have been disregarded for a long time in the economic historiography, but they are certainly more salient before 1800.

The British industrial revolution did in fact inaugurate a period of growth, as it unfettered economic activity from the limitations of humans and animals as sources of force and enhanced the use of technological innovation. Yet it may be that economic growth is more than societies moving through stages of development, from stagnation or immobility, through a revolutionary change, and towards sustained growth. Continuities are certainly more relevant than disruptions and stagnation and growth may occur contemporarily in different sectors in the economy. Thus, "we should move away from models of linearity and convergence toward a mixed pattern of discontinuous growth and deceleration". (de Vries 2001: 186). That perspective becomes quite clear when we look in depth to economic change across a long period of time, as we have done in this book. If we need a single model that explains the different phases of long-term development, instead of a "stages of growth" model, the study of the evolution of the agricultural sector in Portugal pretends to be a contribution on that direction.[2] Our findings reinforce the conclusion that it was not necessary an industrial revolution or other type of breakthrough to step up from a neo-Malthusian environment of growth constrained by the pressure of population on resources, with no role for efficiency enhancements and technological change, to an environment of "modern economic growth" and structural change.

The chapter proceeds in two sections. In the next section we describe the main phases of growth of Portuguese agriculture since the 11th century, looking both at qualitative and quantitative evidence. We will discuss constraints and opportunities for agricultural growth in Portugal and sort out which have dominated in different periods. The second section sets the evolution of Portugal's agriculture in the European context and discusses the overall causes of growth and fluctuations for the sector. We discuss here whether there is enough evidence to conclude for a "little divergence" in Europe, and whether growth can be better described by continuities, spread across time and space, or by short-term "revolutionary" discontinuities. The final section concludes.

Lessons from the European Frontier

The evolution of the institutional setting that concerned Portugal's agricultural sector shows continuities that clearly appear as more important than the discontinuities, the fact that the country suffered a number of important institutional and political changes. Those changes range from the Muslim occupation to the Christian *Reconquista*, four royal dynasties, the dual Monarchy with Spain, a liberal revolution in the early 19th century and, in the 20th century,

2 See Rostow (1960), Kuznets (1966) and Landes (1969).

a number of military coups that put in place a Republic, then a dictatorship and finally a democratic regime. The institutional setting was closely linked to the natural conditions of the country but it was also adaptive and the conjugation of those two elements is probably the source of its strength. The evolution of the agricultural sector was also marked by the adoption of new crops in different parts of the country, some of them imported from overseas. The interruptions in the institutional setting and agricultural developments stem mostly from war and disease, but even in these cases, periods of death and destruction were followed by recovery and often by change in the trend of developments. It is however hard to grasp the intensity of change and that can only be revealed by a structured discussion of the available evidence on output and productivity growth.

Malthusian Growth

During the first century of the new millennium, Western Europe went through a period of economic change, marked by demographic growth, the opening of new lands to agriculture, the expansion of the number and size of the urban centres, and the augmentation of trade across the continent. These changes were accompanied by transformations in the methods of production used in agriculture, manufacturing and trade, as well as the spread of new financial instruments, particularly for long-distance trade.[3] Portugal was no exception in this general European trend and in this country it was closely related with protracted and complex process of Christian conquest and settlement of territories which had been held by a succession of Muslim states for centuries (Henriques: Chapter 1 in this book). In 1031, the once mighty Caliphate of Córdoba, which by then occupied about 2/3 of the territory of the Iberian Peninsula, going from the Mediterranean to the Atlantic shores and from the Ebro to the Douro river valleys, collapsed, thus paving the way for the occupation of central and southern Iberia by expansionist northern polities. A few decades later, in 1064, the cities of Viseu, Lamego and Coimbra were conquered. Between 1143 and 1147, Portugal was founded and Lisbon was taken and made the capital of the new kingdom. In 1249, the new kingdom reached the southern region of the Algarve, nearly trebling its extension, to 90,000 km^2, in one century.

The conquered lands were at the periphery of the former Caliphate and were affected by more than two centuries of warfare which ended with

3 For a periodization of the evolution of medieval agriculture in Europe, see van Bath (1963), Duby (1972), Crouzet (2000: 160–166) and Malanima (2009: Chapter 3). For Portugal, see also Costa et al. (2016).

the *Reconquista*. As such, they were scarcely populated and eventually in demographic and economic decline, punctuated with only a small number of urban centres. With the military occupation came a flow of migration from the more densely occupied northern regions and also migrants of Moorish, Galician, French, Flemish and German origins. Notwithstanding the southward movements, the northern and central regions also registered economic dynamism, which is clearly demonstrated by a cycle of creation of new urban settlements towns and the clearance of forests, marshes and heathlands. Military and administrative concerns led to the establishment of urban centres in the conquered peripheral lands, but proper cities were scarce and most urban centres remained "agro-towns" dominated by the landowning elites and with little connection and trade between them. This was a period of expansion, occupation, conquest of new lands, but that dynamism remained local as the population remained dependent on the land and long distance trade remained insignificant because the surplus was meagre and transport was difficult and expensive. Yet a consequence of the southern military expansion was the securing of the sea route linking the three major ports along the Portuguese coast, namely, Porto, Lisbon and Setúbal, and the increase in trade flows between the different parts of the new kingdom, making that route for the first time relevant in the context of the European sea trade.

The *Reconquista* can thus be depicted as the confrontation between two agrarian societies with contrasting levels of demographic pressure. The eventual victory of the more dynamic contender led to the overall intensification of occupation of the soil. The occupation of south Portugal brought significant changes in the distribution of land, the definition of land ownership, as well as patterns of population settlement. The system of property rights that had developed in the north of the kingdom was transposed to the southern territories, creating an institutional framework that, with little change, would last for centuries to come and would influence to a considerable extent the shape of Portugal's agricultural sector. Even if the crown and the clergy presided over a large patrimony, the fact is that the number of freeholders remained important, as the new institutional framework facilitated the access to land. The institutional setting also changed with the conquest, namely in terms of land ownership, the type of contracts between laborers and landlords, as well as higher levels of centralization of military and political power. Moreover, the replacement of one society by the other was accompanied by some elements of continuity, particularly in what concerns agricultural technology, as the Muslim legacy in this respect was considerable and well adapted to the natural conditions of the south. The old political structures were dissolved and replaced by the new institutions and the Muslim population had a significant change in its status. Yet some elements of continuity remained, particularly those embedded in technology and physical capital.

Thus, agriculture in Portugal had a relatively favourable period from before the new millennium to the mid-14th century, when the new territory allowed for an extension of the agricultural occupation of land, an increase in population and at least a permanence of rates of return in terms of output per unit of land and labor, with an increase of the share of more productive crops, such as wine and olive oil, and with a considerable share of animal husbandry in total output. Yet considerable stretches of the southern territories remained to being cleared or reclaimed and thus we may conclude that it was essentially a frontier economy with a relative abundance of land and other resources and relatively scarce population. The high ratio of land in relation to labor ratio contrasts markedly with the situation in other parts of early 14th century Europe, which appears more as a *monde plein* somehow struggling with an over-supply of labor.

Three and half centuries of territorial, demographic and agricultural expansion were interrupted in 1348 by the Black Death that reached Portugal at the peak of its dissemination throughout Western Europe, and killed in the first years between half and 1/3 of the population, which was reduced to about 700,000 inhabitants (Rodrigues: Chapter 2 in this book). The contraction of the work force happened in a context where the pressure of the population on natural resources was already relatively low and that had a negative effect on the occupation of the land, and eased the pressure for transformation and thus was not globally advantageous. This situation contrasts notably with what was happening in other parts of Europe where the availability of land per unit of labor was smaller. The negative scenario of lower number of workers per unit of land was somehow compensated by the fact that livestock most probably increased, as land for pasturage became more available. The decline in population also led to a rise of wages in relation to rents, thus affecting positively the income of wage earners, but negatively that of landlords, leading to the abandonment of a number of estates, particularly in the areas with lower levels of land productivity. There was also some land redistribution, as manors and free holdings that belonged to proprietors that were killed by the plague were handed over by the king to the Church.

These changes prompted new laws and reforms to protect lay and monastic lords. In 1349, an ordinance compelled landless men and vagrants to work for land owners and established limits on wages. In 1375, another ordinance, the *Sesmarias Law*, introduced regulations on prices of working animals, and compelled landowners to keep the holdings cultivated in order to avoid being seized by the King. As a consequence, a number of landlords had to abandon direct farming of their demesnes and leased their properties, in some cases granting better conditions to their tenants. During the last third of the 14th century, the economic crisis was severely prolonged by a succession of dynastic wars with Castile and the resurgence of outbursts of the plague. Yet by the end of the century as the military, political and demographic situation improved, the country joined a new period of economic stability, and that in an institutional context with relevant changes. The economic implications of the plague were of varied form, but the society proved resilient and adaptive and that is probably one of the most relevant lessons from this period of hardship and yet also of recovery.

New Horizons

Slow but persistent economic recovery from the late 14th century onwards served as the background for the ensuing epoch of the overseas conquest and exploration, which commenced with the conquest of Ceuta, in northern Africa, in 1415. This first expedition ended up being very costly both in human and financial terms, but the following expeditions would prove highly profitable. The century saw the continued overseas expansion and the conquest of trade posts, increasing trade along the coasts of north and northwest Africa, and the agricultural and commercial exploration of the Atlantic islands of Azores and Madeira. But the emerging empire had little impact on Portugal's agrarian sector and would become relevant only in the following centuries. The new era however led to expansion of international trade and to its increasing relevance to the Portuguese economy. These changes ultimately opened new opportunities to the agricultural sector, in the centuries to come, not only because it led to the introducing of new agricultural products, such as maize, but also because it led to the acquisition of colonies and the signing of international alliances that favoured external trade (Miranda: Chapter 3 in this book).

Agriculture in Portugal continued to improve during this period, both of the extension of land put in use and possibly in terms of output, as well as domestic trade. By the end of the 15th century, the level of the population existent before the outbreak of the Black Death was attained once more and eventually superseded. Such recovery accompanied the overall European recovery, or "Europe's second logistic" (Neal and Cameron 2016), which lasted down to the mid-17th century. These changes may also be detected at other levels apart from demography and the extension of land occupation. In fact, farmers increasingly turned to agrarian activities with higher levels of productivity, such as animal husbandry, which increased during the period, and the production of wine and olive oil, implying an alteration in the structure of output, albeit probably marginal and not "revolutionary", and to upsurges in labor productivity. Trade with Europe also expanded, with grain imports becoming increasingly more important, and exports of staples such as cork, sugar-cane from Madeira, and dyestuffs increased in moderate terms. Wheat from northern Africa and the Azores was imported mainly into Lisbon but only in limited quantities.

Following a century of recovery from the Black Death, positive agrarian developments and overseas expansion, most of the 16th century was one of slow if any transformation. Portuguese agriculture continued in its trend of change and adaptation, yet with a lower level of intensity as compared to what was happening in the northern European regions. Trade across the Atlantic Ocean expanded fast, complementing the growth of trade in the axis that crossed England, France, the Low Countries, and Northern Italy. Population and the labor force continued the path of growth that had commenced in the previous century which put pressure on natural resources and the supply of agricultural products. As an immediate consequence, prices and rents increased and ultimately new land was put into use, reaching terrains which were left vacant after the Black Death and leading to the clearance of woods and forests, the diminution of pasturage, and the drainage of marshlands (Neto: Chapter 4 in this book). These changes occurred in a complex institutional setting which was far from stagnant. The variety of types of landownership was considerable and some of the forms were "imperfect", and landowners were separated in different classes with different sets of rights, including the crown and the nobility, the church and the religious orders.[4] Property rights were paramount for the way the land was put under use but there were also many instances of adaptation. In fact, short-term leases could be successively prolonged. The institutional setting ultimately adapted to different natural conditions and historical heritages and changes in demand for agricultural produce and variations in the pressure on the resources. The observed institutional flexibility implied that it was probably not a barrier to changes in the sector, although it probably was neither a source for economic transformation.

Historians have for a long time detected signs of expansion of the agricultural output for most of the 16th century and of changes in its composition towards products with higher land and probably labor productivity, such as wine and corn, as well as animal husbandry, and which catered increasingly to the markets. Trade within the country increased as cities expanded in size and routes were safer and seaborne exports to northern European exports of the expanding agricultural sectors also increased. The evidence regarding these trends is scarce and sometimes rather indirect, so it is difficult still to

4 See Santos (2006).

put figures and particularly to identify major fluctuations. The new quantitative evidence that we gather in the present book points to the conclusion that there was a severe downturn in output in the first decades of the period followed by a recovery to about 1560 and a flat trend down to the middle of the following century. During this period, the divergence of levels of agricultural productivity between the northern and southern regions of Europe began to develop (Allen 2000). The territory remained far from fully exploited in the 16th century. According to Magalhães (1970: 233–243), for example, the stretch of territory of western Alentejo down to the north of the Algarve had no urban centres and was virtually deserted, which implied that the commercial links of the southern coast were stronger with Andalusia then with the rest of Portugal. That scenario would change only much later on.

The 17th century was marked by uneven patterns of European development, as the expansion of agriculture in the North contrasted with slower growth or depression in the Mediterranean region, and it was probably the first century of divergence of national productivity levels across the Continent. The analysis of developments in Portugal confirms that conclusion, as may be shown by looking at the available information on agricultural production, markets, as well as institutional and social developments. The information regarding agricultural output is based on the analysis of tax collections, namely the *dízimo,* as well as prices, wages and rents. Agricultural rents fell in the first three quarters of the 17th century and increased following the end of the Restoration wars with Castile (1640–1668), which affected agricultural and livestock output and greatly disrupted markets. The areas most affected were the territories of Beira Interior and Alentejo, closer to the Spanish border and particularly hit by the war. Moreover, these regions did not benefit from agricultural conversion associated with the diffusion of maize which happened in the rest of the country, including, possibly, also in the southern province of the Algarve. With the end of the war, agriculture recovered throughout the country. The sector was marked by the predomination of grains, particularly barley and wheat, but also maize which was having a large diffusion, and a considerably smaller part for olive oil, wine, fruits and vegetables, as well as livestock. Fisheries and salt extraction were important complementary activities in the coastal areas, particularly to the north of Lisbon and in the southern province of the Algarve. Trade was limited in scope either domestically and internationally, with few exceptions around urban areas. This was in clear contrast with other parts of northern Europe, where grain production was proportionally smaller, livestock and dairy had a large relevance, as did trade.

Throughout the 17th century agriculture remained the main source of revenue for a complex set of taxes, collected by landlords, religious institutions, local authorities (*concelhos*) and the central state. Yet the competition for tax income between the central state and the regional powers led the landlords to increase the control over their estates. Similarly, tax revenues from the colonial trade in the Eastern empire fell and it took some time before the rise of revenues from gold and diamond mining in Brazil surged, what happened from the last two decades of the century onwards. Such changes occurred in a context of the increase in taxation due to the wars of the *Restauração*, after independence from Spain was re-established. The wars and a number of crop failures led to a contraction in agricultural output and to an increase in the number and intensity of famines. Population also fell or increased only very slowly and that contributed to the abandonment of many cultivated lands. On the positive side, the introduction and the expansion of new plant varieties continued, which provided a stimulus for output and productivity growth. There was a progression in the plantation of vines and olive groves in the northern regions, as well as the expansion of the cultivation of maize, which was no longer confined to the centre and north-western regions and progressed to centre and inland territories. These trends were consolidated in the following centuries.

Over the course of the 18th century, transformation in the agricultural sector continued and the most significant element of that was the increase in the level of commercialization of the output, as farmers catered with an increasing intensity to the domestic market and, to a lesser extent, to foreign markets. According to an estimate, by 1770, two thirds of output produced in the Lisbon region was sold in the market and this is indicative of the changes, even though it was probably lower in the rest of the country, as Lisbon was the largest and a fast growing city (Serrão: Chapter 5 in this book). In 1703, Portugal signed a commercial treaty with which granted protection for wine exports to England, mainly from the Douro valley, in exchange for protection for British woollens in Portugal. Agriculture thus responded to changes in the economic environment, which were considerably widespread according to the many sources of qualitative and quantitative information we have about the period. Those changes were in line with what happened in the previous centuries in periods that were not affected by plagues or wars, but it occurred with a higher level of intensity.

The changes in the international setting, particularly in Western Europe, and the abundance of gold imported from Brazil were major drivers of economic dynamism, but the ability of the agricultural sector to expand, invest and introduce and divulge new methods and crops is also part of that story. Such transformations were achieved under high output growth volatility. In fact, agricultural output reached a peak in the mid-18th century, and that upward trend was followed by a sharp downturn to the end of the century which, however, did not cancel all previous gains. Thus, the increase in the population, the expansion of cities and other urban settlements all over the country, new export markets and on top of all that the gold from Brazil, leading to an increase in wealth, enrichment of the elites, provided an economic environment that fostered investment and consumption in the agricultural sector.

Although the market stimuli actually combined a mix of opportunities and risks, it is possible to say that Portuguese agriculture, despite some variations by regions and by sectors, succeeded fairly well in adapting to the new conditions (Santos 2006). The farm sector became more market-oriented and market-integrated. Less profitable crops were discarded and replaced by a specialization in those agricultural products that presented the best comparative advantages, both in the domestic and in the international context. Along with these changes, farming also experienced some innovation and became a more attractive investment. Thus, Portuguese agriculture experienced a "silent revolution" over the course of the 18th century, enriching and reconfiguring the rural landscape and agricultural production, driving the expansion of cultivated land, encouraging the adoption of new technologies and investments, and improving the articulation with other economic sectors. As a final outcome, Portugal became self-sufficient in fruit, olive oil, vegetables, meat, wool and wine, and a major exporter of some of these products in the European context. Furthermore, agriculture was able to supply the majority of the raw materials needed in the domestic wool, silk and hemp industries.

The overall conclusion regarding the performance of the agricultural sector in Portugal in the three centuries to the end of the *Ancien Régime* is that it went through a slow but persistent transformation in many aspects, from the institutional setting, to the structure of output, the expansion of the area under cultivation, the introduction of new crops and methods of production, and the increase in the share of the output that was commercialized and exported. The institutional setting and the structure of property rights did not suffer major changes demonstrating that it was compatible with agrarian change and not an impediment, as it has been already concluded for previous centuries.

The agricultural landscape was thus substantially transformed in terms of the extension of the area cultivated, the organization of the fields, eventually the type of products cultivated, particularly around the expanding urban centres. These changes impacted on the welfare of both peasants and urban dwellers, and both landowners and laborers. These transformations were interrupted by the emergence of wars, particularly in the 17th century, but as conflicts ended the dynamic of change resumed.

An evaluation of the evolution of output growth since the Middle Ages is now possible, based on newly assembled indices for agricultural wages, income and prices. The new indices show that in the period from around 1500 to 1850, total output doubled in real terms, implying an average annual growth rate of 0.2 percent (Reis: Chapter 6 in this book). That upward trend was marked by swings of growth, stagnation and depression of different time lengths. The 16th century started with a sharp output contraction, from 1500 to 1520, followed by recovery and in 1560 the previous peak was already recovered. From then on a century of relative stagnation ensued, ending in 1660. In the following hundred years, to 1750, output increased by 50 percent, and a period of stagnation followed, ending only in the first decade of the 19th century, and output increase again by 30 percent again between 1800 and 1850. The most striking aspect of these estimates is the fact that agricultural output measured in constant prices suffered sharp downturns and outbursts. Clearly the pre-industrial agricultural sector was prone to the vagaries of climate and plant diseases and that may have contributed to such variations. Yet the fact is that no major famines were registered across these three and half centuries, except during the wars with Spain, and also that population increased steadily throughout this long period. Population growth put pressure on agricultural land which however increased, as new fields were occupied, forests cleared and swamps dried. These results question the presence of Malthusian forces in determining the long-term evolution of Portuguese per capita agricultural output in the pre-industrial period, as output declined also in periods of population growth, and we need thus to determine what were the main driving forces behind growth and fluctuations in output, which were the main constraints of growth, whether domestic or international, and whether related to the structure of natural resources or to the social capability of farmers and policy agents.[5]

Growth, Structural Change and Demise

During the 19th century, Portuguese agriculture progressed further and suffered a few fundamental transformations, as output expanded, its composition changed, and labor and capital productivity increased. By the end of the century, the use of artificial fertilizers and mechanization, as the transport to cities by railways, began to rise, particularly in the large cereal production estates of the south (Reis 1982). The transformation of 19th century agriculture in Portugal was also supported by institutional changes that favoured an increase in the commercialization of the land and the occupation of new soils (Branco and Silva: Chapter 7 in this book). Most of the transformations followed the pattern that has been detected since the earlier periods, although there was again some intensification. Growth was to some extent dependent on the expansion of the cultivated area. The overall context remained relatively backward in regard to the more advance economies in northern Europe and even

5 See Simpson (1995: 16–20) and Fonseca and Reis (2011).

to some regions in neighbouring Spain. Two major features of the lag in the transformation in the sector is the fact that the *levels* of labor and land productivity of most crops remained among the lowest in all of Europe, and the share of animal products in total agricultural output was also well beyond that of most other countries, which hampered most of all the fertilization of soils.[6] These factors are of course closely related.

For most of the country the rural landscape was poor and disconnected from the advances in the industrial and urban areas. The detachment between these two worlds of the economic activity also implied that agriculture did not receive a sufficient amount of capital from other economic sectors, and sold few products for industrial transformation, thus contrasting with what happened in the more advanced areas of Europe (Federico 2005). Importantly, virtually all the raw cork produced in the country was sent to be transformed in industrial units abroad, and agricultural products of better quality, such as wood for urban construction, had to be largely imported. Agricultural developments are of course highly dependent from what happens in the rest of the domestic economy and the international environment and the fact that Portugal entered a period of industrialization, however late, and that the international economy was expanding fast meant an overall positive context for agriculture.[7]

Clearly there was a growing perception of the long bridge Portuguese agriculture had to cross in order to reach the level of advance in other parts of Europe and many efforts were in fact conducted by the state and by farmers, including large and small landowners. Yet the degree of backwardness was too large to be surpassed in just half a century. The advances that were felt in the sector also meant that the level of labor and presumably capital productivity was not too far away from the level in some manufacturing sectors which to some extent hindered further shift of resources from agriculture to industry. By the end of the 19th century, wheat protectionism reversed structural transformation of agriculture, as output expanded and more labor and capital was employed in the sector. In fact, total labor force in agriculture did not decline and continued to increase in the next three decades. Yet we may conclude that globally the agricultural sector did not hinder economic growth in 19th century Portugal and that it had changes that were of great relevance for the years to come into the next century.[8]

---

6 See Bairoch (1989), van Zanden (1991), O'Brien and Prados (1992) and Simpson (1995).

7 See Lains (2009) and Pinilla and Ayuda (2009).

8 See for a contrasting view, for example, Neal and Cameron (2016). In the international literature on this topic backwardness is often too hastily associated with absence of change, which indeed occurred although far from fast as fast as wide in more advanced countries such as Sweden. See Myrdal and Morell (2011).

The impact of World War I was mostly of a financial and monetary character, as well as demographic, but economic activity quickly resumed after the end of the conflict, following the pattern elsewhere in Western Europe at least down to the Great Depression. Due to the disequilibria in the public and external deficits, protectionism was reinforced and the level of state intervention also increased significantly and agricultural output increased in response. That expansion led to a small increase in the share of the labor force occupied in the sector and was not followed by major changes in the levels of labor and capital productivity. Yet public and private investment in agriculture increased, as the use of fertilizers, machinery and irrigation increased. Despite the war, the first decades of the twentieth century were relatively favourable for the Portuguese agricultural sector.

During the period from 1929 to 2000, Portugal's agricultural sector experienced three distinct phases, which were somehow marked by changes in the attention it received from national governments and, later on, from the European Union (Amaral and Freire: Chapter 8 in this book). From 1929 up to the early 1960s, tariff protectionism, domestic prices controls and public investment led to the continuation of a period of output and employment growth in the sector.[9] That trend was reversed when industrialization gained momentum and private and public resources were diverted to manufacturing and the shares of agriculture lost ground for the first time in this century. These transformations were certainly inevitable due to the dimension of industrialization, as manufacturing expanded at very fast rates. Agriculture changed considerably in many aspects, including an increase intensive exploitation of natural resources; the intensification of the use of fertilizers, automation, irrigation; the use of research in animal species and plant varieties.

The transformation of the sector was only helped to a relatively low extent by state intervention, particularly during the heyday of fast industrialization, contrarily to what happened in the of western Europe, where industrialization was also fast, but where governments generally introduced policies to promote investment in the sector and compensate the incomes of the farmers. Portugal's accession to the European Communities and the Common Agricultural Policy, in 1986, brought relevant changes as the remaining domestic protectionist policies were lifted, which was particularly important in what trade with Spain was concerned. From its start, the Common Agricultural Policy aimed at increasing the profitability of the sector, by controlling prices of output. Yet as surpluses piled up the policies were changed and farmers' revenues were instead directly subsidized, regardless of output and productivity

9 See Bairoch (1989).

levels. These policies largely contributed to a balanced demise of the sector, as its weight in total labor force and total domestic output converged to the western European average. That decline was however accompanied in a first moment by a relatively small convergence of factor productivity levels, except for a few sectors that benefited either from infrastructure investment or from shifts in consumer preferences. All in all, the agricultural sector ended up being relevant for the preservation of better conditions in the rural areas and a more balanced distribution of population and income across the country, thus contributing very little directly to the growth of output, investment and trade.

Convergence in European Agriculture

The picture we have been able to draw regarding the long term evolution of the Portuguese agricultural sector can now be inserted in the general study of trends and fluctuations of European agriculture, but with great care as the information is rather scant and with many grey areas. We have achieved some degree of comparability but in a context where quantitative data based mostly on indirect indicators of only parts of the sector, in terms of output and regional coverage, needs to be interpreted together with qualitative information on how the sector evolved in Europe. This is truly a large and complex sector in a large and complex territorial entity study across a period where many changes occurred. Despite these limitations, we still can reach some relevant conclusions regarding two crucial aspects. The first is to understand whether agriculture in Portugal followed trends mostly determined by domestic or external factors, whether the cycles in this country were or not common to those of the rest of Europe. The second aspect that the present study can highlight is about the discussion of whether long-term European growth fits better a *Rostowian* typology of stages of growth, with bounds and leaps of ingenuity and productivity, or rather by a more continuous evolution, where productivity advancements take long to spread and consolidate across sectors and the domestic and the international economies.

Patterns of Growth

The study of the development of Portugal's agrarian sector that we carried here is necessarily a relevant contribution to the study of European agriculture and economy, particularly in what concerns its diversity and its relations with the outside world. Europe is diverse continent either because of geography and natural resources, or because of human occupation, and it is a continent inserted in a global context, with connections across Asia, the Mediterranean and the Baltic seas, or the Atlantic Ocean. Such diversity can only be understood if we study with some detail its different components, including the regions on its periphery. Being in one of its frontiers, the study of Portuguese agriculture ends up being also a good field to understand the conditions under which economic activity took place in regions with, on the one hand, poorer soils and lower levels of rainfall and, one the other hand, more closely connected to the Mediterranean, the Northern African and the Atlantic worlds. The diversity of European ecosystem implies the diversity in methods and structures of the economic activity, particularly in what agriculture is concerned.[10] But we may also identify to a certain extent an European "agricultural system" or "European-style agriculture", in opposition to other systems in Asia, Africa or the Americas (Myrdal 2011: 265–264). Portugal is part of it and our closer look of the sector from the early Middle Ages to the present time certainly corroborates that general conclusion.

The diversity of the European economy is also reflected in its relationships with the rest of the world (Crouzet 2000: 11–18). Portugal, since its early days as a state, was open to the outside world despite natural barriers of different sorts. The northern regions of the country were inserted in an area of relatively dense population, but the western and southern frontiers faced for a long period of time areas of the world with little economic activity, either because of the difficulties of seaborne navigation or because of the poverty of the soils and scarcity of rainfall to the south. But the potential that there was for trade and commerce was eventually exploited with military and demographic occupation of the south of the country. The maritime expansion from the earlier 15th century on made Portugal a "sea State" on the cross lines between the Mediterranean and the Atlantic trade (Darwin 2008: 51–52, 96–97). The importance of the empire and of external trade would however remain small for Portugal's agriculture well into the late modern or contemporary periods. The relevance of external relations will however be felt in a number of structural features that directly or indirectly affected the sector. One such feature was the adoption of cultures imported from the Americas or other parts of the world, such as potatoes, maize and rice (and sugar and tea in the Atlantic islands of Madeira and the Azores). One other was the fact that the empire provided a source of state revenue which rendered state finances less dependent on taxing land, labor and domestic trade. Lastly and certainly less importantly, the empire also had demographic impacts which affected the supply of agrarian labor. From the 18th century onwards external relations slowly became more relevant to the agrarian sector, following similar trends elsewhere

10 See Lains and Pinilla (2009).

in Europe, as economic integration within the continent gained momentum, thus reaching its geographical peripheries.

As it happens with the case of Portugal, the evaluation of the long term development of the European economy is very scant and it is necessary to look into many different directions in order to assess the main trends and fluctuations for population, output, and inland or overseas trade. We need thus to resort to conjectures in order to have a broad picture of the main developments. One such conjecture based on many different sources of evidence point to resurgent economic activity in the beginning of the second millennium up to the Black Death in the late 14th century. Our periodization for the Portuguese agrarian economy coincides with that perspective. Trends in the evolution of the agricultural sector in Portugal mirror to a large extent what happened in other countries in Europe and maybe also a general trend for Europe. According to Malanima (2009: 156–157). European agriculture had three large waves in the second millennium, namely, from about 900 to 1300, a period of "slow progress" when "no true revolution took place, but there were modes, short-lived progress both in energy supply and agrarian technology", and climatic conditions were favourable; a "second age" from the 14th century to the beginning of the 19th century, with a "stationary or declining trend in Europe in general despite the success of both English and Dutch agriculture, especially in the 17th century"; and a "third age" of modern growth after 1820.

The changes that we have described here occurred elsewhere and in many parts, particularly in the northern regions of Europe, they were more significant. This was a process of overall transformation of European agriculture of which Portuguese farmers were part, but probably still lagging behind the forerunners in many respects. The fact that there were important transformations in the Portuguese agricultural sector does not imply that backwardness was fully overcome as change occurred elsewhere and in some cases at faster rates of growth. According to Crouzet (2000: 160 and 166) European agriculture suffered from two long depressions since the Middle Ages due to the composite or separated effect of demographic stagnation and climate changes, but since the 18th century if followed mostly an upward trend.

Table 9.1 summarizes the available estimates for the evolution of agricultural output in Portugal and four other countries, since the 14th or the 16th century. Table 9.2 reproduces the estimates for the evolution of agricultural output, population and output per capita from Allen (2000) for nine countries to which we have added data for Portugal and an estimate for "Europe", which was computed as a simple average of growth rates of the countries represented in the table. The comparison of the data in the two tables for England, Holland, Italy and Spain provides a measure of the error of these exercises in computing

TABLE 9.1 *Growth of agricultural output, 1300–1800 (annual rates; percent)*

	England (1)		Holland (2)		Italy (Centre & North) (3)		Spain (4)		Portugal (5)	
	Total	Arable								
1300–1400	-0.44	-0.60	1512–1565	0.37	1360–1430	0.16	1500–1600	0.06	1500–1600	0,03
1400–1470	-0.16	-0.37	1565–1620	0.86	1430–1600	-0.24	1600–1700	0.21	1600–1700	0.06
1470–1600	0.28	0.44	1620–1665	-0.02	1600–1760	0.11	1700–1790	0.32	1800–1840	0.12
1600–1700	0.27	0.29	1665–1720	-0.11	1760–1855	-0.17	1700–1750	0.56	1700–1750	0.52
1700–1800	0.74	0.41	1720–1800	0.28			1750–1790	0.03	1750–1800	-0.13

Sources: Broadberry et al. (2015: Table 3.16), van Zanden and van Leeuwen (2012: Table 4), Malanima (2011: Table 4), Álvarez et al. (2016: Table A-3. column 1) and Reis (Chapter 6 in this book: Statistical Appendix). (1) Rates calculated on decadal averages starting in the years shown; "Total" includes the arable and the livestock sectors. (2) Rates calculated on 5-years averages centred in the years shown. (3) Rates calculated in the source with a Hodrick-Prescott trend. (4) Based on the index of output per head estimated through tithes (upper bound); rates calculated on decadal averages starting in the years shown. (5) Rates calculated from decadal averages starting in the years shown.

long-term growth and we may conclude that there are relevant differences but within acceptable margins of error. For England, for example, the estimates by Broadberry et al. (2015) for the growth of arable product in Table 9.1 are quite similar to those by Allen (2000) in Table 9.2, regarding the periods 1300–1400 (−0.60 vs. −0.58 percent), 1600–1700 (0.29 vs. 0.37 percent) and 1700–1800 (0.41 vs. 0.33 percent), but not for 1400–1500 (−0.37 vs. 0.08 percent) and 1500–1600 (0.44 vs. 0.20 percent). For Holland, the data in Table 9.1 point to a decline in output in 1620–1665 and 1665–1720 (−0.02 and −0.11 percent), whereas the data in Table 9.2 point to a positive trend during the period from 1600–1700 (0.23 percent), but during the 18th century the difference between the two sets of estimates is considerably reduced (0.28 vs. 0.25 percent). Estimates for Italy show a flat or slightly negative trend for the whole period covered, from 1360 to 1800 in both tables, and the differences for Spain are more relevant for the 18th century, as Allen (2000) estimates a decline for 1700–1750 and growth for 1750–1800, whereas Álvarez et al. (2016) estimate growth in the first period and a flat trend in the second. This exercise shows that we need to take with some care the evaluation of output growth but also that some general conclusions can be made.

The overall trends that we can depict from Table 9.2 is that agricultural output growth was overall positive in the 15th century and that it recovered from the 14th century depression, at least in the two cases for which data is shown, namely, England and Italy. During the 16th century the rate of growth turned into negative in six out of the nine countries, following probably the damages of warfare. Then growth resurged in the countries more affected by the previous regression, with the notable exception of Italy. Finally, the 18th century was globally a period of growth and in many cases at rates above those from the previous periods, again with the exception of Italy. The average for Europe depicts in a clear way the trends just identified. Portugal stands out in this table in three aspects. Firstly, output did not decline in the 16th century although it evolved along a flat trend; secondly, the 17th century shows a negative trend, following Italy and unlike the rest of the countries represented; and thirdly the fact that the 18th century had to very distinct phases unlike most of the rest of Europe, as in Portugal output expanded fast until 1750 and contracted afterwards.

Table 9.2 also shows population growth and the growth of output per capita. As to population, the overall trend for Europe is clearly positive, with very few occurrences of negative rates of growth. Moreover, the increase in the intensity of growth of output from the 17th to the 18th century was accompanied by an increase in the intensity of growth of population. If we take the average for Europe in the last column of the table, we may see that during the 17th and the

TABLE 9.2 *Growth of agricultural output, population and output per capita in Europe, 1300–1800*

	England	Netherlands	Belgium	Germany	Poland	Austria	France	Italy	Spain	Portugal	Europe (1)
A – Agricultural output growth (annual growth rates; percent)											
1300–1400	−0.58							−0.15			
1400–1500	0.08		0.16	0.29	0.27	0.10	0.48	0.13	0.21		0.21
1500–1600	0.21	0.30	−0.02	−0.13	−0.05	−0.33	−0.12	0.31	−0.06	0.01	0.02
1600–1700	0.37	0.23	−0.02	−0.09	0.30	0.30	0.12	−0.05	0.13	−0.01	0.11
1700–1800	0.33	0.25	0.52	0.65	0.42	0.39	0.29	−0.06	0.20	0.20	0.30
1700–1750	0.47	0.38	0.81	0.40	0.17	0.60	0.29	−0.01	−0.01	0.67	0.37
1750–1800	0.19	0.12	0.22	0.90	0.68	0.18	0.29	−0.11	0.41	−0.28	0.23
B – Population growth (annual growth rates; percent) (2)											
1300–1400	−0.69							−0.29			
1400–1500	0.00		0.22	0.43	0.40	0.21	0.33	0.19	0.21		0.18
1500–1600	0.56	0.48	0.17	0.14	0.20	0.19	0.13	0.31	0.15	0.62	0.31
1600–1700	0.18	0.23	0.14	0.05	0.19	0.13	0.15	0.00	−0.01	0.32	0.13
1700–1800	0.55	0.12	0.58	0.51	0.39	0.42	0.26	0.31	0.42	0.24	0.38
1700–1750	0.31	−0.01	0.62	0.4	0.30	0.31	0.23	0.32	0.28	0.19	0.29
1750–1800	0.79	0.24	0.55	0.62	0.48	0.53	0.29	0.29	0.57	0.29	0.46

C – Agricultural output per capita growth (annual growth rates; percent)

1300–1400	0.10							0.13			
1400–1500	0.08		–0.06	–0.14	–0.13	–0.11	0.14	–0.06	0.00		0.03
1500–1600	–0.36	–0.17	–0.18	–0.27	–0.25	–0.52	–0.25	0.00	–0.21	–0.62	–0.30
1600–1700	0.19	0.00	–0.16	–0.14	0.11	0.17	–0.03	–0.05	0.14	–0.33	–0.02
1700–1800	–0.22	0.13	–0.07	0.14	0.04	–0.03	0.03	–0.37	–0.22	–0.04	–0.07
1700–1750	0.16	0.39	0.20	0.00	–0.13	0.29	0.06	–0.33	–0.29	0.48	0.08
1750–1800	–0.60	–0.12	–0.33	0.27	0.20	–0.35	0.00	–0.40	–0.16	–0.57	–0.23

Sources and notes: Computed from Allen (2000: Tables 6 and 7) and Reis (Chapter 6 in this volume: Statistical Appendix), for Portugal. For all countries except Portugal, values computed from indices with the base England in 1500 = 100; for Portugal the dates are the centre of 5-year averages, except for 1500, which is average for 1500–1502; For Portugal, consumption per capita. (1) Simple average of all indices. (2) Rates computed as difference between Section A and C.

18th centuries, the rate of growth of output increased alongside with that of population (from 0.11 to 0.30 percent and from 0.13 to 0.38 percent). Probably the most striking conclusion from the same table is that there is a generalization of negative growth rates, which is clearly reflected in the average rates for Europe, in the last column of the table. After 1500, there is a predominance of negative growth rates in southern countries, such as Italy, Spain and Portugal, but we may also find decline in output per capita in Belgium, throughout most of the period, as well as Germany, from 1500 to 1700, and Austria and France. Thus, the data does not show a "little divergence" between northern and southern Europe, and the dividing lines are more chronological than geographical. Álvarez et al. (2016) conclude for such a divergence but only between Holland and England and Spain and Italy. During the second half of the 18th century, growth rates were negative in all countries except Germany and Poland. According to Allen (2000: 20–21), the source of the data for Table 9.2, except for Portugal, agricultural output per worker or productivity followed the same negative trend in all countries but England and the Netherlands, the only two "success stories" of European agriculture, during the 17th and 18th centuries (see also Álvarez et al. 2016).

Thus, the analysis of this data should be taken with great care. The efforts to measure growth in the long-run have just started and although there were considerable advances in the last years, there is still a lot of uncertainty about the main trends of growth and decline. The fact that we are at the start of a new wave of data also implies that we have to be careful in developing theoretical models for causes of growth or decline, as very few information exists about trends of productivity of factors and no information at all on capital and investment. Moreover, it should be taken into account that the dynamics of growth may have changed before and after the industrial revolution as, for instance, it happens with the fact that pre-modern structural transformation is less associated to growth then during the heyday of industrialization. This is so probably because before 1800 structural change could be more strongly associated with changes in demand rather than with technological change and the supply side.

Looking now at the contemporary period, as shown in Table 9.3, from 1870–1913, Portugal's agricultural output increased at 1.2 percent per year, which was slightly above the European average growth rates for the same period. Contrarily, the interwar output depression was felt less intensely in Portugal, as compared to most countries shown in the same table. From 1920 to 1938 output expanded at 3.8 percent per year, which places the sector among the fastest growing in Europe.[11] Growth of output proceeded to the end of the 20th century

11 See also Federico (2005: Chapter 3).

TABLE 9.3 *Growth of agricultural output in Europe, 1870–1938 (annual growth rates; percent)*

	1870–1913	1913–1920	1920–1938	1870–1938
Austria	1.44	-3.55	2.06	1.09
Hungary	2.72	-1.90	0.88	1.79
Belgium	0.70	-0.57	1.60	0.61
Denmark	1.62	-0.60	3.46	1.93
France	0.62	-0.66	1.13	0.69
Germany	1.56	-4.32	3.11	0.91
Greece	2.12	-5.90	4.27	1.53
Italy	0.86	-0.27	0.29	0.82
Netherlands	0.86	1.69	1.96	0.68
Portugal	1.19	-1.56	3.83	0.86
Spain	0.76	1.41	0.77	0.97
Sweden	1.62	-2.64	3.01	1.14
UK	0.00	-0.74	1.87	0.26
Western Europe[1]	0.96	-2.09	1.08	0.75
Southern Europe[2]	0.87	0.19	0.76	0.89

[1]UK, France, Sweden, Denmark, Belgium, Netherlands, Germany and Finland.
[2]Italy, Greece, Spain and Portugal.
SOURCE: LAINS (2003: 49). SEE ALSO FEDERICO (2005: TABLES 3.1 AND 3.3)

but labor also increased down to 1960 and thus its productivity expanded at slower pace. Labor productivity in agriculture expanded fast between 1960 and 1990, at 4.4 percent per year, which was due to the substitution of capital for labor and the increase in total factor productivity.[12] A comparison with output and productivity trends in Portugal, between 1865 and 1951, and Spain, France and Denmark lead to the main conclusion that the performance of the sector was relatively positive.[13] Yet Portuguese agricultural sector had to cope with the increase in agricultural labor force, contrary to what happened in the other countries in the table. Also, land under use in agriculture was still growing in the second half of the 19th century, contrarily to the other countries in the table. After 1902, agricultural land either increased at a lower rate or decrease but, as the labor force kept increasing, the land-labor ratio

12 See Branco and Silva (Chapter 7 in this book) and Lains (2009: 336–340).

13 For agricultural productivity trends in Europe during 1870–1913, see also van Zanden (1991).

declined, which happened only in Spain in a period that comprises the 1936–1939 Spanish Civil War.

Historical analysis that stretches across a millennium necessarily has a large degree of uncertainty. This is particularly true for studies that depend in a crucial way on the evaluation of quantity trends, such as the assessment of levels and trends of output, labor land and capital productivity, as well as the impact of changes in methods of production and technology. The analysis of the evolution of agrarian history as we carry in this study asks questions that medieval, modern and even contemporary sources are not fully prepared to answer, as the concepts of growth or convergence derive from theoretical frameworks which have developed only in recent decades. Historical sources, however, can be pushed a bit further than they have been traditionally, particularly in what concerns variables which also got attention in the past, such as population, prices, wages, foreign trade, and public revenue and debt. Although the available data does not provide the full picture, the new set of questions may on growth and convergence certainly leads to a new perspective on agrarian developments. But we need to take into account that the achievements in this regard are quite unbalanced in terms of the different historical periods since the early Middle Ages and also in terms of different European regions, nations or states.

Uncertainty regarding quantity trends can somehow be complemented by looking into institutional developments which have been largely explored in the historiography. These include first of all the analysis of transformations in the political landscape at the national level, which is particularly relevant as the period we studied here involved territorial conquest and occupation and state formation. Thus, inferences about agrarian developments also depend on the analysis of the new political setting, as well as on institutional developments regarding war and peace, the rule of law, property rights, labor contracts and taxation. The study of institutional factors is crucial to evaluate the context in which economic activity takes place. That is necessarily done by inference as the evidence is indirect. As such, for example, the occurrence of regular armed conflicts is hardly compatible with economic prosperity, whereas changes in labor contracts can be associated with changes in labor demand or supply, as new territorial concessions may be linked to the expansion of agricultural land. Possibly the most relevant contribution of institutions analysis to the present context is the dynamic perspective it necessarily introduces. The picture we provide in the present book has thus many frailties and it cannot be otherwise, due to the extent of the period and the lack of data. But it is on the other hand an enlightening exercise as it involves putting together information that lies on the background of many historical studies, sometimes in an incoherent or inconclusive fashion.

The brighter side of the present study is based on three grounds. Firstly, since earlier periods of time, historians and economic historians have been concerned with the need to measure economic performance, and that is translated into a considerable number of studies on specific parts of the economic activity. It is important to stress that the availability of such studies does not necessarily increase over time, as the earlier medieval period is in certain aspects better covered than the late medieval or the modern eras, and there is quantitative information for the late 18th century which is not available for the earlier 19th century. Secondly, there have been great advances in the elaboration of indices for macroeconomic variables, based on direct historical quantitative evidence, and which can be linked to data for the contemporary period, based on direct estimates or national accounts. Thirdly, the study of institutional aspects of Portugal's history is well connected across themes and epochs and that implies that it is possible to follow closely the path of the institutions that presumably matter the most for the analysis of the agrarian sector. To a certain extent, the exercise carried here undercovers the trends and fluctuations of agrarian output and factor productivity hidden in the relevant historiographical works. The outcome has of course many setbacks but the fact of the matter is that we hope to show that it was possible to step up a degree in our knowledge about agrarian performance, as well as in the set of questions that need to be answered.

Sources of Growth

In the beginning of the period here studied, Portugal was clearly a frontier economy, as land was abundant in relation to labor, thus providing the necessary conditions for increases in output and population.[14] But the advantage of the country in relation to the rest of Western Europe would eventually disappear in the aftermath of the demographic shock imposed by the Black Death, as growth became more dependent on improving methods of production and technological advances. Slowly the sector responded to the changes in the overall conditions, taking advantage of the expansion of the domestic and to a much smaller extension, international markets. By the end of the 18th century, the scenario for Portugal's agricultural sector had changed considerably as it was then "a densely populate country with a population and a low land-labor

14 For the discussion of levels of per capita income in the long-run, see Campbell (2013), Palma and Reis (2016), Broadberry et al. (2015: 422–428) and Fouquet and Broadberry (2015).

ratio which continued to grow at a fast pace".[15] From then on, agricultural change depended to a larger extent on structural transformations and less on the extension of demand conditions.

European agriculture before 1800 had a globally positive trend of growth although certainly slow and with quite sharp fluctuations.[16] Growth was not constrained by a Malthusian trap as it was mostly accompanied by population growth. There was however a number of cases where agriculture apparently suffered from long depressions, as it happen in Italy after 1500 but it does not support "the pessimistic view of the European pre-modern economy" (Malanima 2011: 190). Spain followed closely the Italian trend but for different sets of reasons. Yet according to Cipolla (1981: 250), "the decline of Spain in the 17th century is not difficult to understand. The fundamental fact is that Spain never developed to begin with". As Portugal, Spain can be described as a "frontier economy", as output per worker and labor force evolved in parallel in the long-run, due to the availability of extra land to be put into use.[17]

On the other hand, Holland was the fastest growing economy. More importantly, Holland and the Netherlands, according to van Zanden and van Leeuwen (2012: 128) show that the rate of growth of output in the pre-1800 period hides the intensity of structural change of labor and output. In fact, Holland was the "first modern economy". That implies, according to de Vries and van der Woude (1997: 693), that product and factor markets were "free and pervasive", agricultural productivity was "adequate to support a complex social and occupational structure that makes possible a far-reaching division of labor the state was "attentive to property rights, to freedom of movement and contract" and not "indifferent to the material conditions of life of most inhabitants", and the level of technology was "capable of sustained development and of supporting a material culture of sufficient variety to sustain market-oriented consumer behaviour". But Holland also had the first "modern decline", in the century from the 1660s to the dawn of the British industrial revolution. Yet if we take population trends as an indicator of macroeconomic development it can be concluded that in the Netherlands, contrarily to most of the rest of Europe, demographic change was no longer dependent on the supply of land and food or mortality crises, and instead "the relevant factors became: jobs, urbanization, migration, marriage age and modern expectations". The "modern decline" was not about of a technological or supply backlash, but rather "it

15 Reis and Henriques (2016: 16). See also Henriques (Chapter 1 in this book and 2015).
16 See van Zanden and van Leeuwen (2012) and Fouquet and Broadberry (2015).
17 See Álvarez et al. (2016: 5 and 28). See also Álvarez and Prados (2013).

was a demand-delayed transition from a first to a second cycle of modern economic growth", due to the clash of empires in the international markets.[18]

During the 18th century, the transformation of agriculture was fastest in Britain than in the rest of the continent, but the main distinction was of degree and not so much of the kind of transformation. Simpson (2004) surveys the transformations in British and a number of other European countries' farming during the 17th and 18th centuries, namely, in terms of livestock specialization, increase in labor intensive-farming and changes in the use of land, to find that the main distinction was in the higher ability of British farmers to invest, due to the larger size of British farms.[19] Countries that lay closer to the technological and productivity frontier, such as the Netherlands, first, and Britain, afterwards, were simply better endowed in terms of every kind of resources. Thus, the patterns of development in pre-industrial Europe ranged considerably from cases such as Spain and the Netherlands and England and this can certainly not be grasped only by the evolution of the productivity and technological frontier. Demand shocks could vary in intensity according to differences in the imperial might, or differences in the efficiency of the capacity of the central state to provide the right conditions to growth.[20]

Advances in Portugal, particularly in the 19th and 20th centuries, show that agricultural producers were working below the production frontier allowed by the countries shape of natural resources. In fact, the expansion of the utilization of arable land increased significantly since at least the 1860s, and the peak was reached only a century after. Moreover, the capacity to irrigate the land also increased substantially in the 20th century. These changes needed considerable amounts of investment, both private and public, and that was the true limitation. Hardly can we then argue that Portuguese agricultural expansion was limited by a "neo-Malthusian model": instead, the major constraint was the ability to save and invest or the profitability of investments in the sector as compared to the rest of the economy, either when it was closed or when it became increasingly open.[21] For most of the 19th century and the first half of the 20th, agricultural output expanded through the increase in the use of land and labor. Labor productivity increased in the century that ended in 1960 due to changes in the product mix fostered by changes in the demand pattern,

18 See de Vries and van der Woude (1997: 689 and 720). See also O'Brien (2000 and 2014). The British "climacteric" in the late 19th century is another case of "modern decline" (de Vries 2001: 186).

19 See also Olmstead and Rhode (2009: 32–33).

20 For the Spain, in contrast to the Netherlands and England, see Grafe (2012).

21 See on this issue de Vries (2001: 180–182 and 190).

as overall income growth led to higher levels of consumption of meat, dairy, fruits and vegetables, which had higher levels of factor productivity.[22]

Thus, Portuguese agricultural output growth benefitted first from a large supply of land that allowed high labor-land ratios, then by changes in innovation, particularly in the product mix, which was patent in the increasing share of output of wine, then corn, and then animal products and perishables products. In the mist of these trends, there were was considerable room for the impact of changes in the use of land. Such transformations were accompanied by an increasing role of commercialization and the markets, both domestic and foreign, and this were also forces of structural change and productivity increase, as the products that reached the markets were more advanced in the chain of complexity and had higher levels of land and labor productivity. Our conclusions stem directly from the evidence that we could not identify bottlenecks for growth in Portugal across the period studied, and that when conditions were right institutions evolved, technology developed and was implemented, and public policies were applied.[23] Yet the development of the markets was a crucial factor as well, although it is harder to assess their contribution (Pomeranz 2000: 107). The advantage of pursuing a long term study is that we may observe institutional, technological or policy developments under different economic circumstances, providing an excellent field to observe causes and effects at work.

It is possible to reach an overall conclusion that contributes to a better knowledge of long run economic growth in Europe and eventually to new research questions. And it is important to look to the agricultural sector in particular, because of its dominant presence in economic activity until very recently, and because we need to define better what were the main changes across time and space, which can only be detected through a detailed investigation such as the one we carried here. In fact, agricultural transformations should not be seen as a universal and homogeneous process. Moreover, although Europe can be taken in its unity since the beginning of the period here study, the fact is that it is regionally very diverse and that diversity is a matter of utmost relevance for the study of agriculture.[24] The overall conclusion we have reached here regarding Portugal's agricultural history and which can be extended to other cases is that economic growth, particularly in agriculture, is most of all determined by domestic factors related to natural endowments, demographic patterns, and ability to save and invest. Globalization and the growth of the international

22 See Branco and Silva (Chapter 7 in this book) and Lains (2009).

23 See Fonseca and Reis (2011).

24 See Duby (1972) and Hillbom and Svensson (2013).

economic relations did matter, but to a lesser extent, and its impact had to be felt only during more recent times.[25]

Because of its size, the development of agriculture is crucial for economic growth, particularly before globalization gained momentum well into the 19th century. Agriculture was the source of food and heating for the population, as well as of inputs for construction and manufacturing and because of its sheer size overall factor productivity and income growth depended on changes within agriculture. The transformation was however a slow process conditioned on technological change, the availability of capital to invest in new crops, cattle or infrastructures. The transformations of the agricultural sector were crucial for the success of economic growth and industrialization. Southern European economies were not all the same, but there is a general pattern of lower levels of development across the century in that part of the Continent and it is very important to look at the roots of such different outcomes. The agricultural sector was certainly responsible for it but as we look closely to what happened across such a long period in Portugal, we need to question how much backwardness was a consequence of what happened in the sector, as we may observe that farmers and other relevant agents, such as governments, more often than not showed capacity to adapt to changing conditions.

Conclusion

In the span of 1,000 years studied here, the Portuguese economy was transformed from an essentially rural base, closed to the outside world, to one where services and manufacturing prevailed with borders wide open. That transformation was akin to what happened elsewhere in Europe during the same period, notwithstanding a significant number of differences. Our main conclusions is that the agricultural transformation proceed along a pattern where continuities appear more relevant than discontinuities, where progress is more visible than backwardness, and where the speed of transformation depended more on the domestic conditions than on the international environment, despite the fact that both dimensions did matter. Agriculture in Portugal thus responded to the natural environment and demographic patterns and changes were certainly not revolutionary and their impact took a long time to materialize. That interpretation leads to the conclusion that agricultural transformation in Portugal replicate to a large extent what occurred elsewhere

25 See the discussion of this in Hillbom and Svensson (2013), Timer (2013). For the impact of the empire on the Portuguese economy, see Costa, Palma and Reis (2015).

in Western Europe, as far as our knowledge of both developments through such a long time span can tell, albeit at a distance, both in terms of timing and intensity. If this general conclusion holds, as we believe to demonstrate with the present work, than the broader conclusion that there is an overall process of European agrarian transformation, at least in the Western part of the continent, is globally reinforced, and that the international transformation was the sum of different national transformations. This conclusion goes well with the general idea that European economic history – at least on the western front – can be singled out from the global economic history, although with relevant and well identified global linkages.

The investigation into the causes of economic growth and backwardness leads necessarily to open answers, as theories, qualitative and quantitative knowledge about past economic performance are constantly changing and are rarely unanimous. As we gain better knowledge of factors of growth and economic performance, in an increasing number of regions across the world, we extend our perception on how economies function and how inputs, output, demography and the institutional framework interact, thus gaining new insights on the causes of economic development and backwardness. By going a long way back in time, far from the heyday of the British Industrial Revolution and subsequent European industrialization, and looking at other sectors besides manufacturing, we widen considerably the scope of analysis and improve the chances of better understanding why economies succeed or fail.

Portugal, a nation situated at the southwestern frontier of Europe, founded in the 12th century, with borders established in the 13th century, well open to the outside world and the seat of an empire for most of its history, and where agriculture remained as the largest sector in terms of output and employment well into the 20th century, is a most relevant case to investigate the mechanisms of economic growth and stagnation over the long-run. Due to an historiographical heritage rooted deep in the past, and supported on archival resources mostly untainted by international wars or changes in borders, that study can be carry out with a considerably degree of coherence across such a large period. There is a lot to learn about the investigation of the evolution of the agricultural sector in any European country over the long-term, as it provides a unique perspective to analyse the connections between natural resources, demographic conditions, institutional development, foreign trade, or market integration, and economic growth. Conclusions drawn from any such national study are intimately connected to its particular features within the Continent, and thus provide further insights into the reasons behind patterns of growth and convergence or divergence across its different regions.

This study provides an insight on the constraints and possibilities of agricultural transformation in Portugal, on the periphery of the European Continent, and a new interpretation on conditions for the cycles in growth of agricultural output, population, land use, and factor productivity. In agricultural studies, we still need to understand better how the processes of change unfolded in peripheral and backward regions. Yet in order to test the validity of thesis about the basis of economic success, we need to ascertain not only why some countries succeeded, but also why some countries failed or, at least, took a long time to converge. As we approach the period where international exchanges of goods, capital and people gained an increasing role, we can easily detect the effects of those transformations in agriculture and positive responses, albeit at varying degrees of intensity. Trade with Europe and the empire led to new products, new markets, and the intensification of domestic activity, as well as to institutional adaptations, in what the role of the state and the markets was concerned. Domestic changes culminated in the 18th century but that was also the period where divergence towards the rest of Western Europe increased. The interpretation of the changes that we have detected still needs to be better integrated with the new data for trends in output and productivity in the long-run, as both sources of information are complemented. But the conclusion remains that the Portuguese agricultural sector is shaped both by elements of growth and backwardness, of progress and stagnation and that is the most important common trait that remains for the whole period covered in the present book. Such a conclusion needs to be extended as a general interpretation for economies that lay in the periphery of more advanced areas, of areas where the technological frontier is shaped.

The Portuguese agricultural sector was hardly ever *on* the European technological frontier, was probably only rarely close to it and the distance from it varied substantially in the centuries we have covered here. Yet that did not mean that the sector was not transformed or that it was an impediment to economic change in history. Rather, if we look at periods such as the medieval *Reconquista,* the recovery from the Black Death, the response to the Renaissance globalization, the performance during the enlightened economy period, or the substantial labor and capital (but not land) productivity increases during industrialization, we may conclude after all that the sector was probably as dynamic and adaptive as its ecological and geographical limitations allowed. European economic history is about diversity of paths of development and the agricultural sector in Portugal certainly did not fail.

References

Allen, R.C. (2000). "Economic structure and agricultural productivity in Europe, 1300–1800". *European Review of Economic History*, 4 (1), pp. 1–25.

Álvarez-Nogal, C. and Prados de la Escosura, L. (2013). “The rise and fall of Spain, 1270–1850”. *Economic History Review*, 66 (1), pp. 1–37.

Álvarez-Nogal, C., Prados de la Escosura, L. and Santiago-Caballero, C. (2016). “Spanish agriculture in the Little Divergence”. University of Warwick, *CAGE Working Paper* no. 270.

Bairoch, Paul (1989). “Les trois révolutions agricoles du monde développé: rendements et productivité de 1800 a 1985”. *Annales economies, sociétés, civilisations*, 44 (2), pp. 317–353.

Bath, D.H. Slicher van (1963). *The Agrarian History of Western Europe A.D. 500–1850*. London: Edward Arnold.

Broadberry, S., Campbell, B.M.S., Klein, A., Overton, M. and Leeuwen, B. van (2015). *British Economic Growth, 1270–1870*. Cambridge: Cambridge University Press.

Campbell, Bruce M.S. (2013). “National incomes and economic growth in pre-industrial Europe: insights from recent research”. *Quaestiones Medii Aevi Novae*, 18, pp. 167–196.

Cipolla, Carlo M. (1981). *Before de Industrial Revolution. European Society and Economy, 1000–1700*, 2nd Ed. London: Routledge.

Costa, Leonor Freire, Lains, Pedro and Miranda, Susana Münch (2016). *An Economic History of Portugal, 1142–2010*. Cambridge: Cambridge University Press.

Costa, Leonor Freire, Palma, Nuno and Reis, Jaime (2015). “The great escape? The contribution of the empire to Portugal's economic growth, 1500–1800”. *European Review of Economic History*, 19 (1), pp. 1–22.

Crouzet, François (2000). *Histoire de l'économie européenne, 1000–2000*. Paris: Albin Michel.

Darwin, John (2008). *After Tamerlane. The Rise and Fall of Global Empires, 1400–2000*. London: Penguin Books.

Duby, Georges (1972). *Medieval Agriculture, 900–1500*. In Carlo M. Cipolla (Ed.). *The Fontana Economic History of Europe. The Middle Ages*. London: Collins/Fontana Books, vol. 1, pp. 175–220.

Federico, G. (2005). *Feeding the World: an Economic History of Agriculture 1800–2000*. Princeton: Princeton University Press.

Fonseca, Hélder A. and Reis, Jaime (2011). “The limits of agricultural growth in a fragile eco-system. Total factor productivity in Alentejo, 1750–1850”. In M. Olsson and P. Svensson (Eds.). *Growth and Stagnation in European Historical Agriculture*. Turnhout: Brepols, pp. 167–194.

Fouquet, Roger and Broadberry, Stephen (2015). “Seven centuries of European economic growth and decline”. *Journal of Economic Perspectives*, 29 (4), pp. 227–244.

Grafe, Regina (2012). *Distant Tyranny. Markets, Power, and Backwardness in Spain, 1650–1800*. Princeton: Princeton University Press.

Henriques, António (2015). "Plenty of land, land of plenty: The agrarian output of Portugal (1311–20)". *European Review of Economic History*, 19 (2), pp. 149–170.

Hillbom, E. and Svensson, P. (2013). "Introduction". In E. Hillbom and P. Svensson (Eds.) *Agricultural Transformation in a Global History Perspective*. London: Routledge, pp. 1–25.

Hoffman, Philip T. (2000). *Growth in a Traditional Society. The French Countryside, 1450–1815*. Princeton: Princeton University Press.

Kuznets, Simon (1966). *Modem Economic Growth: Rate, Structure, and Spread*. New Haven: Yale University Press.

Lains, Pedro (2003). "New wine in old bottles: output and productivity trends in Portuguese agriculture, 1850–1950". *European Review of Economic History*, 7, pp. 43–72.

Lains, Pedro (2009). "Agriculture and economic development in Portugal, 1870–1973". In Pedro Lains and Vicente Pinilla (Eds.). *Agriculture and Economic Development in Europe since 1870*. London: Routledge, pp. 333–352.

Lains, Pedro and Pinilla, Vicente (2009). "Introduction". In Pedro Lains and Vicente Pinilla (Eds.) *Agriculture and Economic Development in Europe since 1870*. London: Routledge, pp. 1–24.

Landes, D. (1969). *The Unbound Prometheus: Technological Change and Industrial Development in Western Europe from 1750 to the Present*. New York: Cambridge.

Magalhães, Joaquim Romero de (1970). *Para o Estudo do Algarve Económico durante o Século XVI*. Lisboa: Cosmos.

Malanima, P. (2009). *Pre-Modern European Economy. One Thousand Years (10th–19th Centuries)*. Leiden: Brill.

Malanima, P. (2011). "The long decline of a leading economy: GDP in central and northern Italy, 1300–1913". *European Review of Economic History*, 15 (2), 169–219.

Myrdal, Janken (2011). "Swedish agrarian history – the wider view". In Janken Myrdal and Mats Morell (Eds.). *The Agrarian History of Sweden. From 4000 bc to ad 2000*. Lund: Nordic Academic Press, pp. 257–270.

Myrdal, J. and Morell, M. (Eds.) (2011). *The Agrarian History of Sweden: from 4000 BC to AD 2000*. Lund: Nordic Academic Press.

Neal, Larry and Cameron, Rondo (2016). *A Concise Economic History of the World. From Paleolithic Times to the Present*, 5th Ed. Oxford: Oxford University Press.

O'Brien, Patrick K. (2000). "Mercantilism and imperialism in the rise and decline of the Dutch and British economies 1585–1815". *De Economist*, 148 (4), pp. 469–501.

O'Brien, Patrick K. (2014). "The formation of states and transitions to modern economies: England, Europe and Asia compared". In Larry Neal and Jeffrey G. Williamson (Eds.). *The Cambridge History of Capitalism. The Rise of Capitalism: from Ancient Origins to 1848*. Cambridge: Cambridge University Press, vol. 1, pp. 357–402.

O'Brien, Patrick K. and Prados de la Escosura, Leandro (1992). "Agricultural productivity and European industrialization". *Economic History Review*, 45 (3), pp. 514–536.

Olmstead, Alan L. and Rhode, Paul W. (2009). "Conceptual issues for the comparative study of agricultural development". In Pedro Lains and Vicente Pinilla (Eds.) *Agriculture and Economic Development in Europe since 1870*. London: Routledge, pp. 27–51.

Palma, Nuno and Reis, Jaime (2016). "From convergence to divergence: Portuguese demography and economic growth". Groningen Growth and Development Centre Research Memorandum, 161.

Pinilla, V. and Ayuda, M.I. (2009). "Foreign markets, globalization and agricultural change in Spain, 1850–1935". In V. Pinilla (Ed.) *Markets and Agricultural Change in Europe from the 13th to the 20th Century*. Turnhout: Brepols, pp. 173–208.

Pomeranz, Kenneth (2000). *The Great Divergence. China, Europe, and the Making of the Modern World Economy*. Princeton: Princeton University Press.

Reis, Jaime (1982). "Latifúndio e progresso técnico: a difusão da debulha mecânica no Alentejo, 1860–1913". *Análise Social*, 71, XVIII (1), pp. 371–433.

Reis, Jaime and Henriques, António (2016). "From horn to corn: the two regimes of Portuguese agriculture, 1250–1850". Paper presented at the V Conference of Rural Report, Lisbon: ISCTE-Instituto Universitário de Lisboa.

Rostow, W.W. (1960). *The Stages of Economic Growth: A Non-Communist Manifesto*. Cambridge: Cambridge University Press.

Santos, Rui (2006). "Risk-sharing and social differentiation of demand in land-tenancy markets in southern Portugal, seventeenth-nineteenth centuries". *Continuity and Change*, 21 (2), pp. 287–312.

Simpson, James (1995). *Spanish Agriculture. The Long Siesta, 1765–1965*. Cambridge: Cambridge University Press.

Simpson, James (2004). "European farmers and the British 'agricultural revolution'". In Leandro Prados de la Escosura (Ed.) *Exceptionalism and Industrialization. Britain and its European Rivals, 1688–1815*. Cambridge: Cambridge University Press, pp. 69–85.

Timer, C.P. (2013). "Reflections on the role of agriculture in the structural transformation. A macro-micro perspective". In Janken Myrdal and Mats Morell (Eds.) *Agricultural Transformation in a Global History Perspective*. London: Routledge, pp. 311–326.

Vries, Jan de (2001). "Economic growth before and after the Industrial Revolution: a modest proposal". In Maarten Prak (Ed.). *Early Modern Capitalism. Economic and Social Change in Europe, 1400–1800*. London: Routledge, pp. 177–194.

Vries, Jan de and Woude, Ad van der (1997). *The First Modern Economy. Success, Failure, and Perseverance of the Dutch Economy, 1500–1815*. Cambridge: Cambridge University Press.

Zanden, J.L. van (1991). "The first green revolution: the growth of production and productivity in European agriculture, 1870–1914". *Economic History Review*, 44 (2), pp. 215–239.

Zanden, J.L. van and Leeuwen, Bas van (2012). "Persistent but not consistent: The growth of national income in Holland, 1347–1807". *Explorations in Economic History*, 49, pp. 119–130.

APPENDIX

Maps and Figures for Climate, Relief, Administrative Divisions, Urbanization, and Infrastructures in Portugal, 1000–2000

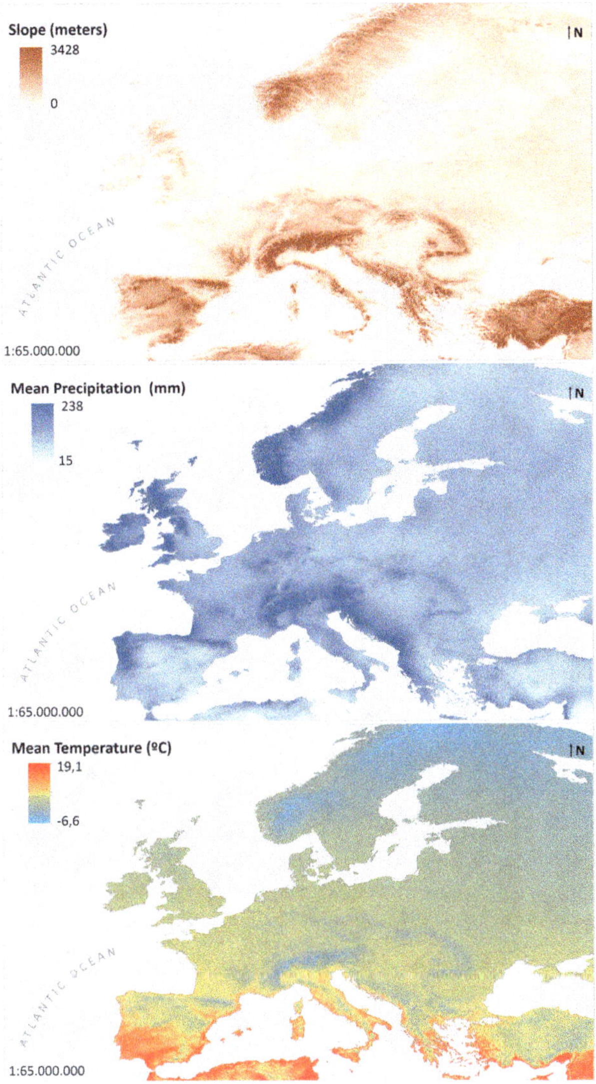

MAP A1 *Slope, precipitation and temperature in Europe (meters; temperature averages for 1960–1990)*

 | DOI 10.1163/9789004311527_012

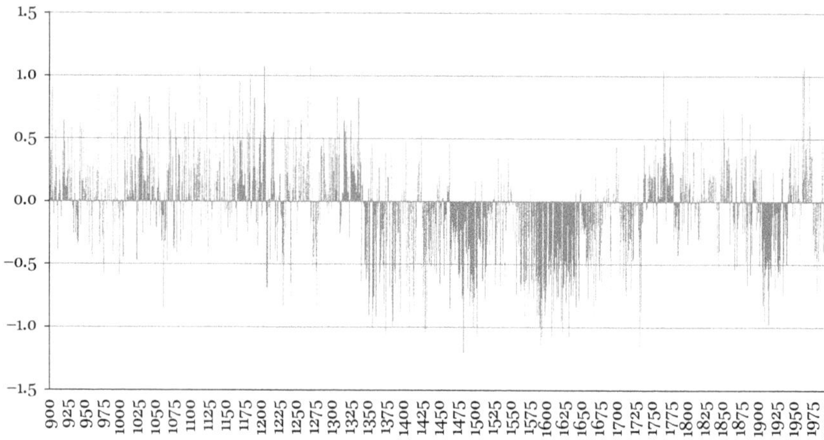

FIGURE A1 *Average temperatures in Portugal, 900–2000 (Celsius)*
Note: Differences of temperature to the average for the 20th century
SOURCE: J. GUIOT AND C. CORONA (2010). "GROWING SEASON TEMPERATURES IN EUROPE AND CLIMATE FORCINGS OVER THE PAST 1400 YEARS". *PLOS ONE*. 5(4), PP. 1–15

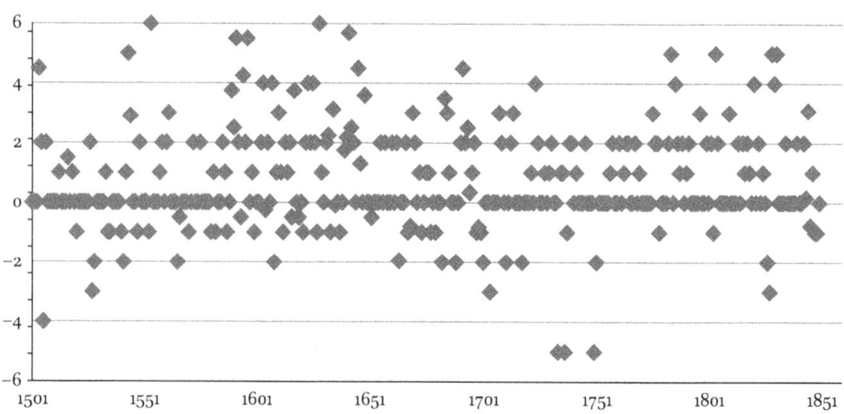

FIGURE A2 *Average rainfall in Portugal, 1501–1851 (<0 = dry years; 0< = wet years)*
SOURCE: A. PAULING, J. LUTERBACHER, C. CASTYAND AND H. WANNER (2006). "FIVE HUNDRED YEARS OF GRIDDED HIGH-RESOLUTION PRECIPITATION RECONSTRUCTIONS OVER EUROPE AND THE CONNECTION TO LARGE-SCALE CIRCULATION". *CLIMATE DYNAMICS*. 26 (4), PP. 387–405

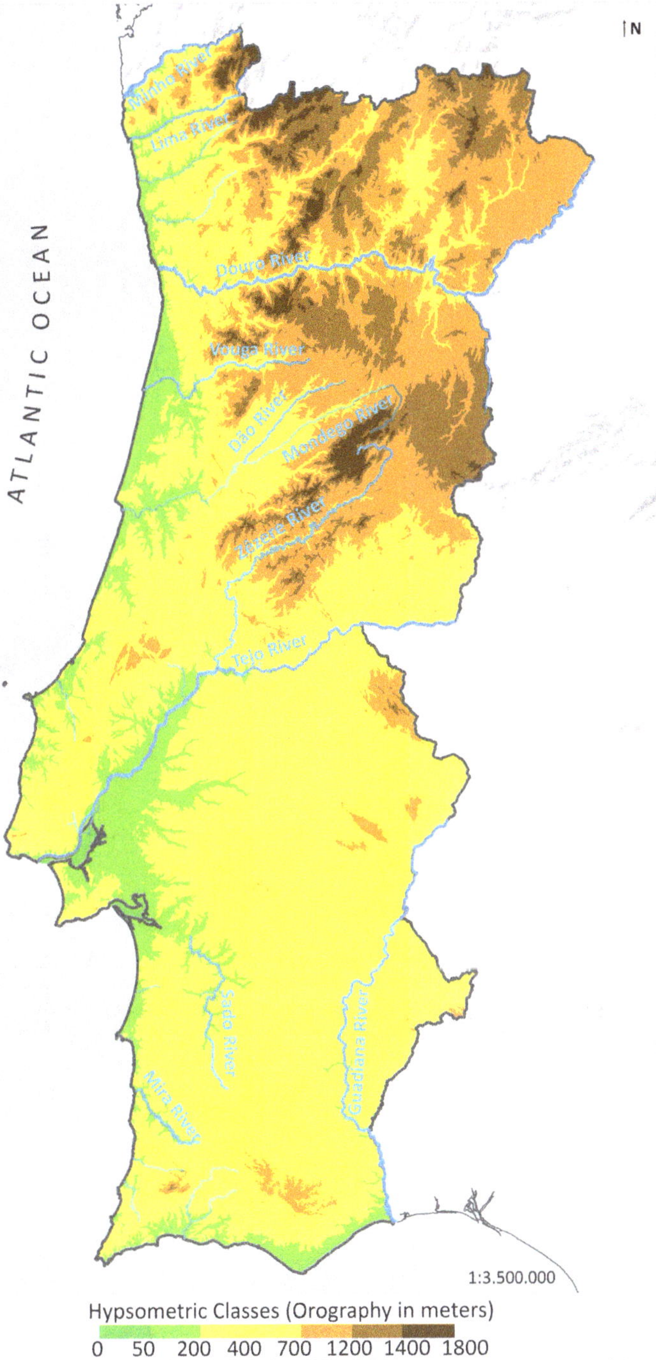

MAP A2 *Hypsometric and orographic map of Portugal*

SOURCE: OWN ELABORATION BASED ON AGÊNCIA PORTUGUESA DO AMBIENTE, SISTEMA NACIONAL DE INFORMAÇÃO DO AMBIENTE

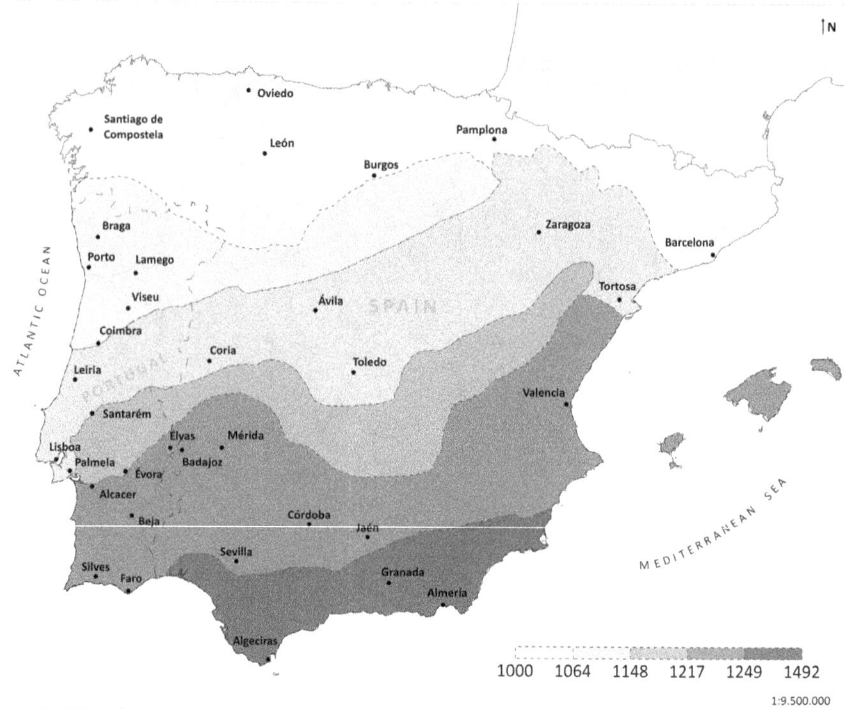

MAP A3 *Phases of the Christian Reconquest in the Iberian Peninsula (in years)*
SOURCE: OWN ELABORATION BASED ON A.H. DE OLIVEIRA MARQUES AND J.J. ALVES DIAS (2003). ATLAS HISTÓRICO DE PORTUGAL E DO ULTRAMAR PORTUGUÊS. LISBON: CENTRO DE ESTUDOS HISTÓRICOS, P. 51

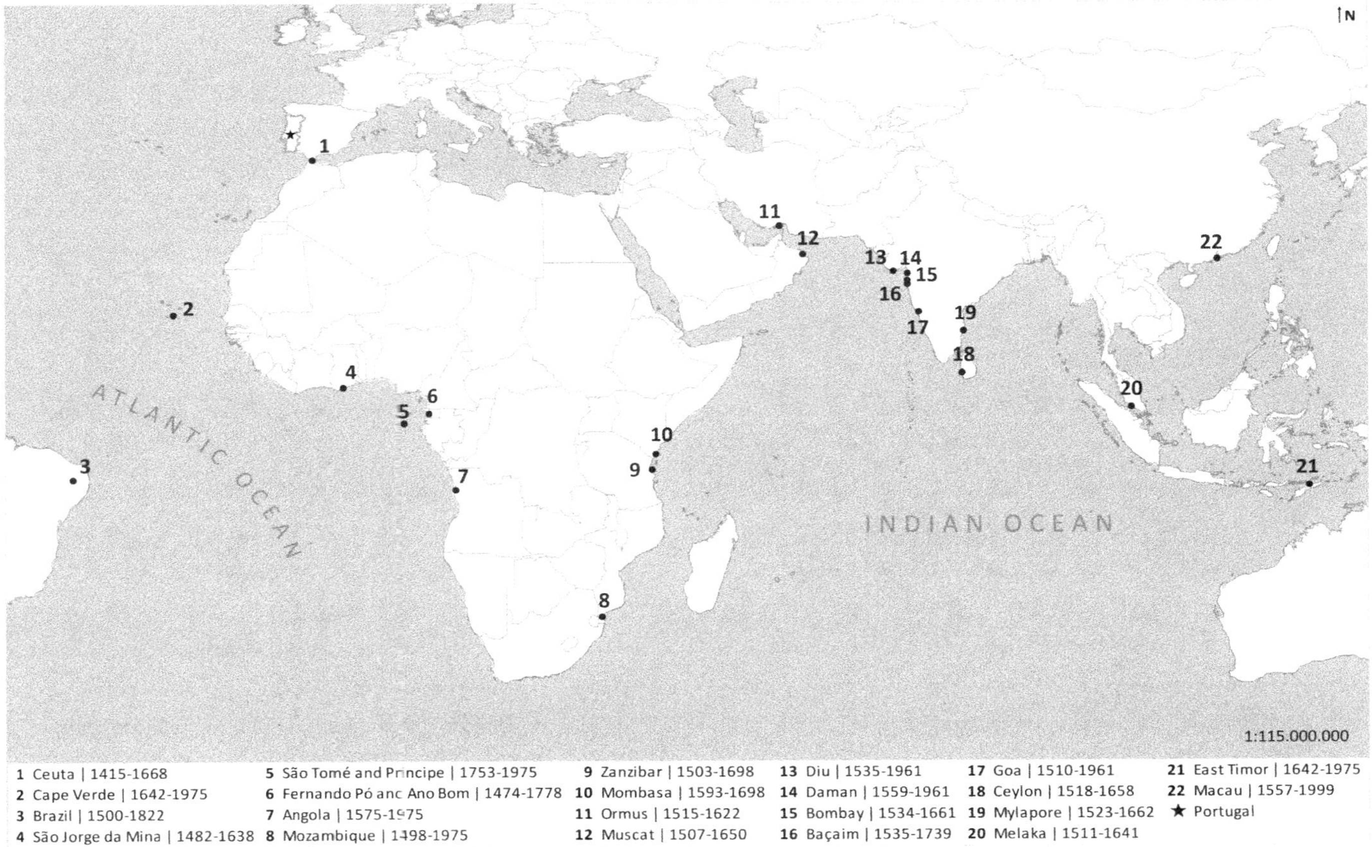

MAP A4 *Portuguese Colonial Empire in the 16th Century*

SOURCE: OWN ELABORATION BASED ON F. BETHENCOURT AND K. CHAUDHURI (1997–1999). *DO ÍNDICO AO ATLÂNTICO (1570–1697)*. IN *HISTÓRIA DA EXPANSÃO PORTUGUESA*. LISBON: CÍRCULO DE LEITORES, VOL. 2

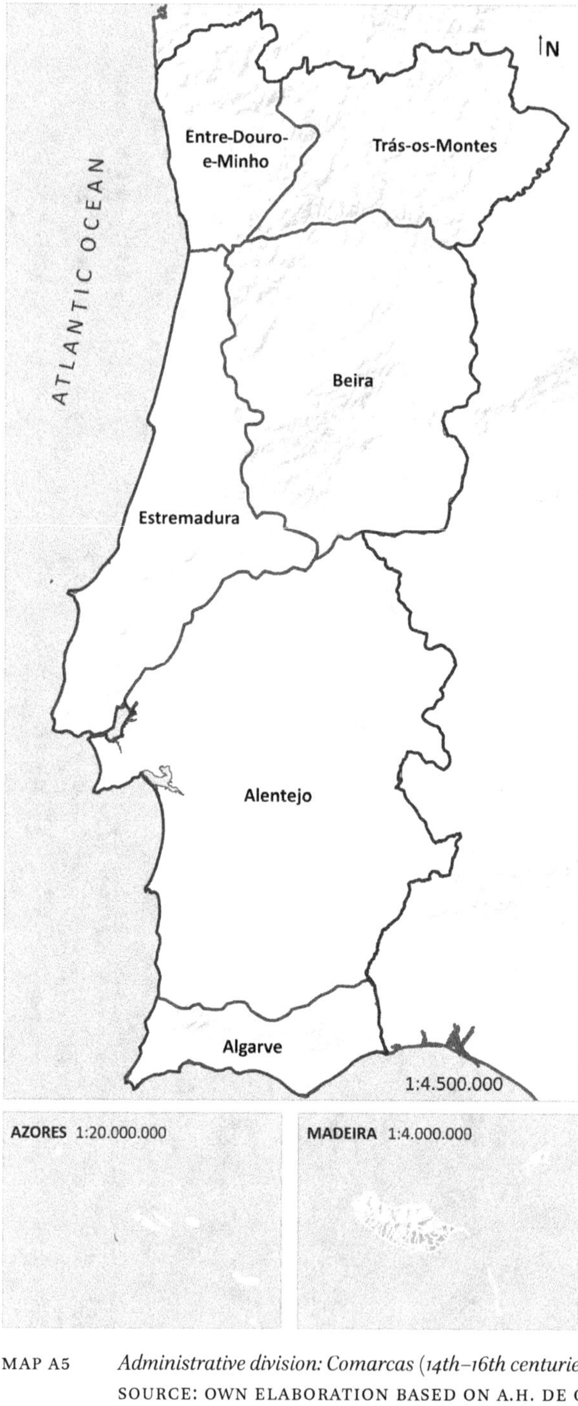

MAP A5 *Administrative division: Comarcas (14th–16th centuries)*
SOURCE: OWN ELABORATION BASED ON A.H. DE OLIVEIRA MARQUES AND J.J. ALVES DIAS (2003). *ATLAS HISTÓRICO DE PORTUGAL E DO ULTRAMAR PORTUGUÊS*. LISBON: CENTRO DE ESTUDOS HISTÓRICOS, PP. 108–332

MAP A6 *Administrative division: Provinces (18th century)*
SOURCE: OWN ELABORATION BASED ON A.H. DE OLIVEIRA MARQUES AND J.J. ALVES DIAS (2003). *ATLAS HISTÓRICO DE PORTUGAL E DO ULTRAMAR PORTUGUÊS*. LISBON: CENTRO DE ESTUDOS HISTÓRICOS, PP. 108–332

MAP A7 *Administrative division: Districts (in 1926)*

SOURCE: OWN ELABORATION BASED ON A.H. DE OLIVEIRA MARQUES AND J.J. ALVES DIAS (2003). *ATLAS HISTÓRICO DE PORTUGAL E DO ULTRAMAR PORTUGUÊS*. LISBON: CENTRO DE ESTUDOS HISTÓRICOS, P. 489

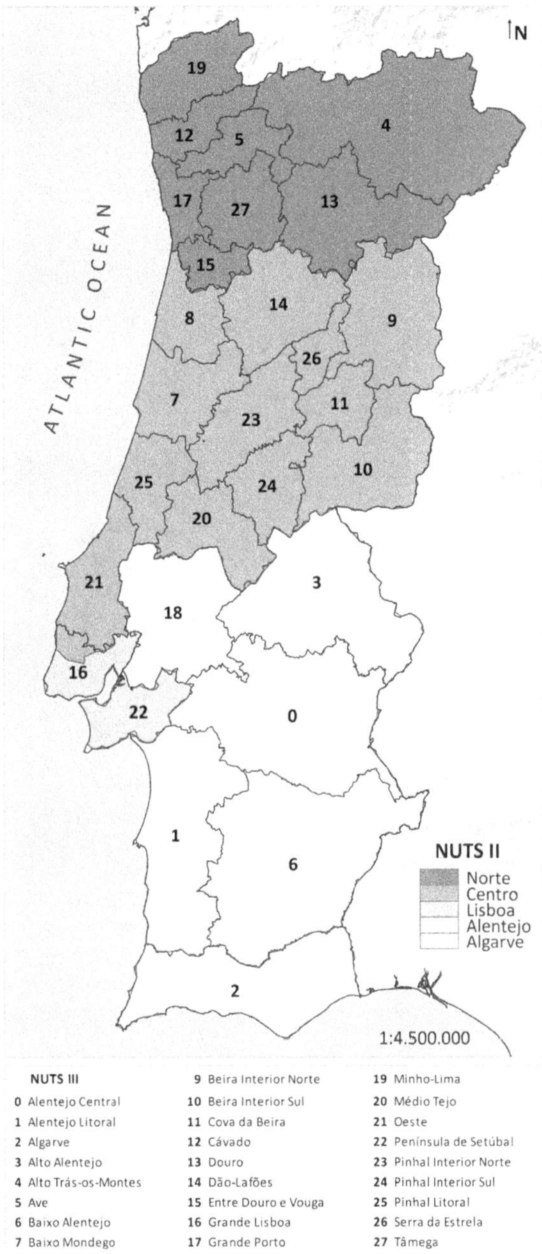

MAP A8 *Administrative division: NUTS (level II and III)*
SOURCE: OWN ELABORATION BASED ON A.H. DE OLIVEIRA MARQUES AND J.J. ALVES DIAS (2003). *ATLAS HISTÓRICO DE PORTUGAL E DO ULTRAMAR PORTUGUÊS*. LISBON: CENTRO DE ESTUDOS HISTÓRICOS, P. 489

MAP A9 *Main towns (c. 1300)*

SOURCE: OWN ELABORATION BASED ON J. VICENTE SERRÃO (1996). "POPULAÇÃO E REDE URBANA NOS SÉCULOS XVI–XVIII". IN C. OLIVEIRA (ED.). *HISTÓRIA DOS MUNICÍPIOS E DO PODER LOCAL (DOS FINAIS DA IDADE MÉDIA À UNIÃO EUROPEIA)*. LISBON: CÍRCULO DE LEITORES, PP. 63–77

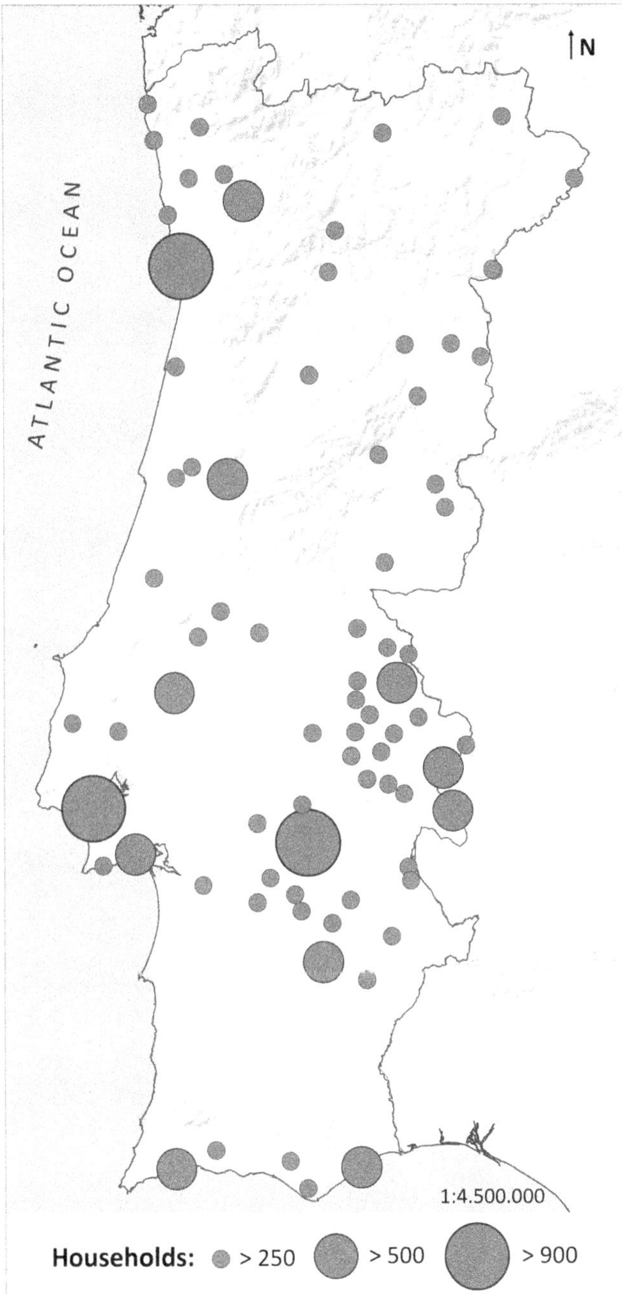

MAP A10 *Urban areas (1527–1532)*

SOURCE: OWN ELABORATION BASED ON J. VICENTE SERRÃO (1996). "POPULAÇÃO E REDE URBANA NOS SÉCULOS XVI–XVIII". IN C. OLIVEIRA (ED.). *HISTÓRIA DOS MUNICÍPIOS E DO PODER LOCAL (DOS FINAIS DA IDADE MÉDIA À UNIÃO EUROPEIA)*. LISBON: CÍRCULO DE LEITORES, PP. 63–77

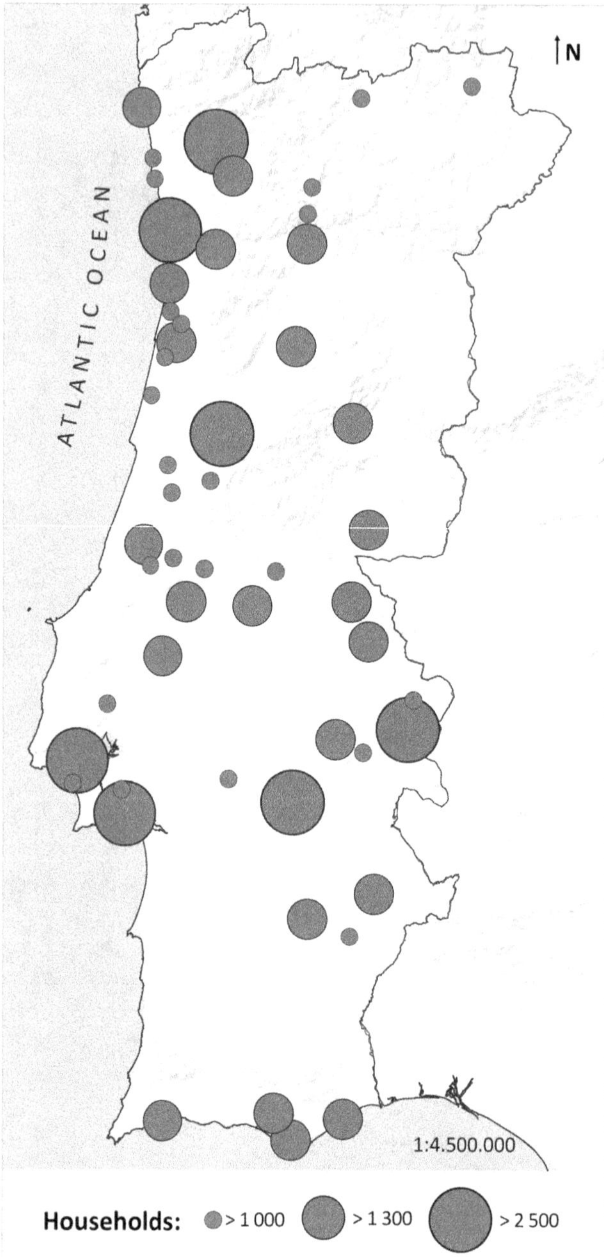

MAP A11 *Urban agglomerations* (*c. 1800*)

SOURCE: OWN ELABORATION BASED ON J. VICENTE SERRÃO (1996). "POPULAÇÃO E REDE URBANA NOS SÉCULOS XVI–XVIII". IN C. OLIVEIRA (ED.). *HISTÓRIA DOS MUNICÍPIOS E DO PODER LOCAL* (*DOS FINAIS DA IDADE MÉDIA À UNIÃO EUROPEIA*). LISBON: CÍRCULO DE LEITORES, PP. 63–77; INSTITUTO NACIONAL DE ESTATÍSTICA (INE) (2011). *RECENSEAMENTOS DA POPULAÇÃO*

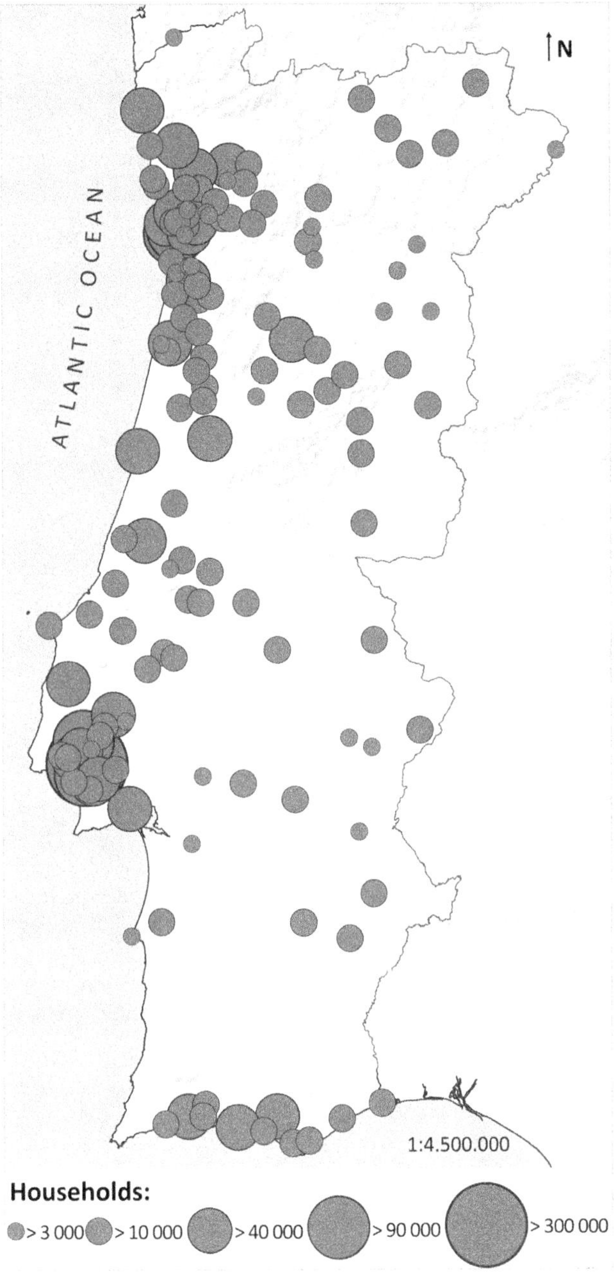

MAP A12 *Portuguese cities (2011)*
SOURCE: OWN ELABORATION BASED ON J. VICENTE SERRÃO (1996). "POPULAÇÃO E REDE URBANA NOS SÉCULOS XVI–XVIII". IN C. OLIVEIRA (ED.). *HISTÓRIA DOS MUNICÍPIOS E DO PODER LOCAL (DOS FINAIS DA IDADE MÉDIA À UNIÃO EUROPEIA)*. LISBON: CÍRCULO DE LEITORES, PP. 63–77; INSTITUTO NACIONAL DE ESTATÍSTICA (INE) (2011). *RECENSEAMENTOS DA POPULAÇÃO*

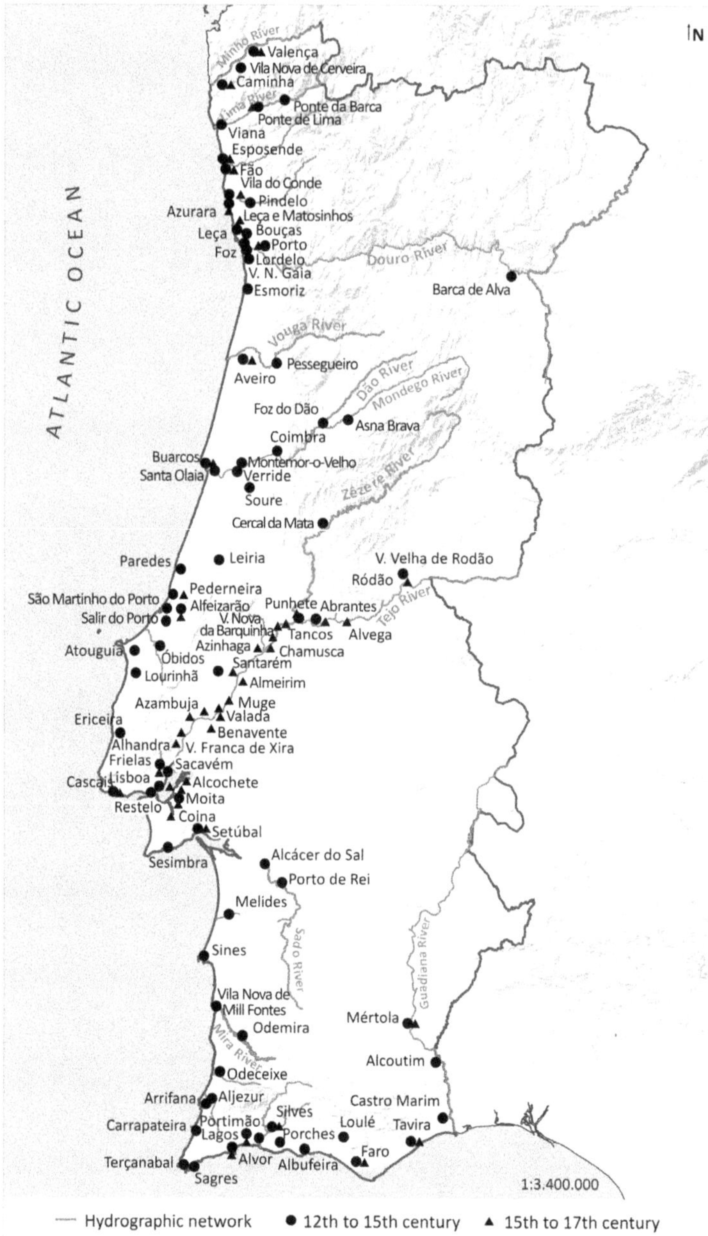

MAP A13 *Major sea and river ports (12th–17th centuries)*

SOURCES: OWN ELABORATION BASED ON JORGE GASPAR (1970). "OS PORTOS FLUVIAIS DO TEJO". *FINISTERRA, REVISTA PORTUGUESA DE GEOGRAFIA* VOL 5, PP. 153–204.; J.R. MAGALHÃES (2002). "A ESTRUTURA DAS TROCAS". IN J. MATTOSO (ED.). *HISTÓRIA DE PORTUGAL*. LISBON: CÍRCULO DE LEITORES, VOL. 3, P. 333; A.H DE OLIVEIRA MARQUES AND J.J. ALVES DIAS (2003). *ATLAS HISTÓRICO DE PORTUGAL E DO ULTRAMAR PORTUGUÊS*. LISBON: CENTRO DE ESTUDOS HISTÓRICOS, P. 92

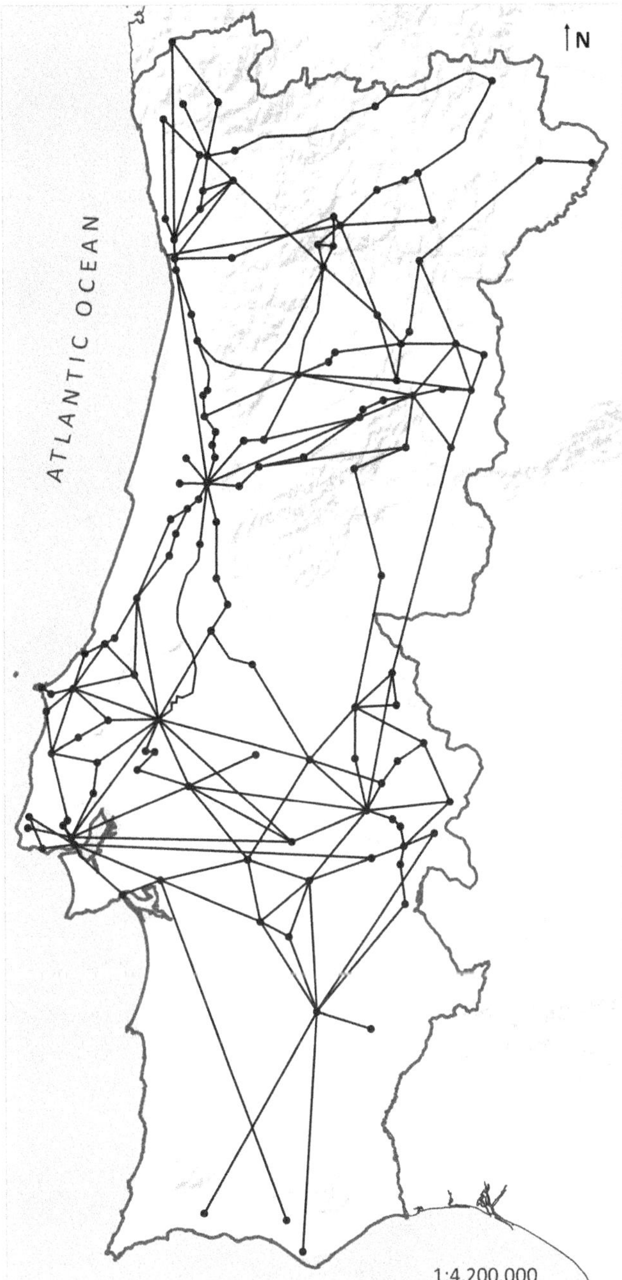

MAP A14 *Medieval royal itineraries (13th and 14th centuries)*
SOURCE: OWN ELABORATION BASED ON J. BAPTISTA DE CASTRO (1870). *MAPPA DE PORTUGAL ANTIGO E MODERNO*, P. 322; A.H. DE OLIVEIRA MARQUES AND J.J. ALVES DIAS (2003). *ATLAS HISTÓRICO DE PORTUGAL E DO ULTRAMAR PORTUGUÊS*. LISBON: CENTRO DE ESTUDOS HISTÓRICOS, PP. 101–103

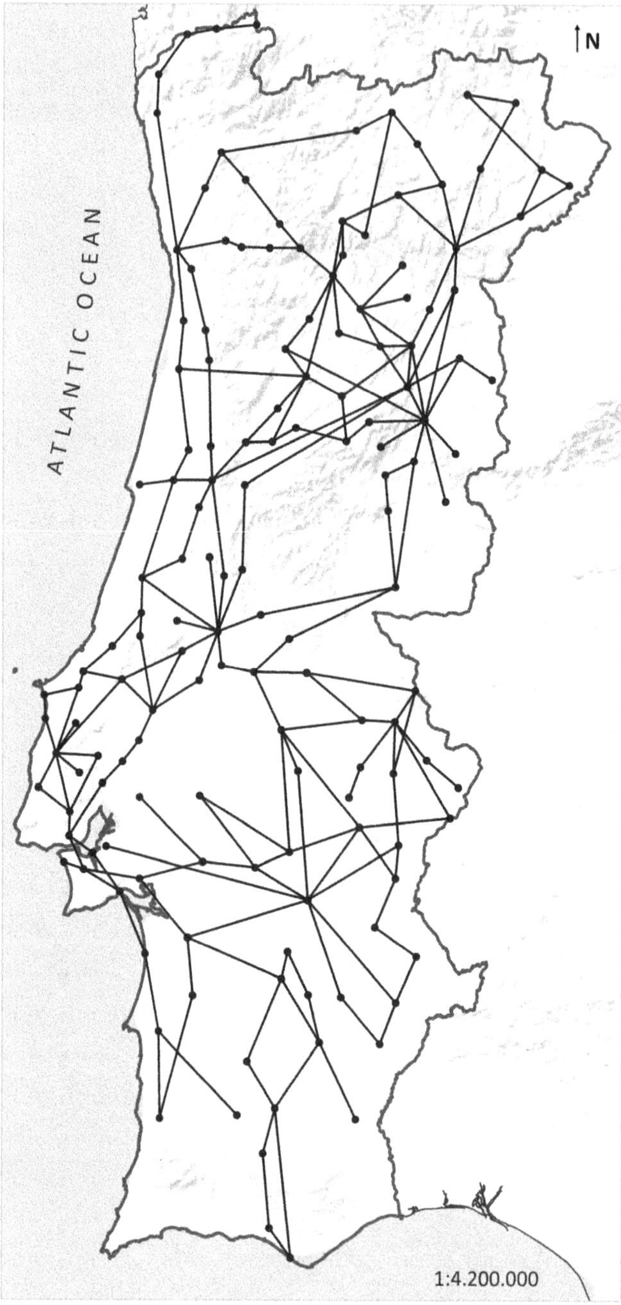

MAP A15 *Road connections between population centres of the kingdom (mid-18th century)*
SOURCE: OWN ELABORATION BASED ON J. BAPTISTA DE CASTRO (1870). *MAPPA DE PORTUGAL ANTIGO E MODERNO*, P. 322; A.H. DE OLIVEIRA MARQUES AND J.J. ALVES DIAS (2003). *ATLAS HISTÓRICO DE PORTUGAL E DO ULTRAMAR PORTUGUÊS*. LISBON: CENTRO DE ESTUDOS HISTÓRICOS, PP. 101–103

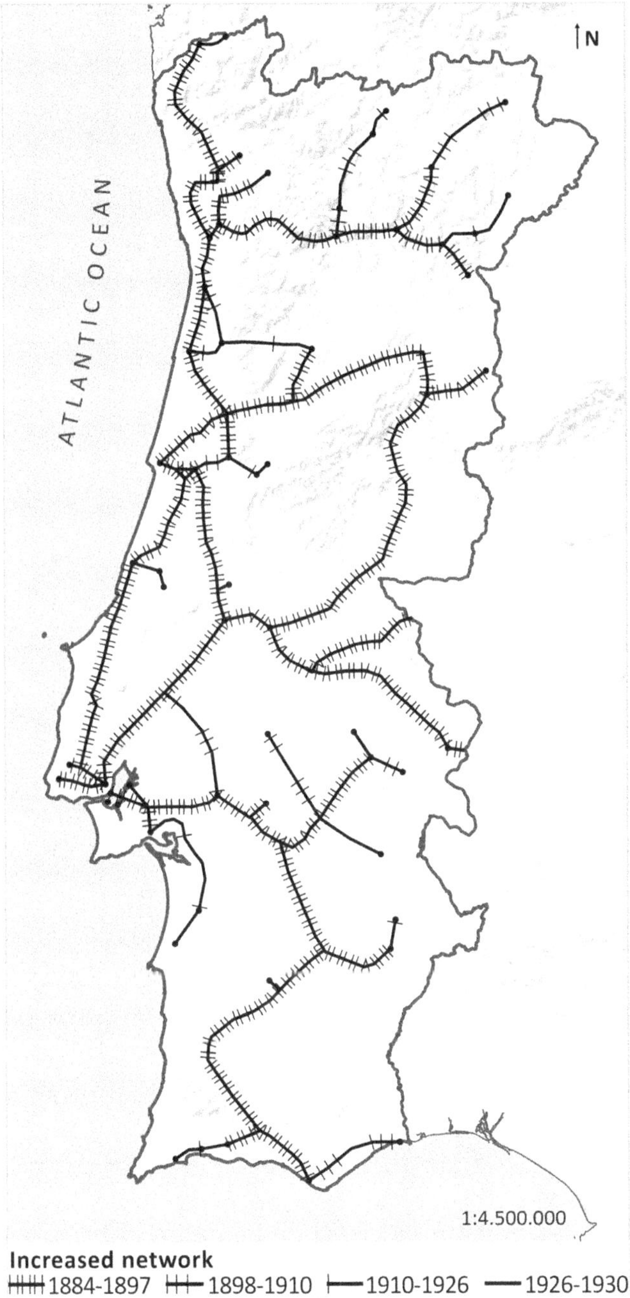

MAP A16 *Railways (19th and 20th centuries)*

SOURCES: OWN ELABORATION BASED ON A.H. DE OLIVEIRA MARQUES AND J.J. ALVES DIAS (2003). *ATLAS HISTÓRICO DE PORTUGAL E DO ULTRAMAR PORTUGUÊS*. LISBON: CENTRO DE ESTUDOS HISTÓRICOS, P. 483; GEOFABRIK GMBH AND OPEN STREET MAP CONTRIBUTORS

MAP A17 *Network of main roads (20th century)*

SOURCES: OWN ELABORATION BASED ON A.H. DE OLIVEIRA MARQUES AND J.J. ALVES DIAS (2003). *ATLAS HISTÓRICO DE PORTUGAL E DO ULTRAMAR PORTUGUÊS*. LISBON: CENTRO DE ESTUDOS HISTÓRICOS, P. 483; GEOFABRIK GMBH AND OPEN STREET MAP CONTRIBUTORS

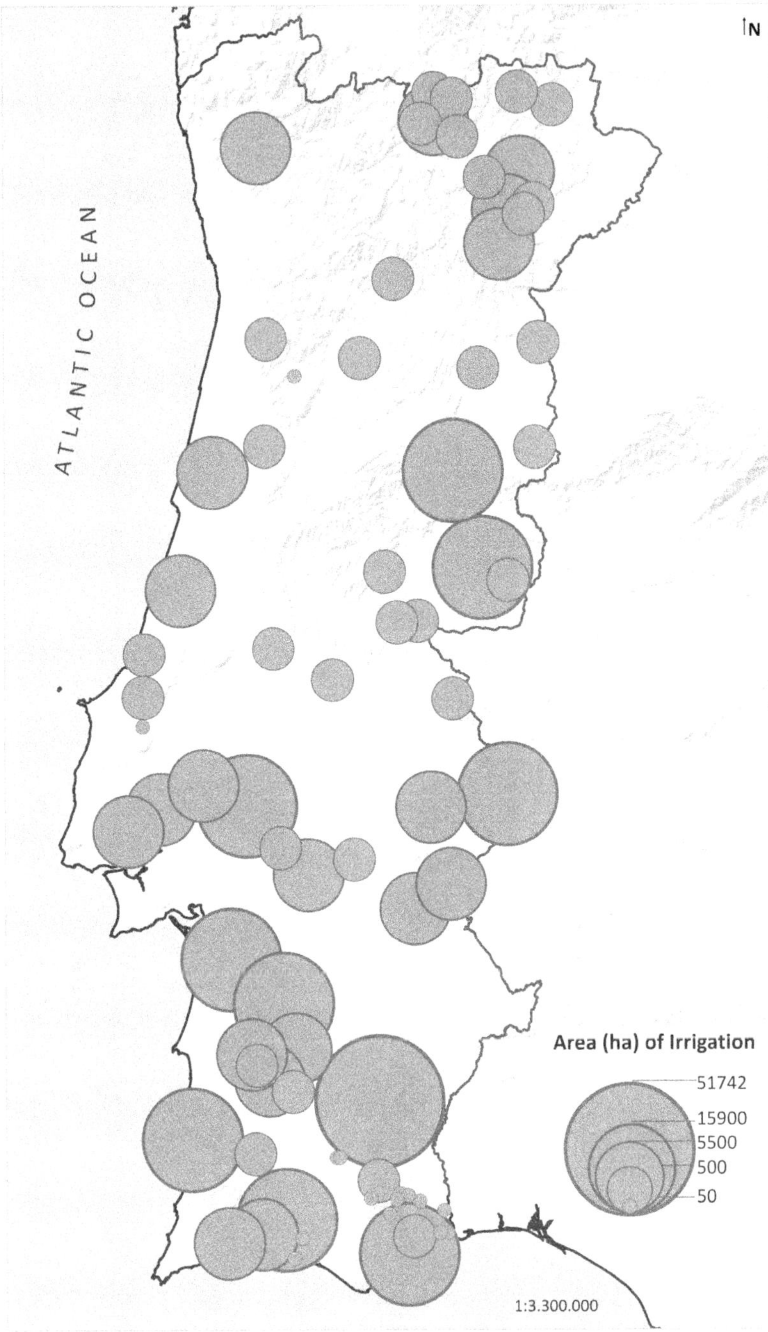

MAP A18 *Dams and irrigation systems build by the state, 1930–2011*
SOURCE: OWN ELABORATION BASED ON SISTEMA DE INFORMAÇÃO DO REGADIO

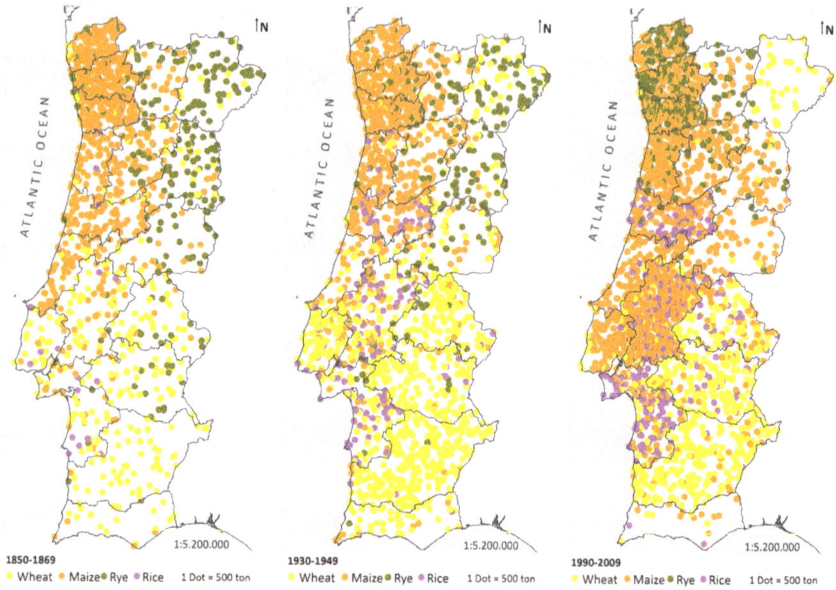

MAP A19 *Cereal production (average per district for 20 years)*
SOURCES: PROJECT «AGRICULTURE IN PORTUGAL: FOOD, DEVELOPMENT AND SUSTAINABILITY (1870–2010)». (FCT-PTDC/HIS-HIS/122589/2010). DATA AVAILABLE AT: WWW.RURALPORTUGAL.ICS.UL.PT.

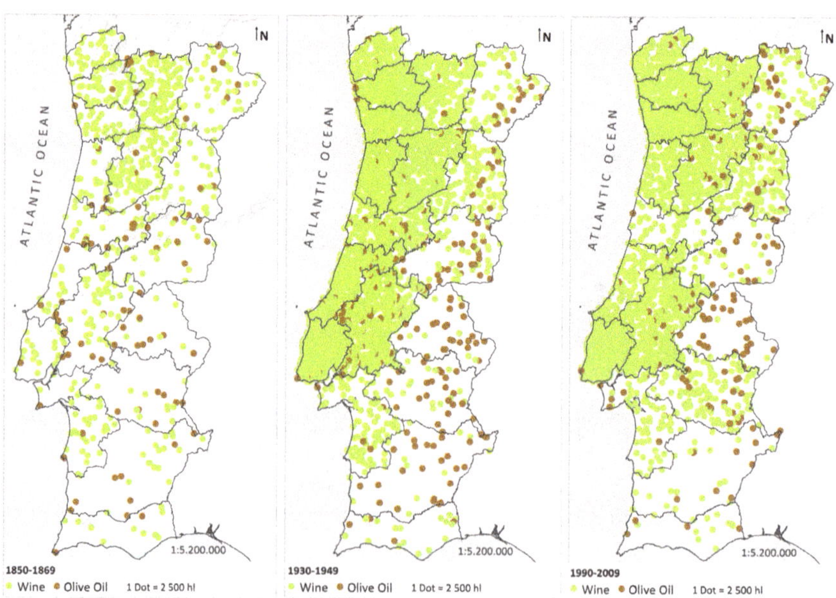

MAP A20 *Wine and olive oil production (average per district for 20 years)*
SOURCES: PROJECT «AGRICULTURE IN PORTUGAL: FOOD, DEVELOPMENT AND SUSTAINABILITY (1870–2010)». (FCT-PTDC/HIS-HIS/122589/2010). DATA AVAILABLE AT: WWW.RURALPORTUGAL.ICS.UL.PT.

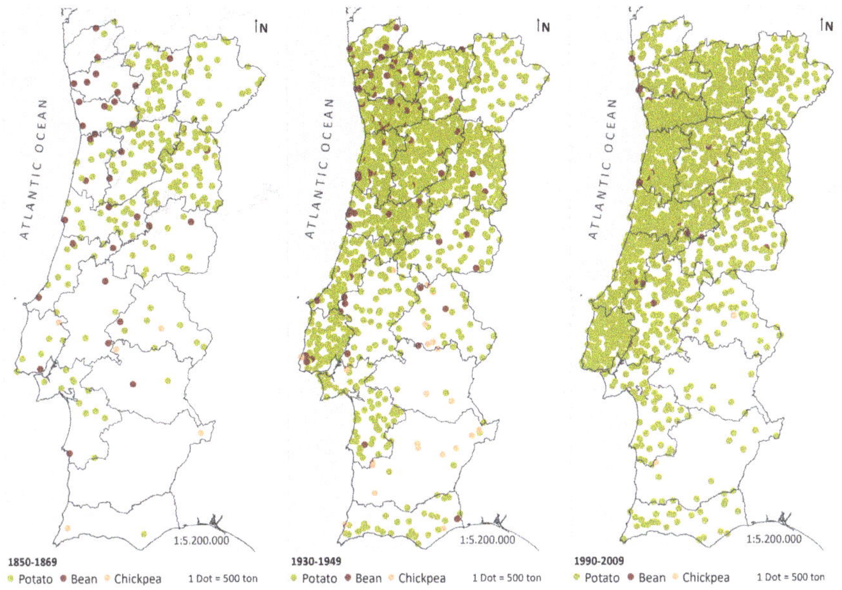

MAP A21 *Potato, bean and grain production (average per district for 20 years)*
SOURCES: PROJECT «AGRICULTURE IN PORTUGAL: FOOD, DEVELOPMENT AND SUSTAINABILITY (1870–2010)». (FCT-PTDC/HIS-HIS/122589/2010). DATA AVAILABLE AT: WWW.RURALPORTUGAL.ICS.UL.PT.

Index

www.ingramcontent.com/pod-product-compliance
Lightning Source LLC
LaVergne TN
LVHW010542300926
849162LV00002B/211